NUMBERS

How Many, How Far, How Long, How Much

NUMBERS

How Many, How Far, How Long, How Much

Edited by Andrea Sutcliffe

A Stonesong Press Book

HarperPerennial

A Division of HarperCollins*Publishers*

Grateful acknowledgment is made for permission to reprint the following:

The Social Readjustment Rating Scale, appearing on page 218, is taken from a paper by Holmes and Rahe in the *Journal of Psychosomatic Research*, vol. II, copyright © 1967 by Elsevier Science Inc. It is reprinted by permission of the publisher.

The table on consumer-product-related injuries around the house, appearing on page 207, was taken from *Accident Facts, 1994 Edition*, page 102, copyright © 1994 by the National Safety Council. It is used by permission of the publisher.

HarperCollins books may be purchased for educational, business, or sales promotional use. For information, please write: Special Markets Department, HarperCollins Publishers, Inc., 10 East 53rd Street, New York, NY 10022.

Text design and copmposition by EEI, Alexandria, VA

FIRST EDITION

A Stonesong Press Book

Library of Congress Cataloging-in-Publication Data

    Numbers : how many, how far, how long, how much / edited by Andrea Sutcliffe. — 1st ed.
          p. cm.
   "A Stonesong Press Book"
   ISBN 0-06-273362-1
 1. Handbooks, vade-mecums, etc. 2. Finance, Personal—Handbooks, manuals, etc. 3. Life skills—Handbooks, manuals, etc. 4. Weights and measures—Handbooks, manuals, etc.   I. Sutcliffe, Andrea.
AG105.N9 1996   95-52658
031.02—dc20   CIP

96 97 98 99 RRD 5 4 3 2 1

Contents

Italicized entries indicate sidebars.

5   Daily Life

7 Time

10   The Universe

11   Conversion Tables and Everyday Math

INTRODUCTION

Numbers are the common denominator of everyday life, as essential to modern society as language. We need numbers constantly, both at home and at work.

If the answer to your question is a number—a date, an amount, a distance, a code, or a formula—we think you'll find it here. This book puts thousands of frequently needed everyday numbers in one convenient volume, making it easy to find the numbers you need to plan a trip, save for college, prepare for retirement, build a deck, address a letter, cook a meal, determine your cholesterol ratio, view an eclipse, or know the time in Tokyo.

In some cases, however, the numbers alone may not be enough. So in many areas—personal finance, for example—we briefly explain the meanings of the numbers as well. In other areas, such as home improvements, we give you formulas to help you calculate rolls of wallpaper, gallons of paint, or number of floor tiles; we even tell you how to determine the right size ceiling fan, air conditioner, and hot water heater to buy.

In the market for a new house? The chapter on money gives the formula that lenders use to qualify you for a loan. You can use the mortgage payment and amortization charts for 15- and 30-year loans to tell you the amount of your monthly payments or the balance left on your loan. Information on interest, taxes, social security, and various investment and retirement plans provides the numbers you need to plan ahead.

On a more enjoyable topic, the chapter on travel puts every number you need in one place; in no other volume can you find these numbers: state and foreign country tourist office phone numbers; international dialing codes; road and air distances for U.S. and international cities; major airline, cruise ship, and hotel reservation numbers; and numbers to call for passport and visa information. We explain how to devise mental tricks for converting foreign currencies to U.S. dollar equivalents and how to convert kilometers to miles. A list showing international electrical requirements will help you decide whether your hair dryer will work without an adapter in Mexico.

The book's other 9 chapters—Food and Nutrition, Health, Daily Life, Transportation, Time, Weather, People and Places, The Universe, and Conversion Tables and Everyday Math—provide equally useful and often hard-to-find numbers, codes, dates, and formulas.

In sidebars throughout the book, you'll find many interesting stories behind the numbers, from how to test for egg freshness to the definition of horsepower.

We made every effort to obtain the most up-to-date information available at the time we went to press, including verification of all phone numbers and addresses. Much of the statistical information, especially in the health chapter, was the most current available; we learned that the process of collecting and analyzing such data can take several years.

The staff at EEI (Editorial Experts, Inc., an editorial and production services firm in Alexandria, VA) researched, wrote, and produced this book. Special thanks go to Patricia Tschirhart-Spangler and Paula Moore, researchers; Denise McLure, word processor; Doreen Jones, editorial proofreader; Connie Moy and her staff of eagle-eyed proofreaders, especially Melissa Barranco and Pat Caudill; Jayne Sutton, production manager; Carla Shaw, Kathryn Hall, and Jean Spencer, desktop designers; and indexer Gloria Peterson.

1
Money

Personal Finance

Perhaps the most important numbers in everyday life are those that affect our pocketbooks: the salary we make, the need to save money (in particular for our children's college costs and our own retirement), and the need to finance our two biggest single expenses—homes and cars.

The chart below shows how the "typical" American family (two-earner married couple, earning $53,354 a year) spent its money in 1994:

How Americans Spend Their Money

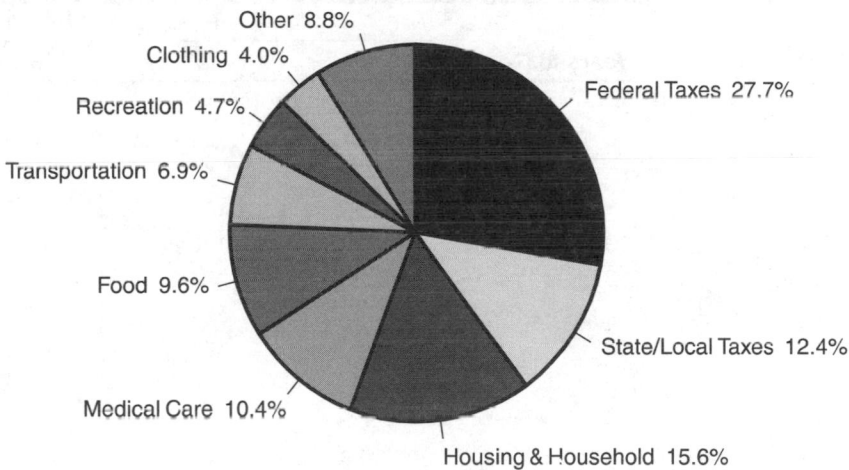

Other 8.8%
Clothing 4.0%
Recreation 4.7%
Transportation 6.9%
Food 9.6%
Medical Care 10.4%
Housing & Household 15.6%
Housing & Household 15.6%
Federal Taxes 27.7%
State/Local Taxes 12.4%

Determining Monthly Savings Needed to Reach Your Goal

If you have a savings goal in mind and want to figure out how much you need to save each month, use this table, which is adjusted for taxes and inflation, and follow these steps:

- Decide how many years you plan to save and the amount of money you need.

- Determine a realistic rate of return—be pessimistic rather than optimistic. Then subtract 2 to 3 percentage points to allow for taxes and inflation.

- Find the point in the body of the table where the "Years to Goal" and "Real Rate of Return" intersect. For example, assume you want to save $15,000 in 4 years for a down payment on a house. You decide that a realistic rate of return is 4 percent; the point where 4 years and 4 percent intersect in the table is 51.9.

- Divide the amount you will need to save by that number (in our example, divide $15,000 by 51.9 = $289.02). The result is the amount you will need to save *each month* to reach your goal.

Years to Goal	Real Rate of Return			
	2%	4%	6%	8%
2	24.5	24.9	25.4	25.9
4	49.9	51.9	54.1	56.4
6	76.5	81.1	86.4	92.1
8	104.2	112.7	122.8	134.1
10	133.0	146.9	163.9	183.4
12	163.0	184.0	210.1	241.2
14	194.2	224.0	262.3	309.0
16	226.8	267.4	321.1	388.7
18	260.7	314.3	387.3	482.2
20	296.1	365.1	462.0	592.0
22	332.8	420.1	546.2	720.8
24	371.2	480.0	641.1	872.0
26	411.1	544.0	748.1	1,049.5
28	452.6	613.7	868.6	1,257.9
30	495.9	689.1	1,004.5	1,502.5

The Cost of Raising a Child

Annual Income	Annual Spending First 3 Years	Total Spending over 18 Years							
		Housing	Food	Trans-portation	Clothing	Health Care	Child Care/ Education	Other*	All Expenses
Less than $32,000	$ 4,960	$30,540	$19,650	$16,530	$ 9,330	$ 6,840	$ 5,790	$ 9,030	$ 97,710
$32,000–$54,100	6,870	43,020	23,700	23,070	10,860	8,460	9,840	13,710	132,660
More than $54,100	10,210	69,780	30,270	27,750	14,370	10,050	16,590	23,970	192,780

Source: U.S. Department of Agriculture, Agricultural Research Service.

Note: Calculations assume two parents and two children in a family; for one child, multiply totals by 1.26. For a family with three children, multiply totals by 0.78 for cost per child.

*Includes personal items, entertainment, books.

Investing Money

How Americans Invest Their Money

Asset Type	Percentage of Households		
	1984	1988	1991
Interest-earning assets at financial institutions	71.8	72.9	73.2
Savings accounts	62.9	61.6	62.4
Money market deposit accounts	15.7	15.2	14.9
Certificates of deposit	19.1	17.7	22.0
Interest-earning checking accounts	24.8	34.3	37.8
Other interest-earning assets	8.5	9.4	9.0
Money market funds	3.8	3.6	4.2
Government securities	1.4	2.2	2.2
Corporate or municipal bonds	2.6	2.8	3.4
Other	2.8	3.3	2.2
Checking accounts	53.9	48.3	46.0
Stocks and mutual fund shares	20.0	21.8	20.7
U.S. Savings Bonds	15.0	17.5	18.1
IRA or Keogh accounts	19.5	24.2	22.9

Source: U.S. Bureau of the Census, *Current Population Reports, P 70–34.*

Value in 1993 of $1 Invested in 1925

This table gives a broad view of different investment approaches over a long period of time. If you had invested $1 in each of the following types of investment at the end of 1925, here is what those dollars would have been worth in 1993:

Small-company stocks	$2,757.15
Large-company stocks	800.08
Long-term government bonds	28.03
Treasury bills	11.73

Investment Results over Time for Various Approaches

Here is the average annual rate of return for the period 1925 through 1994 for four types of investment strategies, as calculated by Ibbotson Associates, plus the average annual rate of inflation for that period:

Investment Strategy	Percent
Diversified portfolio (40% Standard & Poor's 500 stocks, 40% long-term government bonds, 20% U.S. Treasury bills)	7.24
100% stocks	10.19
100% U.S. Treasury bills	3.69
Inflation rate	3.13

The Rule of 72: For Lump-Sum Amounts

If you want to know how long it would take to double your money at various rates of return, try using the rule of 72. If you earn a 1 percent rate of return on $1, you will have to wait 72 years for it to double to $2. To figure how many years it will take for your money to double, divide 72 by the percentage rate you expect to get.

For example, if you expect an average rate of return of 5 percent, you will double your money in 14.4 years:

$$72 \div 5 = 14.4$$

If the expected rate of return is 10 percent, it will take 7.2 years to double; 12 percent, 6 years; 18 percent, 4 years.

The Rule of 144: For Amounts Invested Annually

The rule of 144 lets you determine at what point your money will double when you save a *fixed* amount yearly over a period of time. Divide 144 by the estimated average interest rate to find the number of years it will take to approximately double your investment.

For example, assume you save $2,000 a year at 5 percent.

$$144 \div 5 = 28.8$$

It will take nearly 29 years for your investment of $57,600 ($2,000 deposited yearly) to double to $115,200.

Stock Market Indexes: What They Represent

Index or Average	What It Uses to Track Movements
Dow Jones Industrial Average	30 large NYSE blue-chip stocks
AMEX Market Value Index	800 small- and medium-growth AMEX stocks
NASDAQ Composite Index	All NASDAQ stocks (5,274)
NYSE Composite Index	All NYSE stocks (3,060)
Standard & Poor's Index	500 blue-chip stocks
Wilshire 5000 Equity Index	All NYSE, AMEX, NASDAQ stocks (9,250 total)

Note: NYSE = New York Stock Exchange, AMEX = American Stock Exchange, NASDAQ = National Association of Securities Dealers Automated Quotations (over-the-counter stocks).

The Dow Jones Industrial Average is not a simple average of stock prices. Instead, it takes into account that many stocks have "split" over the years. To keep their stocks affordable, most companies issue split shares. For example, in a 2-for-1 split, 50 shares that had been selling for $100 would become 100 shares selling for $50. To keep the average from going down every time a split happens, a "divisor" or "multiplier" is applied so that the Dow is what it would have been had the stocks never split.

The Dow reached the largely psychological barriers of 100 in 1906, 1,000 in 1972, 2,000 in 1987, 3,000 in 1991, 4,000 in early 1995, and 5,000 in late 1995. Even though the Dow represents the movements of only 30 stocks, it has become a symbolic milestone of stock market growth.

Most individuals buy stocks through mutual funds. There were more than 6,100 mutual funds available to investors in 1995. Fewer than 20 percent of the stocks traded on the New York Stock Exchange are owned by individuals—the vast majority are held by institutional buyers such as mutual funds.

The Presidential Election Year as a Stock Market Predictor

Predicting stock performance is hardly scientific—experts say that market activities cannot be estimated in any rational way. Several prediction schemes have become part of investment folklore. One system is tied to stock market performance in presidential election years: Since 1948, the stock market rose 10 times out of 12, or 83 percent; it rose only 65 percent of the time in nonelection years.

Election Year	President Elected	Percent Change in Standard & Poor's 500
1948	Truman (D)	−0.7
1952	Eisenhower (R)	+11.8
1956	Eisenhower (R)	+2.6
1960	Kennedy (D)	−3.0
1964	Johnson (D)	+13.0
1968	Nixon (R)	+7.7
1972	Nixon (R)	+15.6
1976	Carter (D)	+19.1
1980	Reagan (R)	+25.8
1984	Reagan (R)	+1.4
1988	Bush (R)	+12.4
1992	Clinton (D)	+7.7

Other "predictors" of market action include these:

- Years ending in 7 tend to have an upward trend.
- The first 10 days in January set the tone for the year.
- If General Motors makes a new high or low within 4 months of a previous high or low, the market will also go in that direction.

Darts as a Guide to Picking Stocks

In his book, *A Mathematician Reads the Newspaper*, John Allen Paulos comments on the role of luck in making stock market predictions:

> The sports pages of the January 11, 1994, *Wall Street Journal* reported that the Darts scored a smashing victory over the Pros in the ongoing series. The sport, of course, is stock picking, and the darts are just that—random selections by dart—while the pros are a rotating collection of market experts. The Darts averaged a 42 percent gain in the period July 7, 1993, to December 31, 1993, compared with 8 percent for the Dow Jones and 2.2 percent for the experts.

Fractions-to-Cents Conversion Chart

Stock prices are expressed using fractions—for example, 12³/₄ means $12.75, and 5¹/₈ means $5.12¹/₂. Here's a handy conversion chart:

Fraction	Cents
1/32	3.125
1/16	6.25
1/8	12.5
1/4	25.0
3/8	37.5
1/2	50.0
5/8	62.5
3/4	75.0
7/8	87.5

Fractions-to-Decimals Conversion Chart

Use this table to convert fractions to decimals when entering stock values into your calculator or computer.

Fraction	Decimal	Fraction	Decimal
1/32	0.03125	17/32	0.53125
1/16	0.0625	9/16	0.5625
3/32	0.09375	19/32	0.59375
1/8	0.125	5/8	0.625
5/32	0.15625	21/32	0.65625
3/16	0.1875	11/16	0.6875
7/32	0.21875	23/32	0.71875
1/4	0.25	3/4	0.75
9/32	0.28125	25/32	0.78125
5/16	0.3125	13/16	0.8125
11/32	0.34375	27/32	0.84375
3/8	0.375	7/8	0.875
13/32	0.40625	29/32	0.90625
7/16	0.4375	15/16	0.9375
15/32	0.46875	31/32	0.96875
1/2	0.5		

Types of Bonds and Their Features

There are many different kinds of bonds. All amount to a loan by the investor to the organization issuing the bond. They are purchased at banks for a quoted price, pay a fixed rate of interest, and can be redeemed for their face value on their maturity date. The table on the next page provides a summary of features.

Types of Bonds and Their Features

Type of Bonds	Denominations	Time to Maturity	Tax You Must Pay on Interest	Comments
Series EE Savings Bonds	$50–$5,000	5 years	Federal tax when you cash in; free from state, local	Can be redeemed after 6 months; pay interest from 1st of month; bonds issued after Nov. 1965 earn interest for 30 years. Call (800) US-BONDS for recorded rate information
Series HH Savings Bonds (4% fixed)	$500–$10,000	10 years	Federal tax annually; free from state, local	Bought by trading in Series EE Bonds; pay interest twice annually; bonds issued since 1980 earn interest (4% per year) up to 20 years
Treasury bills (T-bills)	$10,000	3, 6, or 12 months	Federal tax annually; free from state, local	Bought at discount to face value; can be purchased from Federal Reserve banks, branches
Treasury notes	$1,000–$5,000	2, 3, 4, 5, 7, or 10 years	Federal tax annually; free from state, local	Pay interest twice annually; can be purchased through brokers, banks, or directly from government at monthly "auctions." Call (202) 874-4000 for information; call (800) 366-3144 to set up Treasury Direct (direct deposit) account
Treasury bonds	$1,000	10–30 years	Federal tax annually; free from state, local	Same as for Treasury notes (above)
Zero coupon government bonds	$1,000	10 years	Federal tax annually as if tax had been paid; free from state, local	Bought at discount, returns face value
Municipal bonds	$5,000–$10,000	varies	Free from federal, state, local (if in-state); otherwise federal only	Pay interest twice a year
Agency bonds (Freddie Mac, Fannie Mae, etc.)	$1,000–$25,000 and up	20–30 years	Some exempt from state, local	Higher risk, higher interest than Treasury bonds
Corporate bonds	$1,000 (typically)	5–20 years	None	Fixed interest rates

Federal Reserve System Regional Bank Phone Numbers

Regional Bank	Recording	Main Number
Atlanta	(404) 521-8657	(404) 521-8500
Boston	(617) 973-3805	(617) 973-3000
Chicago	(312) 786-1110	(312) 322-5322
Cleveland	(216) 579-2001	(216) 579-2000
Dallas	(214) 651-6177	(214) 922-6000
Kansas City	(816) 881-2364	(816) 881-2000
Minneapolis	—	(612) 340-2345
New York	(212) 720-6693	(212) 720-5000
Philadelphia	(215) 574-6188	(215) 574-6000
Richmond	(804) 697-8355	(804) 697-8000
St. Louis	(314) 444-8602	(314) 444-8444
San Francisco	(415) 974-3491	(415) 974-2000

The main number for the Federal Reserve System in Washington, DC, is (202) 452-3000.

Tax-Free vs. Taxable Investments

The table on the following page will help you decide whether a federal tax-free investment (with its usually lower interest rate) is more beneficial than a taxable investment. The determining factor is your federal tax bracket; see the table on page 107 later in this chapter. Another way to decide is to apply the following formula (which works for federal tax purposes only; your investment may also be free from state and local taxes):

Step	Example
1. Determine your tax bracket (express as a decimal)	0.31
2. Write down the tax-free yield	8%
3. Subtract your tax bracket from 1.00	1.00 − 0.31 = 0.69
4. Divide the tax-free yield by the number in step 3 to find the tax-equivalent yield	8.0% ÷ 0.69 = 11.59%

Tax-Free versus Taxable Bonds: Actual Yields Based on Tax Bracket

	Tax-Free Yield (%) of—			
	4.0	5.0	6.0	7.0
Tax Bracket (%)	Equals a Taxable Yield (%) of —			
15	4.71	5.88	7.06	8.24
28	5.56	6.94	8.33	9.72
31	5.80	7.25	8.70	10.14
36	6.25	7.81	9.38	10.94
39.6	6.62	8.28	9.93	11.59

Recommended Investment Allocations by Age Group

Age	Allocation of Investment
20s–30s	70 percent in aggressive growth or blue-chip stock funds; 30 percent in bonds
40s	50 percent in stocks; 50 percent in bonds, GICs (guaranteed investment contracts), money funds
50s	30 percent in stocks; 70 percent in bonds, GICs, money funds
60s	20 percent in stocks; 80 percent in fixed-income instruments

Liquid Investments: Savings Accounts, Money Markets, and CDs

Savings accounts are insured for up to $100,000 by the Federal Deposit Insurance Corporation (FDIC). If you have accounts in several banks (but not branches of the same bank), each account will be insured to the FDIC limit. If you and your spouse each have individual accounts and also have a joint account at one bank, your total protection would be up to $300,000, because you have three different accounts.

The FDIC's consumer hotline for questions about deposit insurance is (800) 934-3342.

Money market accounts are available through banks and earn the going rate of interest, which of course varies. A minimum balance is usually required, and a limited number of checks may be written each month. As with savings accounts, interest earned is taxable.

Certificates of deposit (CDs) may be purchased through banks, credit unions, and brokers for a minimum amount for a specific period of time, which can be anywhere from 30 days to 5 years. You earn a stated rate of interest and get back the amount you paid for the CD at the end of the investment period, or maturity date. Penalties are often assessed for early withdrawals. CDs are safe investments because you always get back the amount you invest plus interest, and they are insured by the FDIC for up to $100,000. Interest earned is taxable.

Rule of Thumb: Saving for a Rainy Day You should have, in fairly liquid investments such as savings accounts, money market accounts, and CDs, anywhere from 3 to 6 months' income on hand to meet financial emergencies such as job layoffs and unexpected expenses.

Estimating Interest Earned

Most interest is compounded, meaning that the interest you earn (daily, weekly, monthly, quarterly, annually) is added to the amount of your initial investment, and subsequent interest earnings are based on that growing total amount.

Simple interest, on the other hand, merely pays a straight percentage; in other words, if you bought a $1,000 CD for 1 year at 5 percent simple interest, you would earn $50.

The following table lets you determine your earnings on a *lump-sum investment* that is compounded annually; earnings that are compounded daily, monthly, or quarterly will be greater than the amounts shown. Find the number in the body of the table that corresponds with the number of years of your investment and the interest rate it is earning. Then multiply the amount of your investment by that figure to come up with the total amount of your investment for that period.

For example, if you put $1,000 in a savings account that will earn 5 percent interest for 5 years, locate 1.28—where the year 5 row intersects the column of 5 percent—and multiply it by $1,000. Your initial investment after 5 years would be worth $1,280.

Years	Rate of Interest per Year (%)													
	5	6	7	8	9	10	11	12	13	14	15	16	17	18
1	$1.05	$1.06	$1.07	$1.08	$1.09	$1.10	$1.11	$1.12	$ 1.13	$ 1.14	$ 1.15	$ 1.16	$ 1.17	$ 1.18
2	1.10	1.12	1.14	1.17	1.19	1.21	1.23	1.25	1.28	1.30	1.32	1.34	1.37	1.39
3	1.16	1.19	1.22	1.26	1.30	1.33	1.37	1.40	1.44	1.48	1.52	1.56	1.60	1.64
4	1.22	1.26	1.31	1.36	1.41	1.46	1.52	1.57	1.63	1.69	1.75	1.81	1.87	1.94
5	1.28	1.34	1.40	1.47	1.54	1.61	1.68	1.76	1.84	1.93	2.01	2.10	2.19	2.29
6	1.34	1.42	1.50	1.59	1.68	1.77	1.86	1.97	2.08	2.20	2.31	2.44	2.56	2.70
7	1.41	1.50	1.60	1.71	1.83	1.95	2.07	2.21	2.35	2.51	2.66	2.83	3.00	3.19
8	1.48	1.59	1.72	1.86	1.99	2.14	2.30	2.47	2.65	2.86	3.05	3.28	3.51	3.76
9	1.55	1.69	1.84	2.00	2.17	2.36	2.55	2.77	3.00	3.26	3.51	3.80	4.10	4.44
10	1.63	1.79	1.97	2.16	2.37	2.59	2.83	3.10	3.39	3.71	4.04	4.41	4.80	5.24
11	1.71	1.90	2.10	2.33	2.58	2.85	3.14	3.47	3.83	4.24	4.65	5.12	5.62	6.18
12	1.80	2.01	2.25	2.25	2.81	3.14	3.49	3.89	4.33	4.83	5.35	5.94	6.57	7.29
13	1.89	2.13	2.41	2.72	3.07	3.45	3.87	4.35	4.89	5.50	6.15	6.89	7.69	8.60
14	1.98	2.26	2.58	2.94	3.34	3.79	4.29	4.87	5.53	6.27	7.07	7.99	9.00	10.15
15	2.08	2.40	2.76	3.17	3.64	4.17	4.77	5.46	6.25	7.15	8.13	9.26	10.52	11.98
20	2.65	3.21	3.86	4.66	5.60	6.72	8.03	9.63	11.51	13.77	16.35	19.45	23.07	27.30

If you save *a certain amount every month* and want to know what your investment, compounded annually, will be worth in a certain number of years, use the next table. (Results will be slightly higher if compounded more frequently.) Multiply the amount you are putting away each month by 12 to get a yearly total, then find the factor that corresponds with the number of years of your investment and the estimated annual interest rate. Multiply that factor by the number of dollars you invest per year to get a total value at the end of the period.

Estimated Future Value of $1 (Annual Compounding)

Use this table to estimate how much a certain amount invested in 1 year will be worth in future years. For example, $100 invested at 6 percent interest will be worth $1,318 in 10 years (13.18 × $100).

Years	Rate of Interest per Year (%)					
	4	6	8	10	12	14
2	$ 2.04	$ 2.06	$ 2.08	$ 2.10	$ 2.12	$ 2.14
4	4.25	4.37	4.50	4.64	4.77	4.92
6	6.63	6.97	7.33	7.71	8.11	8.53
8	9.21	9.89	10.63	11.43	12.29	13.23
10	12.01	13.18	14.48	15.93	17.54	19.33
12	15.03	16.86	18.97	21.38	24.13	27.27
14	18.29	21.01	24.21	27.97	32.39	37.58
16	21.82	25.67	30.32	35.94	42.75	50.98
18	25.65	30.90	37.45	45.59	55.74	68.39
20	29.78	36.78	45.76	57.27	72.52	91.02

If Only... If you had $1 million in an account that paid interest at 6 percent, you would earn $164.38 *a day*. But how do you get $1 million in the first place? Simply save $1,000—one thousand times.

Borrowing Money

Installment loans are considered short-term loans, with the interest spread out evenly over the life of the loan. Personal and auto loans are usually set up as installment loans.

The next table shows what your monthly payment would be at various interest rates and loan terms for a $1,000 loan; if your loan amount is $10,000, for example, multiply the payment shown by 10.

Monthly Payments on a $1,000 Installment Loan

Interest Rate (%)	12 Months	24 Months	36 Months	48 Months	60 Months
5	$85.63	$43.92	$30.03	$23.08	$18.95
6	86.09	44.37	30.48	23.54	19.41
7	86.55	44.82	30.93	24.00	19.87
8	87.01	45.27	31.38	24.46	20.33
9	87.47	45.72	31.83	24.92	20.79
10	87.92	46.15	32.27	25.37	21.25
11	88.39	46.41	32.74	25.85	21.75
12	88.85	45.08	33.22	26.34	22.25
13	89.32	47.55	33.70	26.83	22.76
14	89.79	48.02	34.18	27.33	23.27
15	90.26	48.49	34.67	27.84	23.79
16	90.74	48.97	35.16	28.35	24.32
17	91.21	49.45	35.66	28.86	24.86
18	91.68	49.93	36.16	29.38	25.40

Loan payments for home mortgages and other long-term loans are based on principal and interest combined. In the beginning years of a mortgage loan, most of the payment is applied toward the interest charges. The longer the period of the loan, the more interest you will pay. The next table can be used to get a rough estimate of a mortgage payment.

Estimating Your Mortgage Payment Amount

The following table provides a simple way to estimate the monthly principal and interest payment on a loan. It gives the factor that lenders use in computing payments.

1. Find your interest rate (or a close approximation) and go to the loan term factor for your type of loan—15 or 30 years.
2. Divide your loan amount by 1,000 and multiply the result by the factor from the chart to get your approximate monthly payment (principal and interest).

	Loan Term Factor	
Interest Rate (%)	15 Years	30 Years
5.0	7.91	5.37
5.5	8.17	5.68
6.0	8.44	6.00
6.5	8.71	6.32
7.0	8.99	6.66
7.5	9.27	7.00
8.0	9.56	7.34
8.5	9.85	7.69
9.0	10.15	8.05
9.5	10.45	8.41
10.0	10.75	8.78
10.5	11.06	9.15
11.0	11.37	9.53
11.5	11.69	9.91
12.0	12.01	10.29
12.5	12.33	10.68
13.0	12.66	11.07
13.5	12.99	11.46
14.0	13.32	11.85

Mortgage Payment Tables

The tables that follow will let you determine your monthly mortgage payment (not including taxes and insurance) for interest rates of 7 to 12 percent and for loan periods of 5 to 40 years. Your loan amount—the number in the first column—is determined by subtracting your down payment from the sales price. Select the table with the interest rate you will be paying and use the column that indicates the number of years of your loan.

Monthly Mortgage Payment
7%

Loan Amount ($)	5 Years	10 Years	15 Years	20 Years	25 Years	30 Years	40 Years
10,000	$ 198.01	$ 116.11	$ 89.88	$ 77.53	$ 70.68	$ 66.53	$ 62.14
15,000	297.02	174.16	134.82	116.29	106.02	99.80	93.21
20,000	396.02	232.22	179.77	155.06	141.36	133.06	124.29
25,000	495.03	290.27	224.71	193.82	176.69	166.33	155.36
30,000	594.04	348.33	269.65	232.59	212.03	199.59	186.43
35,000	693.04	406.38	314.59	271.35	247.37	232.86	217.50
40,000	792.05	464.43	359.53	310.12	282.71	266.12	248.57
45,000	891.05	522.49	404.47	348.88	318.05	299.39	279.64
50,000	990.06	580.54	449.41	387.65	353.39	332.65	310.72
55,000	1089.07	638.60	494.36	426.41	388.73	365.92	341.79
60,000	1188.07	696.65	539.30	465.18	424.07	399.18	372.86
65,000	1287.08	754.71	584.24	503.94	459.41	432.45	403.93
70,000	1386.08	812.76	629.18	542.71	494.75	465.71	435.00
75,000	1485.09	870.81	674.12	581.47	530.08	498.98	466.07
80,000	1584.10	928.87	719.06	620.24	565.42	532.24	497.15
85,000	1683.10	986.92	764.00	659.00	600.76	565.51	528.22
90,000	1782.11	1044.98	808.95	697.77	636.10	598.77	559.29
95,000	1881.11	1103.03	853.89	736.53	671.44	632.04	590.36
100,000	1980.12	1161.08	898.83	775.30	706.78	665.30	621.43
105,000	2079.13	1219.14	943.77	814.06	742.12	698.57	652.50
110,000	2178.13	1277.19	988.71	852.83	777.46	731.83	683.57
120,000	2376.14	1393.30	1078.59	930.36	848.14	798.36	745.72
130,000	2574.16	1509.41	1168.48	1007.89	918.81	864.89	807.86
140,000	2772.17	1625.52	1258.36	1085.42	989.49	931.42	870.00
150,000	2970.18	1741.63	1348.24	1162.95	1060.17	997.95	932.15
175,000	3465.21	2031.90	1572.95	1356.77	1236.86	1164.28	1087.50
200,000	3960.24	2322.17	1797.66	1550.60	1413.56	1330.60	1242.86
225,000	4455.27	2612.44	2022.36	1744.42	1590.25	1496.93	1398.22
250,000	4950.30	2902.71	2247.07	1938.25	1766.95	1663.26	1553.58

Monthly Mortgage Payment (continued)
7-1/2%

Loan Amount ($)	5 Years	10 Years	15 Years	20 Years	25 Years	30 Years	40 Years
10,000	$ 200.38	$ 118.70	$ 92.70	$ 80.56	$ 73.90	$ 69.92	$ 65.81
15,000	300.57	178.05	139.05	120.84	110.85	104.88	98.71
20,000	400.76	237.40	185.40	161.12	147.80	139.84	131.61
25,000	500.95	296.75	231.75	201.40	184.75	174.80	164.52
30,000	601.14	356.11	278.10	241.68	221.70	209.76	197.42
35,000	701.33	415.46	324.45	281.96	258.65	244.73	230.32
40,000	801.52	474.81	370.80	322.24	295.60	279.69	263.23
45,000	901.71	534.16	417.16	362.52	332.55	314.65	296.13
50,000	1001.90	593.51	463.51	402.80	369.50	349.61	329.04
55,000	1102.09	652.86	509.86	443.08	406.45	384.57	361.94
60,000	1202.28	712.21	556.21	483.36	443.39	419.53	394.84
65,000	1302.47	771.56	602.56	523.64	480.34	454.49	427.75
70,000	1402.66	830.91	648.91	563.92	517.29	489.45	460.65
75,000	1502.85	890.26	695.26	604.19	554.24	524.41	493.55
80,000	1603.04	949.61	741.61	644.47	591.19	559.37	526.46
85,000	1703.23	1008.97	787.96	684.75	628.14	594.33	559.36
90,000	1803.42	1068.32	834.31	725.03	665.09	629.29	592.26
95,000	1903.61	1127.67	880.66	765.31	702.04	664.25	625.17
100,000	2003.79	1187.02	927.01	805.59	738.99	699.21	658.07
105,000	2103.98	1246.37	973.36	845.87	775.94	734.18	690.97
110,000	2204.17	1305.72	1019.71	886.15	812.89	769.14	723.88
120,000	2404.55	1424.42	1112.41	966.71	886.79	839.06	789.68
130,000	2604.93	1543.12	1205.12	1047.27	960.69	908.98	855.49
140,000	2805.31	1661.82	1297.82	1127.83	1034.59	978.90	921.30
150,000	3005.69	1780.53	1390.52	1208.39	1108.49	1048.82	987.11
175,000	3506.64	2077.28	1622.27	1409.79	1293.23	1223.63	1151.62
200,000	4007.59	2374.04	1854.02	1611.19	1477.98	1398.43	1316.14
225,000	4508.54	2670.79	2085.78	1812.58	1662.73	1573.23	1480.66
250,000	5009.49	2967.54	2317.53	2013.98	1847.48	1748.04	1645.18

Monthly Mortgage Payment (continued)
8%

Loan Amount ($)	5 Years	10 Years	15 Years	20 Years	25 Years	30 Years	40 Years
10,000	$ 202.76	$ 121.33	$ 95.57	$ 83.64	$ 77.18	$ 73.38	$ 69.53
15,000	304.15	181.99	143.35	125.47	115.77	110.06	104.30
20,000	405.53	242.66	191.13	167.29	154.36	146.75	139.06
25,000	506.91	303.32	238.91	209.11	192.95	183.44	173.83
30,000	608.29	363.98	286.70	250.93	231.54	220.13	208.59
35,000	709.67	424.65	334.48	292.75	270.14	256.82	243.36
40,000	811.06	485.31	382.26	334.58	308.73	293.51	278.12
45,000	912.44	545.97	430.04	376.40	347.32	330.19	312.89
50,000	1013.82	606.64	477.83	418.22	385.91	366.88	347.66
55,000	1115.20	667.30	525.61	460.04	424.50	403.57	382.42
60,000	1216.58	727.97	573.39	501.86	463.09	440.26	417.19
65,000	1317.97	788.63	621.17	543.69	501.68	476.95	451.95
70,000	1419.35	849.29	668.96	585.51	540.27	513.64	486.72
75,000	1520.73	909.96	716.74	627.33	578.86	550.32	521.48
80,000	1622.11	970.62	764.52	669.15	617.45	587.01	556.25
85,000	1723.49	1031.28	812.30	710.97	656.04	623.70	591.01
90,000	1824.88	1091.95	860.09	752.80	694.63	660.39	625.78
95,000	1926.26	1152.61	907.87	794.62	733.23	697.08	660.55
100,000	2027.64	1213.28	955.65	836.44	771.82	733.76	695.31
105,000	2129.02	1273.94	1003.43	878.26	810.41	770.45	730.08
110,000	2230.40	1334.60	1051.22	920.08	849.00	807.14	764.84
120,000	2433.17	1455.93	1146.78	1003.73	926.18	880.52	834.37
130,000	2635.93	1577.26	1242.35	1087.37	1003.36	953.89	903.91
140,000	2838.70	1698.59	1337.91	1171.02	1080.54	1027.27	973.44
150,000	3041.46	1819.91	1433.48	1254.66	1157.72	1100.65	1042.97
175,000	3548.37	2123.23	1672.39	1463.77	1350.68	1284.09	1216.80
200,000	4055.28	2426.55	1911.30	1672.88	1543.63	1467.53	1390.62
225,000	4562.19	2729.87	2150.22	1881.99	1736.59	1650.97	1564.45
250,000	5069.10	3033.19	2389.13	2091.10	1929.54	1834.41	1738.28

Monthly Mortgage Payment (continued)
8-1/2%

Loan Amount ($)	5 Years	10 Years	15 Years	20 Years	25 Years	30 Years	40 Years
10,000	$ 205.17	$ 123.99	$ 98.47	$ 86.78	$ 80.52	$ 76.89	$ 73.31
15,000	307.75	185.98	147.71	130.17	120.78	115.34	109.96
20,000	410.33	247.97	196.95	173.56	161.05	153.78	146.62
25,000	512.91	309.96	246.18	216.96	201.31	192.23	183.27
30,000	615.50	371.96	295.42	260.35	241.57	230.67	219.93
35,000	718.08	433.95	344.66	303.74	281.83	269.12	256.58
40,000	820.66	495.94	393.90	347.13	322.09	307.57	293.24
45,000	923.24	557.94	443.13	390.52	362.35	346.01	329.89
50,000	1025.83	619.93	492.37	433.91	402.61	384.46	366.55
55,000	1128.41	681.92	541.61	477.30	442.87	422.90	403.20
60,000	1230.99	743.91	590.84	520.69	483.14	461.35	439.86
65,000	1333.57	805.91	640.08	564.09	523.40	499.79	476.51
70,000	1436.16	867.90	689.32	607.48	563.66	538.24	513.17
75,000	1538.74	929.89	738.55	650.87	603.92	576.69	549.82
80,000	1641.32	991.89	787.79	694.26	644.18	615.13	586.48
85,000	1743.91	1053.88	837.03	737.65	684.44	653.58	623.13
90,000	1846.49	1115.87	886.27	781.04	724.70	692.02	659.78
95,000	1949.07	1177.86	935.50	824.43	764.97	730.47	696.44
100,000	2051.65	1239.86	984.74	867.82	805.23	768.91	733.09
105,000	2154.24	1301.85	1033.98	911.21	845.49	807.36	769.75
110,000	2256.82	1363.84	1083.21	954.61	885.75	845.80	806.40
120,000	2461.98	1487.83	1181.69	1041.39	966.27	922.70	879.71
130,000	2667.15	1611.81	1280.16	1128.17	1046.80	999.59	953.02
140,000	2872.31	1735.80	1378.64	1214.95	1127.32	1076.48	1026.33
150,000	3077.48	1859.79	1477.11	1301.73	1207.84	1153.37	1099.64
175,000	3590.39	2169.75	1723.29	1518.69	1409.15	1345.60	1282.91
200,000	4103.31	2479.71	1969.48	1735.65	1610.45	1537.83	1466.19
225,000	4616.22	2789.68	2215.66	1952.60	1811.76	1730.06	1649.46
250,000	5129.13	3099.64	2461.85	2169.56	2013.07	1922.28	1832.74

Monthly Mortgage Payment (continued)
9%

Loan Amount ($)	5 Years	10 Years	15 Years	20 Years	25 Years	30 Years	40 Years
10,000	$ 207.58	$ 126.68	$ 101.43	$ 89.97	$ 83.92	$ 80.46	$ 77.14
15,000	311.38	190.01	152.14	134.96	125.88	120.69	115.70
20,000	415.17	253.35	202.85	179.95	167.84	160.92	154.27
25,000	518.96	316.69	253.57	224.93	209.80	201.16	192.84
30,000	622.75	380.03	304.28	269.92	251.76	241.39	231.41
35,000	726.54	443.37	354.99	314.90	293.72	281.62	269.98
40,000	830.33	506.70	405.71	359.89	335.68	321.85	308.54
45,000	934.13	570.04	456.42	404.88	377.64	362.08	347.11
50,000	1037.92	633.38	507.13	449.86	419.60	402.31	385.68
55,000	1141.71	696.72	557.85	494.85	461.56	442.54	424.25
60,000	1245.50	760.05	608.56	539.84	503.52	482.77	462.82
65,000	1349.29	823.39	659.27	584.82	545.48	523.00	501.38
70,000	1453.08	886.73	709.99	629.81	587.44	563.24	539.95
75,000	1556.88	950.07	760.70	674.79	629.40	603.47	578.52
80,000	1660.67	1013.41	811.41	719.78	671.36	643.70	617.09
85,000	1764.46	1076.74	862.13	764.77	713.32	683.93	655.66
90,000	1868.25	1140.08	912.84	809.75	755.28	724.16	694.23
95,000	1972.04	1203.42	963.55	854.74	797.24	764.39	732.79
100,000	2075.84	1266.76	1014.27	899.73	839.20	804.62	771.36
105,000	2179.63	1330.10	1064.98	944.71	881.16	844.85	809.93
110,000	2283.42	1393.43	1115.69	989.70	923.12	885.08	848.50
120,000	2491.00	1520.11	1217.12	1079.67	1007.04	965.55	925.63
130,000	2698.59	1646.79	1318.55	1169.64	1090.96	1046.01	1002.77
140,000	2906.17	1773.46	1419.97	1259.62	1174.87	1126.47	1079.91
150,000	3113.75	1900.14	1521.40	1349.59	1258.79	1206.93	1157.04
175,000	3632.71	2216.83	1774.97	1574.52	1468.59	1408.09	1349.88
200,000	4151.67	2533.52	2028.53	1799.45	1678.39	1609.25	1542.72
225,000	4670.63	2850.20	2282.10	2024.38	1888.19	1810.40	1735.56
250,000	5189.59	3166.89	2535.67	2249.31	2097.99	2011.56	1928.40

Monthly Mortgage Payment (continued)
9-1/2%

Loan Amount ($)	5 Years	10 Years	15 Years	20 Years	25 Years	30 Years	40 Years
10,000	$ 210.02	$ 129.40	$ 104.42	$ 93.21	$ 87.37	$ 84.09	$ 81.01
15,000	315.03	194.10	156.63	139.82	131.05	126.13	121.51
20,000	420.04	258.80	208.84	186.43	174.74	168.17	162.01
25,000	525.05	323.49	261.06	233.03	218.42	210.21	202.52
30,000	630.06	388.19	313.27	279.64	262.11	252.26	243.02
35,000	735.07	452.89	365.48	326.25	305.79	294.30	283.52
40,000	840.07	517.59	417.69	372.85	349.48	336.34	324.02
45,000	945.08	582.29	469.90	419.46	393.16	378.38	364.53
50,000	1050.09	646.99	522.11	466.07	436.85	420.43	405.03
55,000	1155.10	711.69	574.32	512.67	480.53	462.47	445.53
60,000	1260.11	776.39	626.53	559.28	524.22	504.51	486.04
65,000	1365.12	841.08	678.75	605.89	567.90	546.56	526.54
70,000	1470.13	905.78	760.96	652.49	611.59	588.60	567.04
75,000	1575.14	970.48	783.17	699.10	655.27	630.64	607.55
80,000	1680.15	1035.18	835.38	745.70	698.96	672.68	648.05
85,000	1785.16	1099.88	887.59	792.31	742.64	714.73	688.55
90,000	1890.17	1164.58	939.80	838.92	786.33	756.77	729.06
95,000	1995.18	1229.28	992.01	885.52	830.01	798.81	769.56
100,000	2100.19	1293.98	1044.22	932.13	873.70	840.85	810.06
105,000	2205.20	1358.67	1096.44	978.74	917.38	882.90	850.56
110,000	2310.20	1423.37	1148.65	1025.34	961.07	924.94	891.07
120,000	2520.22	1552.77	1253.07	1118.56	1048.44	1009.03	972.07
130,000	2730.24	1682.17	1357.49	1211.77	1135.81	1093.11	1053.08
140,000	2940.26	1811.57	1461.91	1304.98	1223.18	1177.20	1134.09
150,000	3150.28	1940.96	1566.34	1398.20	1310.54	1261.28	1215.09
175,000	3675.33	2264.46	1827.39	1631.23	1528.97	1471.49	1417.61
200,000	4200.37	2587.95	2088.45	1864.26	1747.39	1681.71	1620.12
225,000	4725.42	2911.45	2349.51	2097.30	1965.82	1891.92	1822.64
250,000	5250.47	3234.94	2610.56	2330.33	2184.24	2102.14	2025.15

Monthly Mortgage Payment (continued)
10%

Loan Amount ($)	5 Years	10 Years	15 Years	20 Years	25 Years	30 Years	40 Years
10,000	$ 212.47	$ 132.15	$ 107.46	$ 96.50	$ 90.87	$ 87.76	$ 84.91
15,000	318.71	198.23	161.19	144.75	136.31	131.64	127.37
20,000	424.94	264.30	214.92	193.00	181.74	175.51	169.83
25,000	531.18	330.38	268.65	241.26	227.18	219.39	212.29
30,000	637.41	396.45	322.38	289.51	272.61	263.27	254.74
35,000	743.65	462.53	376.11	337.76	318.05	307.15	297.20
40,000	849.88	528.60	429.84	386.01	363.48	351.03	339.66
45,000	956.12	594.68	483.57	434.26	408.92	394.91	382.12
50,000	1062.35	660.75	537.30	482.51	454.35	438.79	424.57
55,000	1168.59	726.83	591.03	530.76	499.79	482.66	467.03
60,000	1274.82	792.90	644.76	579.01	545.22	526.54	509.49
65,000	1381.06	858.98	698.49	627.26	590.66	570.42	551.94
70,000	1487.29	925.06	752.22	675.52	636.06	614.30	594.40
75,000	1593.53	991.13	805.95	723.77	681.53	658.18	636.86
80,000	1699.76	991.13	859.68	772.02	726.96	702.06	679.32
85,000	1806.00	1123.28	913.41	820.27	772.40	745.94	721.77
90,000	1912.23	1189.36	967.14	868.52	817.83	789.81	764.23
95,000	2018.47	1255.43	1020.87	916.77	863.27	833.69	806.69
100,000	2124.70	1321.51	1074.61	965.02	908.70	877.57	849.15
105,000	2230.94	1387.58	1128.34	1013.27	954.14	921.45	891.60
110,000	2337.17	1453.66	1182.07	1061.52	999.57	965.33	934.06
120,000	2549.65	1585.81	1289.53	1158.03	1090.44	1053.09	1018.98
130,000	2762.12	1717.96	1396.99	1254.53	1181.31	1140.84	1103.89
140,000	2974.59	1850.11	1504.45	1351.03	1272.18	1228.60	1188.80
150,000	3187.06	1982.26	1611.91	1447.53	1363.05	1316.36	1273.72
175,000	3718.23	2312.64	1880.56	1688.79	1590.23	1535.75	1486.01
200,000	4249.41	2643.01	2149.21	1930.04	1817.40	1755.14	1698.29
225,000	4780.59	2973.39	2417.86	2171.30	2044.58	1974.54	1910.58
250,000	5311.76	3303.77	2686.51	2412.55	2271.75	2193.93	2122.86

Monthly Mortgage Payment (continued)
10-1/2%

Loan Amount ($)	5 Years	10 Years	15 Years	20 Years	25 Years	30 Years	40 Years
10,000	$ 214.94	$ 134.93	$ 110.54	$ 99.84	$ 94.42	$ 91.47	$ 88.86
15,000	322.41	202.40	165.81	149.76	141.63	137.21	133.29
20,000	429.88	269.87	221.08	199.68	188.84	182.95	177.71
25,000	537.35	337.34	276.35	249.59	236.05	228.68	222.14
30,000	644.82	404.80	331.62	299.51	283.25	274.42	266.57
35,000	752.29	472.27	386.89	349.43	330.46	320.16	311.00
40,000	859.76	539.74	442.16	399.35	377.67	365.90	355.43
45,000	967.23	607.21	497.43	449.27	424.88	411.63	399.86
50,000	1074.70	674.67	552.70	499.19	472.09	457.37	444.29
55,000	1182.16	742.14	607.97	549.11	519.30	503.11	488.71
60,000	1289.63	809.61	663.24	599.03	566.51	548.84	533.14
65,000	1397.10	877.08	718.51	648.95	613.72	594.58	577.57
70,000	1504.57	944.54	773.78	698.87	660.93	640.32	622.00
75,000	1612.04	1012.01	829.05	748.78	708.14	686.05	666.43
80,000	1719.51	1079.48	884.32	798.70	755.35	731.79	710.86
85,000	1826.98	1146.95	939.59	848.62	802.55	777.53	755.28
90,000	1934.45	1214.41	994.86	898.54	849.76	823.27	799.71
95,000	2041.92	1281.88	1050.13	948.46	896.97	869.00	844.14
100,000	2149.39	1349.35	1105.40	998.38	944.18	914.74	888.57
105,000	2256.86	1416.82	1160.67	1048.30	991.39	960.48	933.00
110,000	2364.33	1484.28	1215.94	1098.22	1038.60	1006.21	977.43
120,000	2579.27	1619.22	1326.48	1198.06	1133.02	1097.69	1066.28
130,000	2794.21	1754.15	1437.02	1297.89	1227.44	1189.16	1155.14
140,000	3009.15	1889.09	1547.56	1397.73	1321.85	1280.64	1244.00
150,000	3224.09	2024.02	1658.10	1497.57	1416.27	1372.11	1332.86
175,000	3761.43	2361.36	1934.45	1747.16	1652.32	1600.79	1555.00
200,000	4298.78	2698.70	2210.80	1996.76	1888.36	1829.48	1777.14
225,000	4836.13	3036.04	2487.15	2246.35	2124.41	2058.16	1999.28
250,000	5373.48	3373.37	2763.50	2495.95	2360.45	2286.85	2221.43

Monthly Mortgage Payment (continued)
11%

Loan Amount ($)	5 Years	10 Years	15 Years	20 Years	25 Years	30 Years	40 Years
10,000	$ 217.42	$ 137.75	$ 113.66	$ 103.22	$ 98.01	$ 95.23	$ 92.83
15,000	326.14	206.63	170.49	154.83	147.02	142.85	139.24
20,000	434.85	275.50	227.32	206.44	196.02	190.46	185.66
25,000	543.56	344.38	284.15	258.05	245.03	238.05	232.07
30,000	652.27	413.25	340.98	309.66	294.03	285.70	278.49
35,000	760.98	482.13	397.81	361.27	343.04	333.31	324.90
40,000	869.70	551.00	454.64	412.88	392.05	380.93	371.32
45,000	978.41	619.88	511.47	464.48	441.05	428.55	417.73
50,000	1087.12	688.75	568.30	516.09	490.06	476.16	464.15
55,000	1195.83	757.63	625.13	567.70	539.06	523.78	510.56
60,000	1304.55	826.50	681.96	619.31	588.07	571.39	556.98
65,000	1413.26	895.38	738.79	670.92	637.07	619.01	603.39
70,000	1521.97	964.25	795.62	722.53	686.08	666.63	649.81
75,000	1630.68	1033.13	852.45	774.14	735.08	714.24	696.22
80,000	1739.39	1102.00	909.28	825.75	784.09	761.86	742.64
85,000	1848.11	1170.88	966.11	877.36	833.10	809.47	789.05
90,000	1956.82	1239.75	1022.94	928.97	882.10	857.09	835.46
95,000	2065.53	1308.63	1079.77	980.58	931.11	904.71	881.88
100,000	2174.24	1377.50	1136.60	1032.19	980.11	952.32	928.29
105,000	2282.95	1446.38	1193.43	1083.80	1029.12	999.94	974.71
110,000	2391.67	1515.25	1250.26	1135.41	1078.12	1047.56	1021.12
120,000	2609.09	1653.00	1363.92	1238.63	1176.14	1142.79	1113.95
130,000	2826.51	1790.75	1477.58	1341.84	1274.15	1238.02	1206.78
140,000	3043.94	1928.50	1591.24	1445.06	1372.16	1333.25	1299.61
150,000	3261.36	2066.25	1704.90	1548.28	1470.17	1428.49	1392.44
175,000	3804.92	2410.63	1989.04	1806.33	1715.20	1666.57	1624.52
200,000	4348.48	2755.00	2273.19	2064.38	1960.23	1904.65	1856.59
225,000	4892.05	3099.38	2557.34	2322.42	2205.25	2142.73	2088.66
250,000	5435.61	3443.75	2841.49	2580.47	2450.28	2380.81	2320.74

Monthly Mortgage Payment (continued)
11-1/2%

Loan Amount ($)	5 Years	10 Years	15 Years	20 Years	25 Years	30 Years	40 Years
10,000	$ 219.93	$ 140.60	$ 116.82	$ 106.64	$ 101.65	$ 99.03	$ 96.83
15,000	329.89	210.89	175.23	159.96	152.47	148.54	145.24
20,000	439.85	281.19	233.64	213.29	203.29	198.06	193.66
25,000	549.82	351.49	292.05	266.61	254.12	247.57	242.07
30,000	659.78	421.79	350.46	319.93	304.94	297.09	290.48
35,000	769.74	492.08	408.87	373.25	355.76	346.60	338.90
40,000	879.70	562.38	467.28	426.57	406.59	396.12	387.31
45,000	989.67	632.68	525.69	479.89	457.41	445.63	435.73
50,000	1099.63	702.98	584.09	533.21	508.23	495.15	484.14
55,000	1209.59	773.27	642.50	586.54	559.06	544.66	532.56
60,000	1319.56	843.57	700.91	639.86	609.88	594.17	580.97
65,000	1429.52	913.87	759.32	693.18	660.70	643.69	629.38
70,000	1539.48	984.17	817.73	746.50	711.53	693.20	677.80
75,000	1649.45	1054.47	876.14	799.82	762.35	742.72	726.21
80,000	1759.41	1124.76	934.55	853.14	813.18	792.23	774.63
85,000	1869.37	1195.06	992.96	906.47	864.00	841.75	823.04
90,000	1979.33	1265.36	1051.37	959.79	914.82	891.26	871.45
95,000	2089.30	1335.66	1109.78	1013.11	965.65	940.78	919.87
100,000	2199.26	1405.95	1168.19	1066.43	1016.47	990.29	968.28
105,000	2309.22	1476.25	1226.60	1119.75	1067.29	1039.81	1016.70
110,000	2419.19	1546.55	1285.01	1173.07	1118.12	1089.32	1065.11
120,000	2639.11	1687.15	1401.83	1279.72	1219.76	1188.35	1161.94
130,000	2859.04	1827.74	1518.65	1386.36	1321.41	1287.38	1258.77
140,000	3078.97	1968.34	1635.47	1493.00	1423.06	1386.41	1355.59
150,000	3298.89	2108.93	1752.29	1599.64	1524.70	1485.44	1452.42
175,000	3848.71	2460.42	2044.33	1866.25	1778.82	1733.01	1694.49
200,000	4398.52	2811.91	2336.38	2132.86	2032.94	1980.58	1936.56
225,000	4948.34	3163.40	2628.43	2399.47	2287.06	2228.16	2178.63
250,000	5498.15	3514.89	2920.47	2660.07	2541.17	2475.73	2420.70

Monthly Mortgage Payment (continued)
12%

Loan Amount ($)	5 Years	10 Years	15 Years	20 Years	25 Years	30 Years	40 Years
10,000	$ 222.44	$ 143.47	$ 120.02	$ 110.11	$ 105.32	$ 102.86	$ 100.85
15,000	333.67	215.21	180.03	165.16	157.98	154.29	151.27
20,000	444.89	286.94	240.03	220.22	210.64	205.72	201.70
25,000	556.11	358.68	300.04	275.27	263.31	257.15	252.12
30,000	667.33	430.41	360.05	330.33	315.97	308.58	302.55
35,000	778.56	502.15	420.06	385.38	368.63	360.01	352.97
40,000	889.78	573.88	480.07	440.43	421.29	411.45	403.40
45,000	1001.00	645.62	540.08	495.49	473.95	462.88	453.82
50,000	1112.22	717.35	600.08	550.54	526.61	514.31	504.25
55,000	1223.44	789.09	660.09	605.60	579.27	565.74	554.67
60,000	1334.67	860.83	720.10	660.65	631.93	617.17	605.10
65,000	1445.89	932.56	780.11	715.71	684.60	668.60	655.52
70,000	1557.11	1004.30	840.12	770.76	737.26	720.03	705.95
75,000	1668.33	1076.03	900.13	825.81	789.92	771.46	756.37
80,000	1779.56	1147.77	960.13	880.87	842.58	822.89	806.80
85,000	1890.78	1219.50	1020.14	935.92	895.24	874.32	857.22
90,000	2002.00	1291.24	1080.15	990.98	947.90	925.75	907.65
95,000	2113.22	1362.97	1140.16	1046.03	1000.56	977.18	958.07
100,000	2224.44	1434.71	1200.17	1101.09	1053.22	1028.61	1008.50
105,000	2335.67	1506.44	1260.18	1156.14	1105.89	1080.04	1058.92
110,000	2446.89	1578.18	1320.18	1211.19	1158.55	1131.47	1109.35
120,000	2669.33	1721.65	1440.20	1321.30	1263.87	1234.34	1210.20
130,000	2891.78	1865.12	1560.22	1431.41	1369.19	1337.20	1311.05
140,000	3114.22	2008.59	1680.24	1541.52	1474.51	1440.06	1411.90
150,000	3336.67	2152.06	1800.25	1651.63	1579.84	1542.92	1512.75
175,000	3892.78	2510.74	2100.29	1926.90	1843.14	1800.07	1764.87
200,000	4448.89	2869.42	2400.34	2202.17	2106.45	2057.23	2017.00
225,000	5005.00	3228.10	2700.38	2477.44	2369.75	2314.38	2269.12
250,000	5561.11	3586.77	3000.42	2752.72	2633.06	2571.53	2521.25

Comparing 30-Year and 15-Year Mortgages

If you have a 30-year mortgage loan at 9 percent interest, you won't reach the halfway point in paying off your loan until its 23rd year. A 15-year loan, because of its shorter term, incurs less interest; at the end of 10 years, the loan is more than half paid.

	30-Year at 10%	15-Year at 9.75%
Monthly payment on $150,000 mortgage	$1,316	$1,589
Total interest	$323,889	$136,028
Principal paid after 5 years	$5,139	$28,486

One way to pay less interest on a 30-year loan is to make extra payments toward principal. One relatively painless way to do this is to pay your mortgage biweekly (in half payments every 2 weeks) instead of one full payment monthly. This means that you will be making an extra month's payment each year (26 half payments equal 13 full monthly payments). Using this method, you'll be able to pay off your mortgage several years earlier and pay less interest than you would have otherwise. Your lender can print out a schedule for you.

Local newspapers usually provide a weekly listing of current area mortgage rates. HSH Associates sells a computerized listing of lenders and their current rates; call (800) 873-2837.

Amortization Tables for 30-Year and 15-Year Loans (per $10,000 of Loan Amount)

The tables that follow will let you determine how much of your monthly mortgage payment goes to paying interest and principal, as well as your projected mortgage balance at the end of each year of the loan. The figures given are for each $10,000 of loan amount, so you must multiply the amounts given by a factor that corresponds to your loan amount. For example, if you have a $164,000 loan, multiply by 16.4; if you have an $85,000 loan, multiply by 8.5.

Amortization Tables for 30-Year Loans
7%

Year	Interest Paid per Year	Principal Paid per Year	Mortgage Balance at End of Year
1	$696.78	$101.58	$9898.42
2	689.44	108.92	9789.49
3	681.56	116.80	9672.70
4	673.12	125.24	9547.45
5	664.06	134.30	9413.16
6	654.36	144.00	9269.16
7	643.95	154.41	9114.74
8	632.78	165.58	8949.17
9	620.81	177.55	8771.62
10	607.98	190.38	8581.24
11	594.22	204.14	8377.09
12	579.46	218.90	8158.19
13	563.63	234.73	7923.47
14	546.67	251.69	7671.78
15	528.47	269.89	7401.89
16	508.96	289.40	7112.49
17	488.04	310.32	6802.17
18	465.61	332.75	6469.42
19	441.55	356.81	6112.61
20	415.76	382.60	5730.01
21	388.10	410.26	5319.75
22	358.44	439.92	4879.83
23	326.64	471.72	4408.11
24	292.54	505.82	3902.29
25	255.98	542.38	3359.91
26	216.77	581.59	2778.32
27	174.72	623.64	2154.68
28	129.64	668.72	1485.96
29	81.30	717.06	768.90
30	29.46	768.90	0.00

Amortization Tables for 30-Year Loans (continued)
7-1/4%

Year	Interest Paid per Year	Principal Paid per Year	Mortgage Balance at End of Year
1	$721.85	$ 96.79	$9903.21
2	714.60	104.04	9799.17
3	706.80	111.84	9687.33
4	698.42	120.22	9567.11
5	689.41	129.23	9437.88
6	679.72	138.92	9298.96
7	669.31	149.33	9149.62
8	658.11	160.53	8989.09
9	646.08	172.56	8816.53
10	633.14	185.50	8631.04
11	619.24	199.40	8431.63
12	604.29	214.35	8217.29
13	588.23	230.41	7986.87
14	570.95	247.69	7739.19
15	552.39	266.25	7472.93
16	532.43	286.21	7186.72
17	510.98	307.66	6879.06
18	487.91	330.73	6548.33
19	463.12	355.52	6192.82
20	436.47	382.17	5810.65
21	407.83	410.81	5399.84
22	377.03	441.61	4958.23
23	343.93	474.71	4483.52
24	308.35	510.29	3973.23
25	270.10	548.54	3424.69
26	228.98	589.66	2835.03
27	184.78	633.86	2201.17
28	137.27	681.37	1519.80
29	86.19	732.45	787.35
30	31.29	787.35	0.00

Amortization Tables for 30-Year Loans (continued)
7-1/2%

Year	Interest Paid per Year	Principal Paid per Year	Mortgage Balance at End of Year
1	$746.86	$ 92.18	$9907.82
2	739.70	99.34	9808.48
3	731.99	107.05	9701.42
4	723.68	115.36	9586.06
5	714.72	124.32	9461.74
6	705.07	133.97	9327.77
7	694.67	144.37	9183.40
8	683.46	155.58	9027.83
9	671.38	167.66	8860.17
10	658.37	180.67	8679.50
11	644.34	194.70	8484.80
12	629.23	209.81	8274.99
13	612.94	226.10	8048.89
14	595.39	243.65	7805.24
15	576.47	262.57	7542.67
16	556.09	282.95	7259.71
17	534.12	304.92	6954.79
18	510.45	328.59	6626.20
19	484.94	354.10	6272.10
20	457.45	381.59	5890.51
21	427.83	411.21	5479.30
22	395.90	443.14	5036.16
23	361.50	477.54	4558.63
24	324.43	514.61	4044.01
25	284.48	554.56	3489.45
26	241.43	597.61	2891.84
27	195.03	644.01	2247.83
28	145.04	694.00	1553.82
29	91.16	747.88	805.94
30	33.10	805.94	0.00

Amortization Tables for 30-Year Loans (continued)
7-3/4%

Year	Interest Paid per Year	Principal Paid per Year	Mortgage Balance at End of Year
1	$771.91	$ 87.77	$9912.23
2	764.86	94.82	9817.41
3	757.25	102.43	9714.98
4	749.02	110.66	9604.32
5	740.13	119.55	9484.77
6	730.53	129.15	9355.63
7	720.16	139.52	9216.11
8	708.95	150.73	9065.38
9	696.85	162.83	8902.55
10	683.77	175.91	8726.64
11	669.64	190.04	8536.60
12	654.38	205.30	8331.31
13	637.89	221.79	8109.52
14	620.08	239.60	7869.92
15	600.84	258.84	7611.08
16	580.05	279.63	7331.45
17	557.59	302.09	7029.36
18	533.33	326.35	6703.01
19	507.12	352.56	6350.45
20	478.80	380.88	5969.57
21	448.22	411.46	5558.11
22	415.17	444.51	5113.60
23	379.47	480.21	4633.39
24	340.90	518.78	4114.61
25	299.24	560.44	3554.17
26	254.23	605.45	2948.72
27	205.60	654.08	2294.64
28	153.07	706.61	1588.03
29	96.32	763.36	824.67
30	35.01	824.67	0.00

Amortization Tables for 30-Year Loans (continued)
8%

Year	Interest Paid per Year	Principal Paid per Year	Mortgage Balance at End of Year
1	$797.02	$ 83.54	$9916.46
2	790.09	90.47	9825.99
3	782.58	97.98	9728.01
4	774.45	106.11	9621.90
5	765.64	114.92	9506.99
6	756.10	124.46	9382.53
7	745.77	134.79	9247.74
8	734.59	145.97	9101.77
9	722.47	158.09	8943.68
10	709.35	171.21	8772.47
11	695.14	185.42	8587.05
12	679.75	200.81	8386.24
13	663.08	217.48	8168.76
14	645.03	235.53	7933.23
15	625.48	255.08	7678.16
16	604.31	276.25	7401.91
17	581.38	299.18	7102.73
18	556.55	324.01	6778.72
19	529.66	350.90	6427.82
20	500.53	380.03	6047.80
21	468.99	411.57	5636.23
22	434.83	445.73	5190.50
23	397.84	482.72	4707.78
24	357.77	522.79	4184.99
25	314.38	566.18	3618.81
26	267.39	613.17	3005.64
27	216.50	664.06	2341.58
28	161.38	719.18	1622.39
29	101.69	778.87	843.52
30	37.05	843.52	0.00

Amortization Tables for 30-Year Loans (continued)
8-1/4%

Year	Interest Paid per Year	Principal Paid per Year	Mortgage Balance at End of Year
1	$822.08	$ 79.48	$9920.52
2	815.27	86.29	9834.23
3	807.87	93.69	9740.54
4	799.85	101.71	9638.83
5	791.13	110.43	9528.40
6	781.67	119.89	9408.50
7	771.39	130.17	9278.34
8	760.24	141.32	9137.02
9	748.13	153.43	8983.58
10	734.98	166.58	8817.00
11	720.71	180.85	8636.15
12	705.21	196.35	8439.80
13	688.38	213.18	8226.62
14	670.11	231.45	7995.17
15	650.28	251.28	7743.90
16	628.75	272.81	7471.08
17	605.37	296.19	7174.90
18	579.99	321.57	6853.33
19	552.43	349.13	6504.20
20	522.52	379.04	6125.16
21	490.04	411.52	5713.63
22	454.77	446.79	5266.84
23	416.49	485.07	4781.77
24	374.92	526.64	4255.13
25	329.79	571.77	3683.36
26	280.79	620.77	3062.59
27	227.60	673.96	2388.63
28	169.84	731.72	1656.91
29	107.14	794.42	862.49
30	39.07	862.49	0.00

Amortization Tables for 30-Year Loans (continued)
8-1/2%

Year	Interest Paid per Year	Principal Paid per Year	Mortgage Balance at End of Year
1	$847.08	$ 75.60	$9924.40
2	840.40	82.28	9842.13
3	833.13	89.55	9752.57
4	825.21	97.47	9655.11
5	816.60	106.08	9549.03
6	807.22	115.46	9433.57
7	797.02	125.66	9307.90
8	785.91	136.77	9171.13
9	773.82	148.86	9022.27
10	760.66	162.02	8860.25
11	746.34	176.34	8683.92
12	730.75	191.93	8491.99
13	713.79	208.89	8283.10
14	695.33	227.35	8055.74
15	675.23	247.45	7808.29
16	653.36	269.32	7538.97
17	629.55	293.13	7245.84
18	603.64	319.04	6926.80
19	575.44	347.24	6579.56
20	544.75	377.93	6201.63
21	511.34	411.34	5790.29
22	474.98	447.70	5342.60
23	435.41	487.27	4855.33
24	392.34	530.34	4324.99
25	345.46	577.22	3747.78
26	294.44	628.24	3119.54
27	238.91	683.77	2435.77
28	178.47	744.21	1691.57
29	112.69	809.99	881.58
30	41.10	881.58	0.00

Amortization Tables for 30-Year Loans (continued)
8-3/4%

Year	Interest Paid per Year	Principal Paid per Year	Mortgage Balance at End of Year
1	$872.16	$ 71.88	$9928.12
2	865.61	78.43	9849.70
3	858.47	85.57	9764.13
4	850.68	93.36	9670.76
5	842.17	101.87	9568.89
6	832.89	111.15	9457.74
7	822.76	121.28	9336.47
8	811.72	132.32	9204.14
9	799.66	144.38	9059.77
10	786.51	157.53	8902.24
11	772.16	171.88	8730.36
12	756.50	187.54	8542.82
13	739.42	204.62	8338.20
14	720.78	223.26	8114.94
15	700.44	243.60	7871.34
16	678.25	265.79	7605.55
17	654.04	290.00	7315.55
18	627.62	316.42	6999.13
19	598.80	345.24	6653.89
20	567.35	376.69	6277.19
21	533.03	411.01	5866.18
22	495.59	448.45	5417.73
23	454.74	489.30	4928.43
24	410.16	533.88	4394.55
25	361.53	582.51	3812.04
26	308.47	635.57	3176.47
27	250.57	693.47	2483.00
28	187.39	756.65	1726.35
29	118.47	825.57	900.78
30	43.26	900.78	0.00

Amortization Tables for 30-Year Loans (continued)
9%

Year	Interest Paid per Year	Principal Paid per Year	Mortgage Balance at End of Year
1	$897.20	$ 68.32	$9931.68
2	890.79	74.73	9856.95
3	883.78	81.74	9775.21
4	876.11	89.41	9685.81
5	867.73	97.79	9588.01
6	858.55	106.97	9481.05
7	848.52	117.00	9364.05
8	837.54	127.98	9236.07
9	825.54	139.98	9096.09
10	812.41	153.11	8942.97
11	798.04	167.48	8775.50
12	782.33	183.19	8592.31
13	765.15	200.37	8391.94
14	746.35	219.17	8172.77
15	725.79	239.73	7933.05
16	703.31	262.21	7670.83
17	678.71	286.81	7384.02
18	651.80	313.72	7070.31
19	622.37	343.15	6727.16
20	590.19	375.33	6351.83
21	554.98	410.54	5941.28
22	516.46	449.06	5492.23
23	474.34	491.18	5001.05
24	428.26	537.26	4463.79
25	377.87	587.65	3876.14
26	322.74	642.78	3233.36
27	262.44	703.08	2530.28
28	196.49	769.03	1761.25
29	124.35	841.17	920.08
30	45.44	920.08	0.00

Amortization Tables for 30-Year Loans (continued)
9-1/4%

Year	Interest Paid per Year	Principal Paid per Year	Mortgage Balance at End of Year
1	$922.32	$ 64.92	$9935.08
2	916.06	71.18	9863.90
3	909.19	78.05	9785.85
4	901.65	85.59	9700.26
5	893.39	93.85	9606.41
6	884.33	102.91	9503.50
7	874.40	112.84	9390.66
8	863.51	123.73	9266.93
9	851.56	135.68	9131.25
10	838.47	148.77	8982.48
11	824.11	163.13	8819.35
12	808.36	178.88	8640.47
13	791.10	196.14	8444.32
14	772.16	215.08	8229.25
15	751.40	235.84	7993.41
16	728.64	258.60	7734.81
17	703.68	283.56	7451.24
18	676.31	310.93	7140.31
19	646.29	340.95	6799.36
20	613.38	373.86	6425.51
21	577.30	409.94	6015.57
22	537.73	449.51	5566.06
23	494.34	492.90	5073.16
24	446.76	540.48	4532.68
25	394.59	592.65	3940.04
26	337.39	649.85	3290.19
27	274.66	712.58	2577.61
28	205.88	781.36	1796.25
29	130.46	856.78	939.48
30	47.76	939.48	0.00

Amortization Tables for 30-Year Loans (continued)
9-1/2%

Year	Interest Paid per Year	Principal Paid per Year	Mortgage Balance at End of Year
1	$947.42	$ 61.66	$9938.34
2	941.30	67.78	9870.55
3	934.57	74.51	9796.04
4	927.17	81.91	9714.13
5	919.04	90.04	9624.10
6	910.11	98.97	9525.13
7	900.29	108.79	9416.33
8	889.49	119.59	9296.74
9	877.62	131.46	9165.28
10	864.57	144.51	9020.77
11	850.23	158.85	8861.92
12	834.46	174.62	8687.31
13	817.13	191.95	8495.36
14	798.08	211.00	8284.36
15	777.14	231.94	8052.43
16	754.12	254.96	7797.47
17	728.82	280.26	7517.21
18	701.00	308.08	7209.13
19	670.43	338.65	6870.48
20	636.82	372.26	6498.22
21	599.87	409.21	6089.02
22	559.26	449.82	5639.20
23	514.62	494.46	5144.73
24	465.54	543.54	4601.20
25	411.00	597.40	4003.71
26	352.30	656.78	3346.93
27	287.11	721.97	2624.97
28	215.46	793.62	1831.35
29	136.70	872.38	958.97
30	50.11	958.97	0.00

Amortization Tables for 30-Year Loans (continued)
9-3/4%

Year	Interest Paid per Year	Principal Paid per Year	Mortgage Balance at End of Year
1	$972.48	$ 58.56	$9941.44
2	966.51	64.53	9876.92
3	959.93	71.11	9805.81
4	952.68	78.36	9727.45
5	944.69	86.35	9641.10
6	935.88	95.16	9545.94
7	926.18	104.86	9441.08
8	915.49	115.55	9325.53
9	903.70	127.34	9198.19
10	890.72	140.32	9057.87
11	876.41	154.63	8903.24
12	860.64	170.40	8732.84
13	843.26	187.78	8545.06
14	824.11	206.93	8338.14
15	803.01	228.03	8110.11
16	779.76	251.28	7858.82
17	754.13	276.91	7581.92
18	725.89	305.15	7276.77
19	694.78	336.26	6940.51
20	660.49	370.55	6569.95
21	622.70	408.34	6161.61
22	581.06	449.98	5711.63
23	535.17	495.87	5215.75
24	484.60	546.44	4669.31
25	428.88	602.16	4067.15
26	367.47	663.57	3403.58
27	299.80	731.24	2672.34
28	225.23	805.81	1866.51
29	143.05	887.99	978.54
30	52.50	978.54	0.00

Amortization Tables for 30-Year Loans (continued)
10%

Year	Interest Paid per Year	Principal Paid per Year	Mortgage Balance at End of Year
1	$997.53	$ 55.59	$9944.41
2	991.71	61.41	9883.00
3	985.28	67.84	9815.16
4	978.18	74.94	9741.22
5	970.33	82.79	9657.43
6	961.66	91.46	9565.97
7	952.08	101.04	9464.94
8	941.50	111.62	9353.32
9	929.82	123.30	9230.02
10	916.90	136.22	9093.80
11	902.64	150.48	8943.32
12	886.88	166.24	8777.09
13	869.48	183.64	8593.44
14	850.25	202.87	8390.57
15	829.00	224.12	8166.46
16	805.54	247.58	7918.87
17	779.61	273.51	7645.36
18	750.97	302.15	7343.21
19	719.33	333.79	7009.43
20	684.38	368.74	6640.69
21	645.77	407.35	6233.33
22	603.11	450.01	5783.33
23	555.99	497.13	5286.20
24	503.94	549.18	4737.01
25	446.43	606.69	4130.32
26	382.90	670.22	3460.10
27	312.72	740.40	2719.70
28	235.19	817.93	1901.77
29	149.54	903.58	998.19
30	54.93	998.19	0.00

Amortization Tables for 30-Year Loans (continued)
10-1/4%

Year	Interest Paid per Year	Principal Paid per Year	Mortgage Balance at End of Year
1	$1022.57	$ 52.75	$9947.25
2	1016.90	58.42	9888.82
3	1010.62	64.70	9824.12
4	1003.67	71.65	9752.47
5	995.97	79.35	9673.12
6	987.44	87.88	9585.24
7	978.00	97.32	9487.91
8	967.54	107.78	9380.13
9	955.96	119.36	9260.77
10	943.13	132.19	9128.58
11	928.93	146.39	8982.19
12	913.20	162.12	8820.07
13	895.78	179.54	8640.52
14	876.48	198.84	8441.69
15	855.12	220.20	8221.48
16	831.46	243.87	7977.62
17	805.25	270.07	7707.55
18	776.23	299.09	7408.46
19	744.09	331.23	7077.23
20	708.50	366.82	6710.41
21	669.08	406.24	6304.17
22	625.43	449.89	5854.28
23	577.09	498.23	5356.05
24	523.55	551.77	4804.28
25	464.26	611.06	4193.22
26	398.60	676.72	3516.49
27	325.88	749.44	2767.05
28	245.35	829.97	1937.08
29	156.16	919.16	1017.92
30	57.40	1017.92	0.00

Amortization Tables for 30-Year Loans (continued)
10-1/2%

Year	Interest Paid per Year	Principal Paid per Year	Mortgage Balance at End of Year
1	$1047.59	$ 50.05	$9949.95
2	1042.07	55.57	9894.38
3	1035.95	61.69	9832.69
4	1029.15	68.49	9764.21
5	1021.60	76.04	9688.17
6	1013.22	84.42	9603.75
7	1003.92	93.72	9510.04
8	993.59	104.05	9405.99
9	982.13	115.51	9290.48
10	969.40	128.24	9162.24
11	955.27	142.37	9019.86
12	939.58	158.06	8861.80
13	922.16	175.48	8686.31
14	902.82	194.82	8491.49
15	881.35	216.29	8275.20
16	857.51	240.13	8035.07
17	831.05	266.59	7768.47
18	801.67	295.97	7472.50
19	769.05	328.59	7143.91
20	732.84	364.80	6779.11
21	692.64	405.00	6374.11
22	648.00	449.64	5924.47
23	598.45	499.19	5425.28
24	543.44	554.20	4871.08
25	482.37	615.27	4255.81
26	414.56	683.08	3572.73
27	339.28	758.36	2814.37
28	255.71	841.93	1972.44
29	162.92	934.72	1037.72
30	59.92	1037.72	0.00

Amortization Tables for 30-Year Loans (continued)
10-3/4%

Year	Interest Paid per Year	Principal Paid per Year	Mortgage Balance at End of Year
1	$1072.73	$ 47.47	$9952.53
2	1067.37	52.83	9899.69
3	1061.40	58.80	9840.89
4	1054.79	65.44	9775.45
5	1047.36	72.84	9702.61
6	1039.14	81.06	9621.55
7	1029.98	90.22	9531.33
8	1019.79	100.41	9430.92
9	1008.45	111.75	9319.16
10	995.82	124.38	9194.79
11	981.77	138.43	9056.36
12	966.14	154.06	8902.30
13	948.73	171.47	8730.83
14	929.37	190.83	8540.00
15	907.81	212.39	8327.61
16	883.82	236.38	8091.22
17	857.12	263.08	7828.14
18	827.40	292.80	7535.34
19	794.33	325.87	7209.47
20	757.52	362.68	6846.78
21	716.55	403.65	6443.13
22	670.95	449.25	5993.88
23	620.21	499.99	5493.89
24	563.73	556.47	4937.41
25	500.87	619.33	4318.08
26	430.91	689.29	3628.79
27	353.05	767.15	2861.64
28	266.39	853.81	2007.84
29	169.95	950.25	1057.59
30	62.61	1057.59	0.00

Amortization Tables for 30-Year Loans (continued)
11%

Year	Interest Paid per Year	Principal Paid per Year	Mortgage Balance at End of Year
1	$1097.75	$ 45.01	$9954.99
2	1092.54	50.22	9904.77
3	1086.73	56.03	9848.73
4	1080.24	62.52	9786.22
5	1073.01	69.75	9716.46
6	1064.94	77.82	9638.64
7	1055.93	86.83	9551.81
8	1045.88	96.88	9454.94
9	1034.67	108.09	9346.85
10	1022.17	120.59	9226.26
11	1008.21	134.55	9091.71
12	992.64	150.12	8941.59
13	975.27	167.49	8774.10
14	955.89	186.87	8587.22
15	934.26	208.50	8378.73
16	910.14	232.62	8146.10
17	883.22	259.54	7886.56
18	853.18	289.58	7596.98
19	819.67	323.09	7273.89
20	782.29	360.47	6913.42
21	740.57	402.19	6511.23
22	694.03	448.73	6062.50
23	642.10	500.66	5561.85
24	584.17	558.59	5003.25
25	519.53	623.23	4380.02
26	447.41	695.35	3684.67
27	366.95	775.81	2908.86
28	277.17	865.59	2043.27
29	177.00	965.76	1077.51
30	65.25	1077.51	0.00

Amortization Tables for 30-Year Loans (continued)
11-1/4%

Year	Interest Paid per Year	Principal Paid per Year	Mortgage Balance at End of Year
1	$1122.89	$ 42.67	$9957.33
2	1117.83	47.73	9909.61
3	1112.18	53.38	9856.23
4	1105.86	59.70	9796.52
5	1098.78	66.78	9729.74
6	1090.87	74.69	9655.05
7	1082.02	83.54	9571.51
8	1072.12	93.44	9478.07
9	1061.05	104.51	9373.56
10	1048.67	116.89	9256.67
11	1034.82	130.74	9125.92
12	1019.32	146.24	8979.69
13	1002.00	163.56	8816.13
14	982.62	182.94	8633.18
15	960.94	204.62	8428.57
16	936.70	228.86	8199.71
17	909.58	255.98	7943.73
18	879.25	286.31	7657.42
19	845.33	320.23	7337.18
20	807.37	358.18	6979.01
21	764.95	400.61	6578.40
22	717.48	448.08	6130.31
23	664.39	501.17	5629.14
24	605.01	560.55	5068.59
25	538.59	626.97	4441.61
26	464.30	701.26	3740.35
27	381.21	784.35	2956.00
28	288.28	877.28	2078.72
29	184.33	981.23	1097.49
30	68.07	1097.49	0.00

Amortization Tables for 30-Year Loans (continued)
11-1/2%

Year	Interest Paid per Year	Principal Paid per Year	Mortgage Balance at End of Year
1	$1147.92	$ 40.44	$9959.56
2	1143.02	45.34	9914.22
3	1137.52	50.84	9863.38
4	1131.36	57.00	9806.38
5	1124.44	63.92	9742.47
6	1116.69	71.67	9670.80
7	1108.00	80.36	9590.44
8	1098.26	90.10	9500.35
9	1087.34	101.02	9399.32
10	1075.08	113.28	9286.05
11	1061.35	127.01	9159.04
12	1045.95	142.41	9016.62
13	1028.68	159.68	8856.94
14	1009.32	179.04	8677.90
15	987.61	200.75	8477.14
16	963.26	225.10	8252.05
17	935.97	252.39	7999.65
18	905.36	283.00	7716.66
19	871.05	317.31	7399.34
20	832.57	355.79	7043.55
21	789.43	398.93	6644.62
22	741.05	447.31	6197.31
23	686.81	501.55	5695.76
24	625.99	562.37	5133.40
25	557.80	630.56	4502.84
26	481.34	707.02	3792.82
27	395.61	792.75	3003.07
28	299.48	888.88	2114.19
29	191.69	996.67	1117.52
30	70.84	1117.52	0.00

Amortization Tables for 30-Year Loans (continued)
11-3/4%

Year	Interest Paid per Year	Principal Paid per Year	Mortgage Balance at End of Year
1	$1172.97	$ 38.31	$9961.69
2	1168.22	43.06	9918.63
3	1162.88	48.41	9870.22
4	1156.87	54.41	9815.81
5	1150.12	61.16	9754.65
6	1142.54	68.74	9685.91
7	1134.01	77.27	9608.64
8	1124.42	86.86	9521.78
9	1113.65	97.63	9424.15
10	1101.54	109.74	9314.42
11	1087.93	123.35	9191.06
12	1072.63	138.65	9052.41
13	1055.43	155.85	8896.56
14	1036.10	175.18	8721.38
15	1014.37	196.91	8524.47
16	989.95	221.33	8303.14
17	962.49	248.79	8054.35
18	931.63	279.65	7774.70
19	896.95	314.33	7460.37
20	857.96	353.32	7107.04
21	814.13	397.15	6709.89
22	764.87	446.41	6263.48
23	709.49	501.79	5761.70
24	647.25	564.03	5197.67
25	577.29	633.99	4563.88
26	498.65	712.63	3851.05
27	410.26	801.02	3050.03
28	310.90	900.68	2149.66
29	199.22	1012.06	1137.60
30	73.68	1137.60	0.00

Amortization Tables for 30-Year Loans (continued)
12%

Year	Interest Paid per Year	Principal Paid per Year	Mortgage Balance at End of Year
1	$1198.03	$ 36.29	$9963.71
2	1193.43	40.89	9922.82
3	1188.24	46.08	9876.75
4	1182.40	51.92	9824.83
5	1175.82	58.50	9766.32
6	1168.40	65.92	9700.40
7	1160.04	74.29	9626.11
8	1150.61	83.71	9542.41
9	1140.00	94.32	9448.08
10	1128.04	106.28	9341.80
11	1114.56	119.76	9222.04
12	1099.37	134.95	9087.08
13	1082.25	152.07	8935.01
14	1062.97	171.35	8763.66
15	1041.23	193.09	8570.57
16	1016.74	217.58	8353.00
17	989.15	245.17	8107.83
18	958.06	276.26	7831.56
19	923.02	311.30	7520.26
20	883.54	350.78	7169.48
21	839.05	395.27	6774.22
22	788.92	445.40	6328.82
23	732.43	501.89	5826.93
24	668.78	565.54	5261.39
25	597.06	637.26	4624.13
26	516.24	718.08	3906.05
27	425.17	809.15	3096.90
28	322.55	911.77	2185.12
29	206.91	1027.41	1157.71
30	76.61	1157.71	0.00

Amortization Tables for 30-Year Loans (continued)
12-1/4%

Year	Interest Paid per Year	Principal Paid per Year	Mortgage Balance at End of Year
1	$1223.12	$ 34.36	$9965.64
2	1218.66	38.82	9926.82
3	1213.63	43.85	9882.97
4	1207.95	49.53	9833.44
5	1201.53	55.95	9777.49
6	1194.28	63.20	9714.29
7	1186.08	71.40	9642.89
8	1176.83	80.65	9562.24
9	1166.38	91.10	9471.14
10	1154.57	102.91	9368.22
11	1141.23	116.25	9251.97
12	1126.16	131.32	9120.65
13	1109.14	148.34	8972.31
14	1089.91	167.57	8804.74
15	1068.19	189.29	8615.45
16	1043.66	213.82	8401.63
17	1015.94	241.54	8160.09
18	984.63	272.85	7887.24
19	949.27	308.21	7579.03
20	909.32	348.16	7230.87
21	864.19	393.29	6837.58
22	813.21	444.27	6393.31
23	755.63	501.85	5891.46
24	690.58	566.90	5324.56
25	617.10	640.38	4684.18
26	534.09	723.39	3960.79
27	440.33	817.15	3143.64
28	334.41	923.07	2220.58
29	214.77	1042.71	1177.86
30	79.62	1177.86	0.00

Amortization Tables for 30-Year Loans (continued)
12-1/2%

Year	Interest Paid per Year	Principal Paid per Year	Mortgage Balance at End of Year
1	$1248.23	$ 32.53	$9967.47
2	1243.92	36.84	9930.63
3	1239.04	41.72	9889.91
4	1233.52	47.24	9841.67
5	1227.26	53.50	9788.18
6	1220.18	60.58	9727.60
7	1212.16	68.60	9658.99
8	1203.07	77.69	9581.31
9	1192.79	87.97	9493.33
10	1181.14	99.62	9393.71
11	1167.95	112.81	9280.90
12	1153.01	127.75	9153.15
13	1136.09	144.67	9008.48
14	1116.94	163.82	8844.65
15	1095.24	185.52	8659.14
16	1070.68	210.08	8449.05
17	1042.86	237.90	8211.15
18	1011.36	269.40	7941.75
19	975.68	305.08	7636.67
20	935.29	345.47	7291.20
21	889.54	391.22	6899.98
22	837.74	443.02	6456.96
23	779.07	501.69	5955.27
24	712.64	568.12	5387.15
25	637.41	643.35	4743.80
26	552.22	728.54	4015.27
27	455.75	825.01	3190.26
28	346.51	934.25	2256.01
29	222.80	1057.96	1198.05
30	82.71	1198.05	0.00

Amortization Tables for 30-Year Loans (continued)
12-3/4%

Year	Interest Paid per Year	Principal Paid per Year	Mortgage Balance at End of Year
1	$1273.25	$ 30.79	$9969.21
2	1269.09	34.95	9934.26
3	1264.36	39.68	9894.58
4	1258.99	45.05	9849.53
5	1252.90	51.14	9798.39
6	1245.99	58.05	9740.34
7	1238.14	65.90	9674.44
8	1229.23	74.81	9599.63
9	1219.11	84.93	9514.70
10	1207.63	96.41	9418.29
11	1194.59	109.45	9308.84
12	1179.79	124.25	9184.59
13	1162.99	141.05	9043.54
14	1143.92	160.12	8883.42
15	1122.26	181.78	8701.64
16	1097.68	206.36	8495.29
17	1069.78	234.26	8261.03
18	1038.10	265.94	7995.09
19	1002.14	301.90	7693.20
20	961.32	342.72	7350.48
21	914.98	389.06	6961.42
22	862.37	441.67	6519.75
23	802.65	501.39	6018.35
24	734.85	569.19	5449.16
25	657.88	646.16	4803.00
26	570.51	733.53	4069.47
27	471.32	832.72	3236.74
28	358.72	945.32	2291.42
29	230.89	1073.15	1218.27
30	85.77	1218.27	0.00

Amortization Tables for 30-Year Loans (continued)
13%

Year	Interest Paid per Year	Principal Paid per Year	Mortgage Balance at End of Year
1	$1298.31	$ 29.13	$9970.87
2	1294.28	33.16	9937.71
3	1289.71	37.73	9899.98
4	1284.50	42.94	9857.03
5	1278.57	48.87	9808.17
6	1271.83	55.61	9752.55
7	1264.15	63.29	9689.26
8	1255.41	72.03	9617.23
9	1245.47	81.97	9535.26
10	1234.16	93.28	9441.98
11	1221.28	106.16	9335.82
12	1206.63	120.81	9215.01
13	1189.95	137.49	9077.52
14	1170.97	156.47	8921.05
15	1149.38	178.06	8742.99
16	1124.80	202.64	8540.34
17	1096.83	230.61	8309.73
18	1064.99	262.45	8047.28
19	1028.77	298.67	7748.61
20	987.54	339.90	7408.71
21	940.62	386.82	7021.89
22	887.23	440.21	6581.68
23	826.47	500.97	6080.70
24	757.32	570.12	5510.58
25	678.62	648.82	4861.76
26	589.06	738.38	4123.38
27	487.14	840.30	3283.08
28	371.15	956.29	2326.79
29	239.15	1088.29	1238.51
30	88.93	1238.51	0.00

Amortization Tables for 30-Year Loans (continued)
13-1/4%

Year	Interest Paid per Year	Principal Paid per Year	Mortgage Balance at End of Year
1	$1323.40	$ 27.56	$9972.44
2	1319.52	31.44	9940.99
3	1315.09	35.87	9905.12
4	1310.03	40.93	9864.19
5	1304.27	46.69	9817.50
6	1297.69	53.27	9764.24
7	1290.19	60.77	9703.47
8	1281.63	69.33	9634.14
9	1271.87	79.09	9555.04
10	1260.73	90.23	9464.81
11	1248.02	102.94	9361.87
12	1233.52	117.44	9244.42
13	1216.97	133.99	9110.44
14	1198.10	152.86	8957.58
15	1176.57	174.39	8783.19
16	1152.01	198.95	8584.24
17	1123.99	226.97	8357.27
18	1092.02	258.94	8098.33
19	1055.55	295.41	7802.91
20	1013.94	337.02	7465.89
21	966.47	384.49	7081.40
22	912.31	438.65	6642.75
23	850.53	500.43	6142.32
24	780.04	570.92	5571.40
25	699.63	651.33	4920.07
26	607.89	743.07	4177.00
27	503.23	847.73	3329.26
28	383.82	967.14	2362.13
29	247.60	1103.36	1258.77
30	92.19	1258.77	0.00

Amortization Tables for 30-Year Loans (continued)
13-1/2%

Year	Interest Paid per Year	Principal Paid per Year	Mortgage Balance at End of Year
1	$1348.41	$ 26.07	$9973.93
2	1344.67	29.81	9944.12
3	1340.38	34.10	9910.02
4	1335.48	39.00	9871.02
5	1329.88	44.60	9826.42
6	1323.47	51.01	9775.42
7	1316.14	58.34	9717.08
8	1307.76	66.72	9650.37
9	1298.18	76.30	9574.06
10	1287.22	87.26	9486.80
11	1274.68	99.80	9387.00
12	1260.34	114.14	9272.86
13	1243.94	130.54	9142.32
14	1225.18	149.30	8993.02
15	1203.73	170.75	8822.27
16	1179.20	195.28	8627.00
17	1151.15	223.33	8403.66
18	1119.06	255.42	8148.24
19	1082.36	292.12	7856.12
20	1040.39	334.09	7522.03
21	992.39	382.09	7139.94
22	937.49	436.99	6702.96
23	874.71	499.77	6203.19
24	802.91	571.57	5631.62
25	720.79	653.69	4977.92
26	626.87	747.61	4230.31
27	519.45	855.03	3375.28
28	396.61	977.87	2397.41
29	256.11	1118.37	1279.05
30	95.43	1279.05	0.00

Amortization Tables for 30-Year Loans (continued)
13-3/4%

Year	Interest Paid per Year	Principal Paid per Year	Mortgage Balance at End of Year
1	$1373.47	$ 24.65	$9975.35
2	1369.86	28.26	9947.09
3	1365.72	32.40	9914.69
4	1360.97	37.15	9877.54
5	1355.53	42.59	9834.95
6	1349.29	48.83	9786.11
7	1342.13	55.99	9730.13
8	1333.93	64.19	9665.94
9	1324.53	73.59	9592.35
10	1313.75	84.37	9507.98
11	1301.39	96.73	9411.24
12	1287.21	110.91	9300.34
13	1270.97	127.15	9173.18
14	1252.34	145.78	9027.40
15	1230.98	167.14	8860.26
16	1206.49	191.63	8668.63
17	1178.42	219.70	8448.93
18	1146.23	251.89	8197.04
19	1109.33	288.79	7908.24
20	1067.02	331.10	7577.14
21	1018.51	379.61	7197.53
22	962.89	435.23	6762.23
23	899.13	498.99	6263.31
24	826.03	572.09	5691.22
25	742.21	655.91	5035.31
26	646.12	752.00	4283.31
27	535.95	862.17	3421.13
28	409.63	988.49	2432.65
29	264.81	1133.31	1299.34
30	98.78	1299.34	0.00

Amortization Tables for 15-Year Loans
(per $10,000 of loan amount)
7%

Year	Interest Paid per Year	Principal Paid per Year	Mortgage Balance at End of Year
1	$687.61	$ 390.98	$9609.02
2	659.35	419.24	9189.78
3	629.04	449.55	8740.23
4	596.54	482.05	8258.18
5	561.70	516.90	7741.28
6	524.33	554.26	7187.02
7	484.26	594.33	6592.69
8	441.30	637.29	5955.39
9	395.23	683.36	5272.03
10	345.83	732.77	4539.26
11	292.86	785.74	3753.52
12	236.06	842.54	2910.99
13	175.15	903.45	2007.54
14	109.84	968.76	1038.79
15	39.81	1038.79	0.00

7-1/4%

Year	Interest Paid per Year	Principal Paid per Year	Mortgage Balance at End of Year
1	$712.44	$ 383.00	$9617.00
2	683.73	411.70	9205.30
3	652.87	442.57	8762.73
4	619.70	475.74	8286.99
5	584.04	511.40	7775.59
6	545.70	549.73	7225.86
7	504.49	590.94	6634.92
8	460.20	635.24	5999.68
9	412.58	682.85	5316.83
10	361.40	734.04	4582.79
11	306.37	789.06	3793.73
12	247.23	848.21	2945.52
13	183.65	911.79	2033.74
14	115.30	960.13	1053.60
15	41.83	1053.60	0.00

Amortization Tables for 15-Year Loans (continued)
7-1/2%

Year	Interest Paid per Year	Principal Paid per Year	Mortgage Balance at End of Year
1	$737.28	$ 375.14	$9624.86
2	708.16	404.26	9220.60
3	676.77	435.64	8784.96
4	642.95	469.46	8315.50
5	606.51	505.91	7809.59
6	567.23	545.18	7264.41
7	524.91	587.51	6676.90
8	479.30	633.12	6043.78
9	430.15	682.27	5361.52
10	377.18	735.23	4626.28
11	320.10	792.31	3833.97
12	258.59	853.82	2980.15
13	192.31	920.11	2060.05
14	120.88	991.54	1068.51
15	43.90	1068.51	0.00

7-3/4%

Year	Interest Paid per Year	Principal Paid per Year	Mortgage Balance at End of Year
1	$762.13	$ 367.40	$9632.60
2	732.62	396.91	9235.69
3	700.75	428.78	8806.91
4	666.31	463.22	8343.69
5	629.11	500.42	7843.27
6	588.92	540.61	7302.66
7	545.50	584.03	6718.63
8	498.59	630.94	6087.69
9	447.92	681.61	5406.08
10	393.18	736.35	4669.73
11	334.04	795.49	3874.24
12	270.15	859.38	3014.86
13	201.13	928.40	2086.47
14	126.57	1002.96	1083.51
15	46.02	1083.51	0.00

Amortization Tables for 15-Year Loans (continued)
8%

Year	Interest Paid per Year	Principal Paid per Year	Mortgage Balance at End of Year
1	$787.00	$ 359.78	$9640.22
2	757.14	389.65	9250.57
3	724.80	421.99	8828.58
4	689.77	457.01	8371.57
5	651.84	494.94	7876.63
6	610.76	536.02	7340.60
7	566.27	580.51	6760.09
8	518.09	628.70	6131.39
9	465.91	680.88	5450.52
10	409.39	737.39	4713.13
11	348.19	798.59	3914.53
12	281.91	864.88	3049.66
13	210.12	936.66	2113.00
14	132.38	1014.40	1098.60
15	48.19	1098.60	0.00

8-1/4%

Year	Interest Paid per Year	Principal Paid per Year	Mortgage Balance at End of Year
1	$811.88	$ 352.29	$9647.71
2	781.69	382.48	9265.23
3	748.91	415.26	8849.97
4	713.33	450.84	8399.13
5	674.69	489.47	7909.66
6	632.75	531.42	7378.24
7	587.21	576.96	6801.28
8	537.77	626.40	6174.89
9	484.09	680.07	5494.81
10	425.82	738.35	4756.46
11	362.55	801.62	3954.84
12	293.85	870.31	3084.53
13	219.28	944.89	2139.63
14	138.31	1025.86	1113.77
15	50.40	1113.77	0.00

Amortization Tables for 15-Year Loans (continued)
8-1/2%

Year	Interest Paid per Year	Principal Paid per Year	Mortgage Balance at End of Year
1	$836.77	$ 344.92	$9655.08
2	806.28	375.41	9279.67
3	773.10	408.59	8871.08
4	736.98	444.71	8426.38
5	697.67	484.01	7942.37
6	654.89	526.80	7415.57
7	608.33	573.36	6842.21
8	557.65	624.04	6218.17
9	502.49	679.20	5538.97
10	442.45	739.23	4799.74
11	377.11	804.58	3995.16
12	305.99	875.69	3119.47
13	228.59	953.10	2166.37
14	144.35	1037.34	1129.03
15	52.66	1129.03	0.00

8-3/4%

Year	Interest Paid per Year	Principal Paid per Year	Mortgage Balance at End of Year
1	$861.67	$ 337.67	$9662.33
2	830.91	368.43	9293.91
3	797.35	401.99	8891.92
4	760.73	438.61	8453.31
5	720.77	478.56	7974.74
6	677.18	522.16	7452.58
7	629.61	569.73	6882.86
8	577.71	621.63	6261.23
9	521.09	678.25	5582.98
10	459.30	740.04	4842.94
11	391.88	807.45	4035.48
12	318.33	881.01	3154.47
13	238.07	961.27	2193.21
14	150.51	1048.83	1144.38
15	54.96	1144.38	0.00

Amortization Tables for 15-Year Loans (continued)
9%

Year	Interest Paid per Year	Principal Paid per Year	Mortgage Balance at End of Year
1	$886.59	$ 330.53	$9669.47
2	855.58	361.54	9307.93
3	821.66	395.46	8912.47
4	784.57	432.55	8479.92
5	743.99	473.13	8006.79
6	699.61	517.51	7489.28
7	651.06	566.06	6923.23
8	597.96	619.16	6304.07
9	539.88	677.24	5626.83
10	476.35	740.77	4886.06
11	406.86	810.26	4075.81
12	330.86	886.26	3189.54
13	247.72	969.40	2220.14
14	156.78	1060.34	1159.80
15	57.31	1159.80	0.00

9-1/4%

Year	Interest Paid per Year	Principal Paid per Year	Mortgage Balance at End of Year
1	$911.51	$ 323.52	$9676.48
2	880.28	354.75	9321.74
3	846.04	388.99	8932.75
4	808.50	426.53	8506.21
5	767.33	467.70	8038.51
6	722.18	512.85	7525.66
7	672.68	562.35	6963.31
8	618.40	616.63	6346.68
9	558.88	676.15	5670.52
10	493.61	741.42	4929.11
11	422.05	812.98	4116.12
12	343.58	891.45	3224.67
13	257.53	977.50	2247.17
14	163.18	1071.85	1175.31
15	59.72	1175.31	0.00

Amortization Tables for 15-Year Loans (continued)
9-1/2%

Year	Interest Paid per Year	Principal Paid per Year	Mortgage Balance at End of Year
1	$936.45	$ 316.62	$9683.38
2	905.03	348.04	9335.34
3	870.48	382.59	8952.75
4	832.51	420.56	8532.19
5	790.77	462.30	8069.89
6	744.89	508.18	7561.72
7	694.46	558.61	7003.10
8	639.01	614.06	6389.05
9	578.07	675.00	5714.05
10	511.08	741.99	4972.06
11	437.44	815.63	4156.43
12	356.49	896.58	3259.85
13	267.51	985.56	2274.28
14	169.69	1083.38	1190.90
15	62.17	1190.90	0.00

9-3/4%

Year	Interest Paid per Year	Principal Paid per Year	Mortgage Balance at End of Year
1	$961.40	$ 309.84	$9690.16
2	929.80	341.43	9348.73
3	894.98	376.25	8972.47
4	856.61	414.62	8557.85
5	814.33	456.90	8100.95
6	767.74	503.50	7597.45
7	716.39	554.84	7042.60
8	659.81	611.43	6431.18
9	597.46	673.78	5757.40
10	528.75	742.49	5014.92
11	453.03	818.20	4196.71
12	369.59	901.64	3295.07
13	277.65	993.59	2301.48
14	176.32	1094.91	1206.57
15	64.67	1206.57	0.00

Amortization Tables for 15-Year Loans (continued)
10%

Year	Interest Paid per Year	Principal Paid per Year	Mortgage Balance at End of Year
1	$986.35	$ 303.17	$9696.83
2	954.61	334.92	9361.91
3	919.54	369.99	8991.92
4	880.80	408.73	8583.19
5	838.00	451.53	8131.66
6	790.72	498.81	7632.85
7	738.48	551.04	7081.81
8	680.78	608.74	6473.06
9	617.04	672.49	5800.58
10	546.62	742.91	5057.67
11	468.83	820.70	4236.97
12	382.89	906.64	3330.33
13	287.95	1001.57	2328.76
14	183.08	1106.45	1222.31
15	67.22	1222.31	0.00

10-1/4%

Year	Interest Paid per Year	Principal Paid per Year	Mortgage Balance at End of Year
1	$1011.32	$ 296.62	$9703.38
2	979.45	328.49	9374.89
3	944.15	363.79	9011.10
4	905.06	402.88	8608.22
5	861.77	446.17	8162.04
6	813.82	494.12	7667.92
7	760.73	547.21	7120.71
8	701.93	606.01	6514.70
9	636.81	671.13	5843.57
10	564.69	743.25	5100.32
11	484.83	823.12	4277.20
12	396.38	911.56	3365.64
13	298.43	1009.52	2356.12
14	189.95	1117.99	1238.13
15	69.81	1238.13	0.00

Amortization Tables for 15-Year Loans (continued)
10-1/2%

Year	Interest Paid per Year	Principal Paid per Year	Mortgage Balance at End of Year
1	$1036.30	$ 290.18	$9709.82
2	1004.32	322.16	9387.66
3	968.82	357.66	9030.00
4	929.40	397.08	8632.92
5	885.64	440.84	8192.08
6	837.06	489.42	7702.67
7	783.12	543.35	7159.31
8	723.25	603.23	6556.08
9	656.77	669.71	5886.37
10	582.96	743.52	5142.85
11	501.02	825.45	4317.40
12	410.06	916.42	3400.97
13	309.06	1017.42	2383.56
14	196.94	1129.54	1254.02
15	72.46	1254.02	0.00

10-3/4%

Year	Interest Paid per Year	Principal Paid per Year	Mortgage Balance at End of Year
1	$1061.28	$ 283.85	$9716.15
2	1029.22	315.92	9400.23
3	993.54	351.60	9048.63
4	953.82	391.32	8657.31
5	909.62	435.52	8221.79
6	860.42	484.72	7737.07
7	805.67	539.47	7197.60
8	744.73	600.41	6597.20
9	676.91	668.23	5928.97
10	601.43	743.71	5185.26
11	517.42	827.72	4357.55
12	423.92	921.21	3436.33
13	319.87	1025.27	2411.06
14	204.05	1141.08	1269.98
15	75.16	1269.98	0.00

Amortization Tables for 15-Year Loans (continued)
11%

Year	Interest Paid per Year	Principal Paid per Year	Mortgage Balance at End of Year
1	$1086.28	$ 277.64	$9722.36
2	1054.15	309.76	9412.60
3	1018.31	345.61	9066.99
4	978.31	385.60	8681.38
5	933.69	430.23	8251.16
6	883.91	480.01	7771.15
7	828.36	535.56	7235.59
8	766.38	597.53	6638.06
9	697.24	666.68	5971.38
10	620.09	743.82	5227.55
11	534.02	829.90	4397.66
12	437.96	925.93	3471.72
13	330.83	1033.08	2438.64
14	211.29	1152.63	1286.01
15	77.91	1286.01	0.00

11-1/4%

Year	Interest Paid per Year	Principal Paid per Year	Mortgage Balance at End of Year
1	$1111.28	$ 271.53	$9728.47
2	1079.11	303.70	9424.76
3	1043.12	339.69	9085.08
4	1002.88	379.94	8705.14
5	957.86	424.95	8280.18
6	907.51	475.31	7804.88
7	851.19	531.62	7273.26
8	788.20	594.61	6678.65
9	717.75	665.07	6013.58
10	638.95	743.87	5269.71
11	550.81	832.00	4437.71
12	452.23	930.58	3507.13
13	341.97	1040.85	2466.28
14	218.64	1164.17	1302.11
15	80.70	1302.11	0.00

Amortization Tables for 15-Year Loans (continued)
11-1/2%

Year	Interest Paid per Year	Principal Paid per Year	Mortgage Balance at End of Year
1	$1136.29	$ 265.53	$9734.47
2	1104.09	297.73	9436.73
3	1067.99	333.84	9102.90
4	1027.51	374.32	8728.58
5	982.12	419.71	8308.87
6	931.23	470.60	7838.27
7	874.16	527.66	7310.61
8	810.18	591.65	6718.96
9	738.44	663.39	6055.57
10	657.99	743.83	5311.74
11	567.80	834.03	4477.71
12	466.66	935.16	3542.55
13	353.27	1048.56	2493.98
14	226.12	1175.71	1318.28
15	83.55	1318.28	0.00

11-3/4%

Year	Interest Paid per Year	Principal Paid per Year	Mortgage Balance at End of Year
1	$1161.31	$ 259.65	$9740.35
2	1129.11	291.85	9448.50
3	1092.90	328.05	9120.45
4	1052.21	368.74	8751.71
5	1006.47	414.48	8337.22
6	955.06	465.89	7871.33
7	897.27	523.68	7347.64
8	832.32	588.64	6759.00
9	759.30	661.66	6097.35
10	677.23	743.73	5353.62
11	584.98	835.98	4517.64
12	481.29	939.67	3577.97
13	364.73	1056.23	2521.74
14	233.72	1187.24	1334.50
15	86.45	1334.50	0.00

Amortization Tables for 15-Year Loans (continued)
12%

Year	Interest Paid per Year	Principal Paid per Year	Mortgage Balance at End of Year
1	$1186.34	$ 253.86	$9746.14
2	1154.14	286.06	9460.08
3	1117.86	322.34	9137.74
4	1076.98	363.22	8774.52
5	1030.92	409.28	8365.23
6	979.01	461.19	7904.04
7	920.52	519.68	7384.36
8	854.61	585.59	6798.77
9	780.34	659.86	6138.91
10	696.66	743.55	5395.36
11	602.35	837.85	4557.51
12	496.09	944.11	3613.41
13	376.36	1063.84	2549.56
14	241.44	1198.77	1350.80
15	89.40	1350.80	0.00

12-1/4%

Year	Interest Paid per Year	Principal Paid per Year	Mortgage Balance at End of Year
1	$1211.37	$ 248.19	$9751.81
2	1179.20	280.36	9471.46
3	1142.86	316.69	9154.76
4	1101.81	357.74	8797.02
5	1055.44	404.11	8392.91
6	1003.06	456.49	7936.41
7	943.90	515.66	7420.75
8	877.06	582.50	6838.25
9	801.55	658.00	6180.24
10	716.26	743.29	5436.95
11	619.92	839.64	4597.31
12	511.09	948.47	3648.84
13	388.15	1071.41	2577.43
14	249.28	1210.28	1367.15
15	92.41	1367.15	0.00

Amortization Tables for 15-Year Loans (continued)
12-1/2%

Year	Interest Paid per Year	Principal Paid per Year	Mortgage Balance at End of Year
1	$1236.41	$ 242.61	$9757.39
2	1204.29	274.74	9482.65
3	1167.91	311.12	9171.53
4	1126.71	352.32	8819.21
5	1080.06	398.97	8420.24
6	1027.23	451.80	7968.44
7	967.40	511.63	7456.81
8	899.65	579.37	6877.44
9	822.93	656.09	6221.35
10	736.06	742.97	5478.38
11	637.68	841.35	4637.03
12	526.27	952.76	3684.27
13	400.11	1078.92	2605.35
14	257.24	1221.78	1383.57
15	95.46	1383.57	0.00

12-3/4%

Year	Interest Paid per Year	Principal Paid per Year	Mortgage Balance at End of Year
1	$1261.46	$ 237.15	$9762.85
2	1229.39	269.21	9493.64
3	1192.99	305.62	9188.03
4	1151.66	346.94	8841.09
5	1104.75	393.85	8447.23
6	1051.49	447.11	8000.12
7	991.03	507.57	7492.55
8	922.40	576.21	6916.34
9	844.48	654.12	6262.22
10	756.03	742.57	5519.65
11	655.62	842.98	4676.66
12	541.63	956.97	3719.69
13	412.23	1086.38	2633.32
14	265.33	1233.28	1400.04
15	98.56	1400.04	0.00

Amortization Tables for 15-Year Loans (continued)
13%

Year	Interest Paid per Year	Principal Paid per Year	Mortgage Balance at End of Year
1	$1286.51	$ 231.78	$9768.22
2	1254.52	263.77	9504.45
3	1218.11	300.18	9204.27
4	1176.68	341.62	8862.65
5	1129.52	388.77	8473.89
6	1075.86	442.43	8031.45
7	1014.79	503.50	7527.95
8	945.29	573.00	6954.95
9	866.20	652.09	6302.86
10	776.19	742.10	5560.75
11	673.75	844.54	4716.21
12	557.18	961.11	3755.10
13	424.51	1093.78	2661.32
14	273.54	1244.75	1416.57
15	101.72	1416.57	0.00

13-1/4%

Year	Interest Paid per Year	Principal Paid per Year	Mortgage Balance at End of Year
1	$1311.57	$ 226.51	$9773.49
2	1279.67	258.42	9515.07
3	1243.27	294.81	9220.26
4	1201.74	336.34	8883.92
5	1154.37	383.71	8500.20
6	1100.32	437.76	8062.44
7	1038.67	499.42	7563.03
8	968.32	569.76	6993.27
9	888.07	650.01	6343.25
10	796.52	741.57	5601.69
11	692.07	846.02	4755.67
12	572.91	965.18	3790.50
13	436.96	1101.12	2689.37
14	281.87	1256.22	1433.16
15	104.93	1433.16	0.00

Amortization Tables for 15-Year Loans (continued)
13-1/2%

Year	Interest Paid per Year	Principal Paid per Year	Mortgage Balance at End of Year
1	$1336.64	$ 221.35	$9778.65
2	1304.83	253.15	9525.51
3	1268.46	289.52	9235.99
4	1226.87	331.12	8904.87
5	1179.29	378.69	8526.18
6	1124.89	433.10	8093.09
7	1062.66	495.32	7597.77
8	991.50	566.49	7031.28
9	910.11	647.88	6383.40
10	817.02	740.96	5642.45
11	710.57	847.42	4795.03
12	588.82	969.17	3825.86
13	449.57	1108.41	2717.45
14	290.32	1267.66	1449.79
15	108.19	1449.79	0.00

13-3/4%

Year	Interest Paid per Year	Principal Paid per Year	Mortgage Balance at End of Year
1	$1361.71	$ 216.28	$9783.72
2	1330.02	247.96	9535.76
3	1293.69	284.29	9251.46
4	1252.04	325.94	8925.52
5	1204.29	373.69	8551.83
6	1149.54	428.44	8123.38
7	1086.77	491.21	7632.17
8	1014.81	563.18	7069.00
9	932.30	645.69	6423.31
10	837.70	740.28	5683.03
11	729.25	848.74	4834.29
12	604.90	973.06	3861.21
13	462.34	1115.64	2745.57
14	298.90	1279.09	1466.48
15	111.50	1466.48	0.00

Installment Loan Debt Guidelines

Financial advisors almost unanimously recommend that you spend no more than 15 to 20 percent of your annual take-home pay to pay off installment loan debt—car loans, student loans, credit cards, lines of credit, and so on (they exclude home mortgage loans from this percentage). This table gives some examples using the 20 percent guideline.

Monthly Take-Home Pay	Maximum Amount of Total Monthly Credit and Loan Payments
$1,500	$ 300
1,750	350
2,000	400
2,250	450
2,500	500
2,750	550
3,000	600
3,250	650
3,500	700
3,750	750
4,000	800
4,250	850
4,500	900
4,750	950
5,000	1,000
5,250	1,050
5,500	1,100
5,750	1,150
6,000	1,200
6,250	1,250
6,500	1,300
6,750	1,350
7,000	1,400
7,250	1,450
7,500	1,500
7,750	1,550
8,000	1,600

Credit Cards

Interest paid on credit card debt is usually called the finance charge. This charge may vary widely among lenders because of differences in the terms of the card—annual fees, lack of grace periods, and other factors. Finance charges may be assessed (1) on the daily unpaid balance, (2) on the adjusted balance (the difference between the amount of your payment and the beginning balance), or (3) on the total amount of your previous balance before you make your payment. Interest rates and annual fees can vary widely among card-issuing institutions, so it pays to shop around.

Your record of paying on credit cards, home loans, personal loans, and car loans is most likely being reported to one of three major credit bureaus. You have a right to see what your credit report says about you, and it's a good idea to check it for errors periodically—especially before you plan to take out a large loan. Here are addresses and phone numbers for the three bureaus; each has files on about 200 million Americans:

Equifax Information Service Center
P.O. Box 740241
Atlanta, GA 30374
(800) 525-6285

Trans Union National Consumer Disclosure Center
P.O. Box 7000
North Olmstead, OH 44070
(800) 680-7289

TRW Consumer Assistance Center
P.O. Box 2350
Chatsworth, CA 91313
(800) 301-7195

Credit Card Debt The average American carries 8 credit cards. In 1994 the typical balance on VISA and MasterCard accounts was $1,680 at an average interest rate of 17.48 percent. This means that the average credit card holder paid $294 in interest that year.

**What Does
APR Mean?** The annual percentage rate (APR) and the periodic interest
rate are not necessarily the same thing. The APR represents all
fees connected with the loan plus a year's worth of interest
charges. Using the APR makes it easier for consumers to
compare the true cost of loans.

Reporting Requirements for Erroneous Billings and Lost or Stolen Bank and Credit Cards

Type	To Report Billing Errors	To Report Loss
Credit	Notify issuer within 60 days of bill receipt	If you call card issuer immediately, you will have no liability; if you don't report loss, your liability is $50
Debit	Same as above	Call bank within 2 to 4 days (depends on your bank); if you wait, your liability may be up to $500; after 60 to 90 days, you may be responsible for entire loss
ATM	Same as above	Notify bank within 2 days; liability is limited to $50; same as debit card (above) for longer periods

Checks and Checking Accounts

Here is a general rule for the earliest times your bank will let you
write checks on deposits you have made:

- Cash deposits, money orders, government checks, local bank
  checks: 2nd business day after deposit

- Out-of-town checks: 5th business day after deposit

In general, banks will not honor checks that are 6 months older than
the dates on them. U.S. government checks issued after October 1,
1989, are void after 1 year.

The various numbers printed on your checks have certain meanings
and purposes:

The first two digits are the bank's American Bankers Association number for the state or area of the bank, followed by the bank's identification number. The number below the rule is the bank's Federal Reserve district, office, and state.

Check number

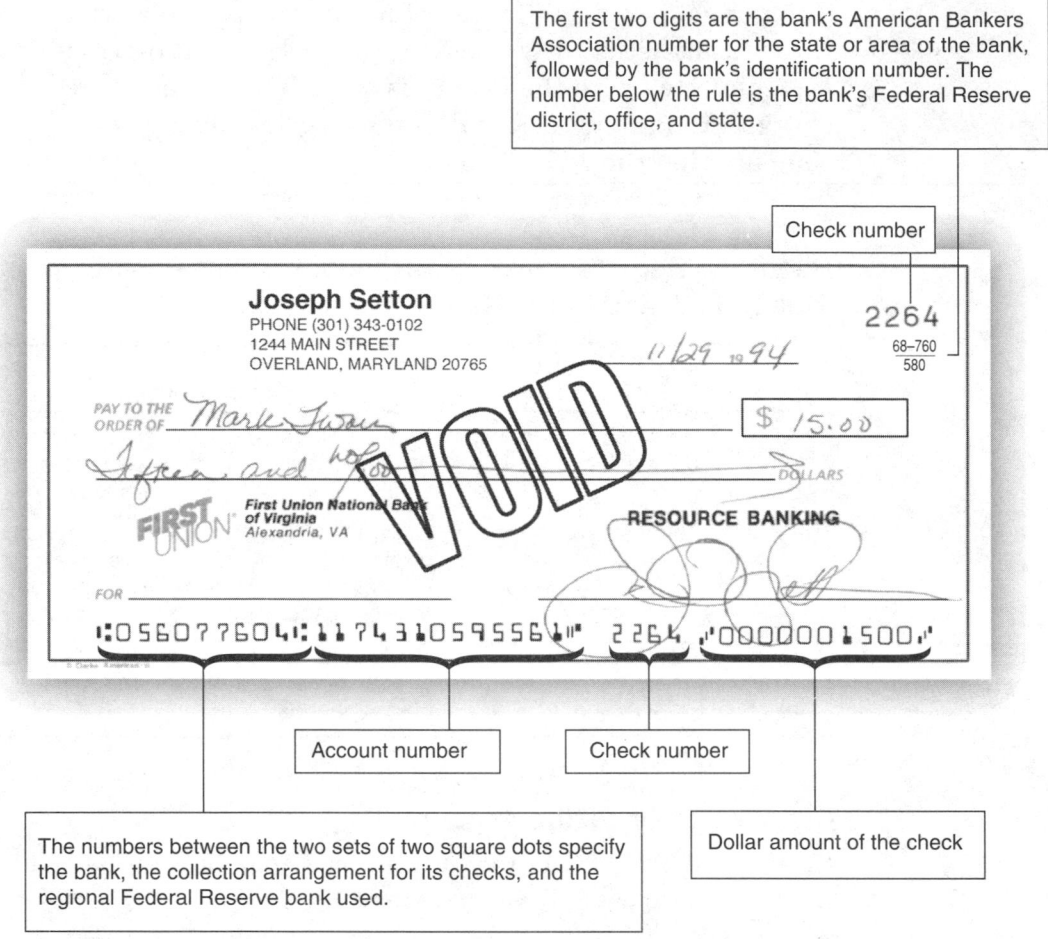

Account number

Check number

The numbers between the two sets of two square dots specify the bank, the collection arrangement for its checks, and the regional Federal Reserve bank used.

Dollar amount of the check

Balancing Your Checkbook: The Power of 9

Balancing your checkbook is never a pleasant task. Here's a trick to use if your numbers won't balance. Take the difference between your bottom line and the bank's and divide it by 9. If the result is evenly divisible by 9 (for example, 81 or 72), you may have simply transposed some numbers (writing 234 instead of 324). This won't solve the problem, but at least you'll have something to look for as you search for the error.

Other important numbers appear on the back of the canceled check:

This date shows the day your bank debited your account.

This number indicates the date the check was deposited; it can be used to prove the date of deposit.

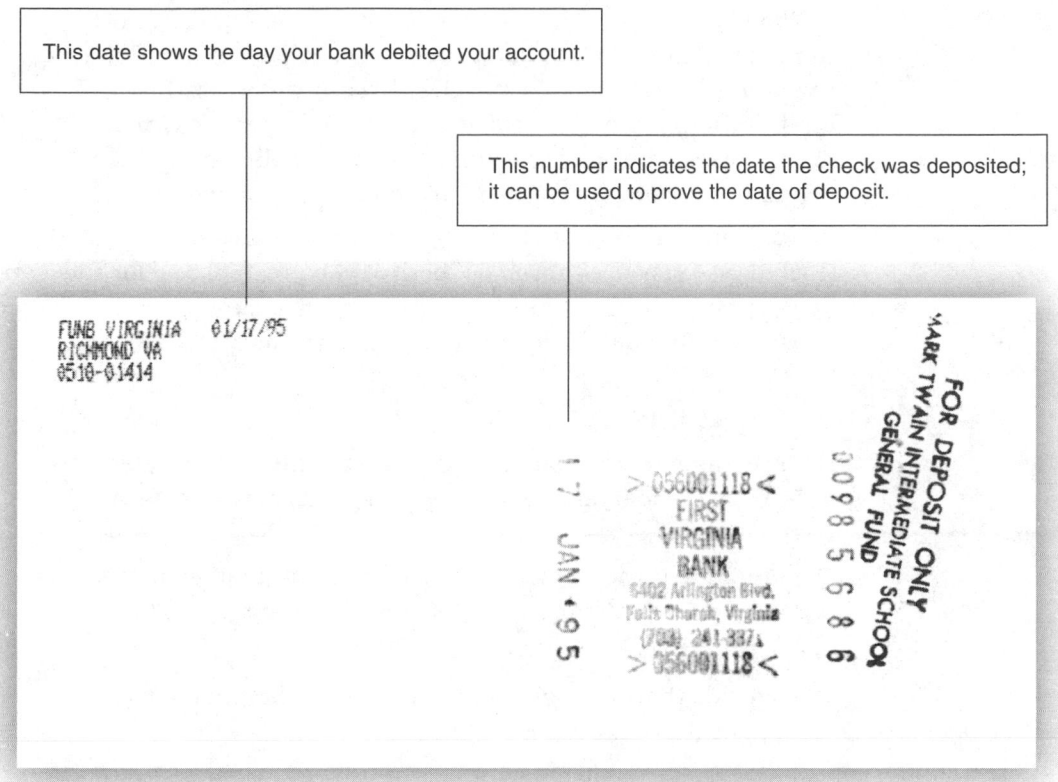

Using Numbers to Stop Check Fraud

A scholar in Nova Scotia is using an obscure mathematical theory to help companies and tax authorities catch check forgers and tax cheats.

The theory, known as Benford's Law, describes the statistical frequency in which the numbers 1 through 9 appear in any set of random numbers. Mark Negrini, who teaches accounting at St. Mary's University in Halifax, uses a computer program to check tax returns and suspicious checks. Legitimate amounts will be truly random, whereas made-up numbers will not.

Buying and Selling Your Home

Most lenders use the 28/36 qualifying ratio to decide whether they will give you a loan. The 28 refers to the percentage of your total income you can afford to spend on mortgage payments, property taxes, and homeowner's insurance. Even if you meet this qualification, you may be turned down if your total payments for mortgage loans and installment loans (credit card debt, car payments, etc.) will exceed 36 percent of your income.

For example, assume that your gross monthly income is $4,000 (include your spouse's income if both of you are applying for the loan). To determine the maximum monthly mortgage payment plus related expenses that will be allowed by your lender, multiply 28 percent by $4,000:

$$0.28 \times \$4,000 = \$1,120$$

Refer to the mortgage payment tables starting on page 19 to determine the loan amount that corresponds to that monthly payment.

To determine if you might be turned down for a home mortgage loan, multiply your total annual income by 36 percent. For example,

$$0.36 \times \$48,000 = \$17,280$$

Then divide that number by 12 to get a monthly amount:

$$\$17,280 \div 12 = \$1,440$$

You may be turned down for a loan if your expected monthly house payment (use the charts on pages 19–29 to estimate monthly mortgage payments, then add 15 percent to cover taxes and insurance), combined with all other monthly payments for loans and credit cards, adds up to more than that number.

Numbers to Know: Buying or Selling Your Home

A *point* is 1 percent of the loan amount.

The minimum down payment you will need is usually 10 to 20 percent of the purchase price.

Equity means the amount of money you have invested in your house—the down payment you made and the principal paid to date on your loan—plus any increase in the appraised value of your house since the time of its purchase. For example, if your house is appraised at $200,000 and your remaining loan amount (see the amortization tables starting on page 31 to estimate this amount, or call your lender) is $106,000, your equity is $94,000.

All mortgage loan interest is deductible on your federal income taxes for loans up to $1 million. You may deduct mortgage interest on a maximum of 2 homes.

Real estate agents and brokers generally charge 6 percent commission on home sales and up to 10 percent on sales of properties in resort or vacation areas.

First-Time Homebuyer Loan Affordability

Year	Starter Home Price	10% Down Payment	Loan Amount	Effective Interest Rate* (%)	Monthly Payment	Prime First-Time Median Income	Qualifying Income
1992	$88,100	$8,810	$79,290	8.11	$602	$23,133	$28,887
1993	90,800	9,080	81,720	7.16	566	23,475	27,186
1994	93,300	9,330	83,970	7.47	600	24,203	28,792
1995	94,400	9,440	84,960	7.98	637	24,735	30,580

Source: National Association of Realtors®.

*Effective rate on loans closed on existing homes from the Federal Housing Finance Board.

Median Sales Prices of Existing Single-Family Homes, 1992 to 1995

Metropolitan Area	1992	1993	1994	1995*
Albany/Schenectady/Troy, NY	$111,400	$112,300	$112,000	$104,400
Albuquerque, NM	92,000	100,400	110,000	116,300
Anaheim/Santa Ana, CA	230,900	217,200	211,000	207,200
Atlanta, GA	89,500	91,800	93,600	97,000
Austin/San Marcos, TX	83,800	91,300	96,200	101,900
Baltimore, MD	113,400	115,700	115,400	109,000
Birmingham, AL	90,900	96,500	100,200	98,500
Boston, MA	171,100	173,200	179,300	179,000
Buffalo/Niagara Falls, NY	81,700	83,500	82,300	81,300
Charlotte/Gastonia/Rock Hill, NC/SC	102,200	106,100	106,500	105,900
Chicago, IL	136,800	142,000	144,100	147,200
Cincinnati, OH/KY/IN	88,600	91,400	96,500	100,600
Cleveland, OH	90,700	95,000	98,500	103,400
Columbus, OH	91,000	91,800	94,800	98,100
Dallas, TX	91,300	94,500	95,000	95,300
Denver, CO	96,200	104,700	116,800	125,500
Detroit, MI	81,300	86,000	87,000	97,100
Ft. Lauderdale/Hollywood/Pompano Beach, FL	99,100	103,100	103,100	103,100
Ft. Worth/Arlington, TX	80,200	82,900	82,500	82,100
Hartford, CT	141,100	135,300	133,400	130,800
Honolulu, HI	349,000	358,500	360,000	340,000
Houston, TX	80,300	80,900	80,500	78,000
Indianapolis, IN	83,700	86,600	90,700	93,700
Jacksonville, FL	76,800	77,100	81,900	80,700
Kansas City, MO/KS	79,500	83,600	87,100	90,300
Las Vegas, NV	104,300	108,200	110,500	110,500
Los Angeles, CA	210,800	195,400	189,100	176,300
Louisville, KY/IN	69,500	74,500	80,500	84,000
Memphis, TN/AR/MS	85,300	87,000	86,300	83,300
Miami/Hialeah, FL	97,100	98,800	103,200	106,300
Milwaukee, WI	97,000	104,100	109,000	115,300
Minneapolis/St. Paul, MN/WI	94,200	98,200	101,500	103,300
Nashville, TN	88,800	90,400	96,500	103,900
New Orleans, LA	73,600	76,800	76,900	79,400

Median Sales Prices of Existing Single-Family Homes, 1992 to 1995 (continued)

Metropolitan Area	1992	1993	1994	1995*
New York/N. New Jersey/ Long Island, NY/NJ/CT	$172,700	$173,200	$173,200	$169,200
Oklahoma City, OK	61,600	64,900	66,700	69,300
Orlando, FL	87,600	90,100	90,700	89,200
Philadelphia, PA/NJ	117,000	118,000	119,500	117,100
Phoenix, AZ	86,800	89,100	91,400	94,800
Pittsburgh, PA	78,600	82,200	80,700	83,700
Portland, OR	97,700	106,000	116,900	127,200
Providence, RI	118,500	116,300	116,400	115,400
Raleigh/Durham, NC	105,900	109,200	115,200	126,400
Richmond/Petersburg, VA	93,900	94,100	95,400	96,500
Riverside/San Bernardino, CA	136,200	134,500	129,100	122,600
Rochester, NY	84,700	84,800	85,600	84,100
Sacramento, CA	134,000	129,200	124,500	118,300
St. Louis, MO/IL	83,200	84,800	85,000	86,400
Salt Lake City/Ogden, UT	76,500	84,900	98,000	111,500
San Antonio, TX	70,400	77,000	78,200	80,000
San Diego, CA	183,100	176,900	176,000	170,000
San Francisco, CA	259,300	254,400	255,600	260,900
Seattle, WA	145,700	150,200	155,900	158,800
Tampa/St. Petersburg/Clearwater, FL	72,600	75,000	76,200	77,300
Washington, DC/MD/VA	157,800	158,300	157,900	156,200
West Palm Beach/Boca Raton/Delray Beach, FL	114,100	114,600	117,600	118,600

Source: National Association of Realtors®.

Note: All areas are metropolitan statistical areas (MSAs) as defined by the U.S. Office of Management and Budget as of 1992. They include the named central city and surrounding areas.

*Figures are for the 2nd quarter and are not seasonally adjusted.

Rule of Thumb: How Much Home Can You Afford? An often-used guideline is that you can afford to buy a home that costs up to $2\frac{1}{2}$ times your gross annual income.

Median Sales Prices of New Single-Family Homes, 1970 to 1994

Year	U.S.	Northeast	Midwest	South	West
1970	$ 23,400	$ 30,300	$ 24,400	$ 20,300	$ 24,000
1971	25,200	30,600	27,200	22,500	25,500
1972	27,600	31,400	29,300	25,800	27,500
1973	32,500	37,100	32,900	30,900	32,400
1974	35,900	40,100	36,100	34,500	35,800
1975	39,300	44,000	39,600	37,300	40,600
1976	44,200	47,300	44,800	40,500	47,200
1977	48,800	51,600	51,500	44,100	53,500
1978	55,700	58,100	59,200	50,300	61,300
1979	62,900	65,500	63,900	57,300	69,600
1980	64,600	69,500	63,400	59,600	72,300
1981	68,900	76,000	65,900	64,400	77,800
1982	69,300	78,200	68,900	66,100	75,000
1983	75,300	82,200	79,500	70,900	80,100
1984	79,900	88,600	85,400	72,000	87,300
1985	84,300	103,300	80,300	75,000	92,600
1986	92,000	125,000	88,300	80,200	95,700
1987	104,500	140,000	95,000	88,000	111,000
1988	112,500	149,000	101,600	92,000	126,500
1989	120,000	159,600	108,800	96,400	139,000
1990	122,900	159,000	107,900	99,000	147,500
1991	120,000	155,900	110,000	100,000	141,100
1992	121,500	169,000	115,600	105,500	130,400
1993	126,500	162,600	125,000	115,000	135,000
1994	130,000	169,000	132,900	116,900	140,400

Source: *Statistical Abstract of the United States, 1995.*

Americans on the Move

More than 43 million Americans—16.7 percent of the population—moved during a 12-month period in 1993 and 1994. Most people—26.6 million—moved within the same county. Eight million moved between counties in the same state, 7 million moved to another state, and 1.2 million moved to the United States from abroad.

These numbers reflect a general decrease in mobility over the past 10 years; 20.2 percent of Americans moved during a similar period in 1984–85.

Features of New Single-Family Homes, Selected Years, 1970 to 1994

Characteristic	1970	1980	1990	1992	1994
Total new houses (in thousands)	793	957	966	964	1,160
Financing (%)					
Mortgage	84	81	82	86	89
FHA-insured	30	16	14	10	8
VA-guaranteed	7	8	4	5	5
Conventional	47	55	62	69	74
Farmers Home Administration	(a)	3	2	1	1
Cash or equivalent	16	18	18	14	11
Floor area					
Under 1,200 sq. ft. (%)	36	21	11	10	9
1,200 to 1,599 sq. ft. (%)	28	29	22	22	21
1,600 to 1,999 sq. ft. (%)	16	22	22	23	24
2,000 to 2,399 sq. ft. (%)	21	13	17	17	18
2,400 sq. ft. and over (%)	(b)	15	29	29	29
Average (sq. ft.)	1,500	1,740	2,080	2,095	2,100
Median (sq. ft.)	1,385	1,595	1,905	1,920	1,940
Number of stories (%)					
1	74	60	46	48	49
2 or more	17	31	49	47	47
Split-level	10	8	4	5	3
Foundation (%)					
Full or partial basement	37	36	38	42	39
Slab	36	45	40	38	41
Crawlspace	27	19	21	20	20
Bedrooms (%)					
2 or fewer	13	17	15	12	12
3	63	63	57	59	58
4 or more	24	20	29	29	30
Bathrooms (%)					
1½ or fewer	20	10	5	13	11
2	32	48	42	40	40
2½ or more	16	25	45	47	49
Heating fuel (%)					
Electricity	28	50	33	29	29
Gas	62	41	59	65	67
Oil	8	3	5	4	3
Other	1	5	3	2	1

Features of New Single-Family Homes, Selected Years, 1970 to 1994 (continued)

Characteristic	1970	1980	1990	1992	1994
Central air conditioning (%)					
With	34	63	76	77	79
Without	66	37	24	23	21
Fireplaces (%)					
None	65	43	34	36	36
1 or more	35	56	66	64	64
Parking facilities (%)					
Garage	58	69	82	83	86
Carport	17	7	2	2	2
No garage or carport	25	24	16	15	13

Source: *Statistical Abstract of the United States, 1995.*

Note: Percentages may not total 100 percent because some categories were not fully reported.

a. Included with conventional financing.

b. Included with floor area of 2,000 to 2,399 square feet.

Refinancing Your Home

When interest rates are low, it may pay to refinance your mortgage. To decide if refinancing makes sense for you, ask your lender to tell you the amount of your new monthly payment and give you an estimate of closing costs on the new loan. Subtract the new payment from the old. Estimate how much longer you plan to live in your house, in terms of months. Then multiply the number of months you plan to stay in your house by the amount you will save monthly in mortgage payments.

If the resulting number is more than the amount of your closing costs (which can be substantial), refinancing makes sense.

Rule of Thumb: Another Guide on When to Refinance

If the new interest rate you could get is 2 or more percentage points lower than your existing rate, refinancing will usually pay off, especially if you plan to stay in your house for several more years.

Rule of Thumb: How Much to Spend on Remodeling

Many people refinance or take out home equity loans to make major improvements to their homes. If you decide to remodel or add on to your existing home, be careful not to increase the value of your house more than 20 percent over the value of other houses in your neighborhood. See Chapter 5, Daily Life, for more numbers about remodeling.

Home Equity Loans and Lines of Credit

If you took out a home equity loan after October 13, 1987, you can deduct the interest incurred on loans up to $100,000, regardless of the purpose of the loan. If you took out a loan before that date, you must be able to show how the proceeds were used. You can usually borrow up to 80 percent of the equity you have in your home.

Renting Your Home

If you rent your vacation home for more than 14 days a year, the IRS considers it to be a rental business, not a residence, for tax purposes.

If you plan to take the one-time capital gains tax exclusion of $125,000 after age 55, you can rent your home before you sell it, but you must have lived in it (as your principal residence) for at least 3 of the previous 5 years before you sell it.

Selling Your Home

Any profit you make when you sell your home is taxable unless you buy a house of equal or greater value within 2 years. There are a few exceptions to this rule, the most common one being the *age 55 exclusion*. If you or your spouse is age 55 or older when you sell, you may exclude up to $125,000 in profits on the sale of your home (the amount is $62,500 for single persons) one time only in a lifetime, even if each spouse owns a house. The home must be your main residence, and you must have lived in it for at least 3 of the past 5 years (or 1 year of the past 5 if the owner is in a nursing home or similar facility). Consult a tax advisor for more information.

To figure the amount of profit when you sell, you first need to determine your home's *cost basis*. The cost basis is your original purchase price minus the costs of all major improvements—for example, a new roof, a remodeled bathroom, a new deck. To determine your profit for tax purposes, total your selling costs (including title insurance, transfer fees, legal fees, and real estate commission). Then subtract the cost basis number and the selling costs number from the selling price. For example, assume you purchased your home for $100,000 and are selling it for $150,000:

Original purchase price	$ 100,000
Improvements	−10,000
Cost basis	90,000
Selling costs	−12,000
Adjusted selling price	$ 78,000

To figure your profit on the sale, subtract $78,000 from your selling price of $150,000:

Selling price	$ 150,000
Adjusted selling price	− 78,000
Profit on sale	$ 72,000

Calculating Your Net Worth

Knowing your net worth will give you a starting point for setting personal finance goals. If you apply for a mortgage or other large loan, your lender may require you to complete a similar form.

Assets

Cash in checking, savings, money market accounts _____

Any debts you are owed _____

Current market value of stocks and bonds _____

Value of other investments _____

Cash value of life insurance (cash surrender value) _____

IRA, Keogh, and 401(k) accounts _____

Vested interest in pension or profit sharing _____

Real estate (current value of your home, other property) _____

Business investments _____

Personal property (furnishings, jewelry, cars, collectibles, etc.) _____

TOTAL ASSETS _____

Liabilities

Balances on—

    Mortgages _____

    Bank loans, including home equity _____

    Car loans _____

    Lines of credit _____

    Charge accounts _____

    Student loans _____

Alimony/child support owed (this year) _____

Taxes owed (income, real estate, etc.) _____

TOTAL LIABILITIES _____

Net Worth (subtract liabilities from assets) _____

Saving for College

Estimating Future College Costs

To estimate future college costs, take the current cost of the college of your choice (or an average from the table on the following page) and multiply by the inflation factor below. The inflation factor assumes a 6 percent yearly rate of inflation in college costs; however, some financial experts are now predicting a drop in college cost increases, to about 5 percent.

Years Until College	Inflation Factor
1	1.06
2	1.12
3	1.19
4	1.26
5	1.34
6	1.42
7	1.51
8	1.60
9	1.70
10	1.80
11	1.91
12	2.02
13	2.14
14	2.27
15	2.41
16	2.55
17	2.70
18	2.87

Average College Costs for Undergraduates, 1993–1994 School Year

Type of Institution/ Item	Public Colleges (in-state student)		Private Colleges	
	Resident	Commuter	Resident	Commuter
4-year colleges, total	$8,562	$6,809	$17,846	$15,200
Tuition and fees	2,527	2,527	11,025	11,025
Books and supplies	552	552	556	556
Room and board	3,680	1,601	4,793	1,722
Transportation	557	870	498	824
Other	1,246	1,259	974	1,073
2-year colleges, total	(a)	5,372	12,142	10,190
Tuition and fees	1,229	1,229	6,175	6,175
Books and supplies	533	533	566	566
Room and board	(a)	1,643	3,980	1,589
Transportation	(a)	923	487	890
Other	(a)	1,044	934	970

Source: The College Board, New York, NY, *Annual Survey of Colleges*.

a. Base too small to meet statistical standards for reliability of a derived figure.

Student Aid Information

Many kinds of student loans, grants, and scholarships are available. Here are a few sources.

- For information about federal government grants and loans, call the Federal Student Aid Information Center at (800) 433-3243.

- The College Board offers a software program called the College Explorer Fund Finder that contains information on thousands of private scholarships and financial aid programs. This software is often available in high school guidance offices; if yours does not have it, call the College Board at (212) 713-8165.

- Private companies also provide scholarship and aid information for a fee. One is the National Scholarship Research Service at (800) 432-3782.

Amount to Save to Meet College Expenses

This table lets you estimate how much you will need to save, per year, to have $10,000 available for your child's college education. If you think you'll need $80,000 to cover 4 years of schooling, multiply the monthly savings amount shown by 8.

If You Start Saving When Your Child Is—	Number of Years of Saving	Monthly Savings	Principal	Interest Earned	Total Savings
4% interest					
Newborn	18	$ 32	$6,912	$3,187	$10,099
Age 4	14	45	7,560	2,552	10,112
Age 8	10	68	8,160	1,853	10,013
Age 12	6	124	8,928	1,144	10,072
Age 16	2	401	9,624	378	10,002
8% interest					
Newborn	18	$ 21	$4,536	$5,546	$10,082
Age 4	14	33	5,544	4,621	10,165
Age 8	10	55	6,600	3,462	10,062
Age 12	6	109	7,848	2,183	10,031
Age 16	2	386	9,264	746	10,010

Source: U.S. Department of Education, "Preparing Your Child for College."

Retirement Planning

Many financial advisors say that in retirement you should expect that you will need 70 percent of your current income (adjusted for inflation) to maintain your current lifestyle. They also warn that company pensions and Social Security may provide only 20 to 40 percent of your total retirement income needs. The table below shows the amount you need to save each month (assuming 6 percent interest) to have a certain amount by age 65.

Monthly Savings Needed to Reach Savings Goal by Age 65

Current Age	Amount Saved by Age 65			
	$200,000	$300,000	$500,000	$1,000,000
	Monthly Savings			
25	$ 102	$ 153	$ 255	$ 510
30	140	210	350	700
35	198	297	495	990
40	286	429	715	1,430
45	426	639	1,065	2,130
50	674	1,011	1,685	3,370
55	1,192	1,788	2,980	5,960

Unisex Life Expectancy Table

Knowing average life expectancies can help in planning for retirement. The U.S. government has developed a unisex life expectancy chart for this purpose:

Current Age	Life Expectancy
55	23 more years
60	20 more years
65	16 more years
70	13 more years

Estimating the Future Value of Assets and Investments

To get a general idea of what your savings and other assets will be worth at retirement, use this worksheet and tables A and B on the following page.

Present Assets/Investments

Type	Current Value		Factor from Table A		Future Value
Savings	$_____	×	_____	=	$_____
Equity in home	_____	×	_____	=	_____
Pension/401(k)	_____	×	_____	=	_____
IRA/Keogh/SEP	_____	×	_____	=	_____
Annuities	_____	×	_____	=	_____
Stocks/mutual funds	_____	×	_____	=	_____
Bonds	_____	×	_____	=	_____
Real estate	_____	×	_____	=	_____
Other investments	_____	×	_____	=	_____
Collectibles	_____	×	_____	=	_____
Cash value of life insurance	_____	×	_____	=	_____
TOTAL					$_____

Future Investments

Type	Annual Investment		Factor from Table B		Future Value
_____	$_____	×	_____	=	$_____
_____	_____	×	_____	=	_____
TOTAL					$_____

Total Future Wealth

Add both totals from above columns	$_____
Subtract expected annual tax payments	−_____
Subtract anticipated inflation*	−_____
ACTUAL TOTAL FUTURE WEALTH	$_____

*Assuming 4 percent annual inflation rate, use the guide on the following page.

Anticipated Inflation

Years to Retirement	Factor
5	0.83
10	0.68
15	0.56
20	0.46
25	0.38
30	0.31

Table A

Expected Return (%)	Years to Retirement				
	5	10	15	20	25
2	1.10	1.22	1.35	1.49	1.64
4	1.22	1.48	1.80	2.19	2.67
6	1.34	1.79	2.40	3.21	4.29
8	1.47	2.16	3.17	4.66	6.85
10	1.61	2.59	4.18	6.73	10.83

Table B

Expected Return (%)	Years to Retirement				
	5	10	15	20	25
2	5.20	10.95	17.29	24.30	32.03
4	5.42	12.01	20.02	29.78	41.65
6	5.64	13.18	23.28	36.79	54.86
8	5.87	14.49	27.15	45.76	73.11
10	6.11	15.94	31.77	57.28	96.35

How Long Will Your Savings Last?

Now that you have an idea of how much money you will have at age 65, the next step is to calculate how long it will last. The table below shows how much you will need to have invested in order to withdraw $100 a month for the periods shown. To get to the monthly amount you can draw from your savings, divide your expected savings (from the worksheet above) by the number in the body of the table where the expected interest rate and number of years intersect.

For example, assume you will have $350,000 in savings by age 65. You expect to live another 20 years, and you estimate that your investments will earn 9 percent annual interest. Go to the point in the body of the table where 20 years and 9 percent intersect, to get $11,114. Divide that number into $350,000; the result will be 31.49. Multiply 31.49 by $100 to get $3,149, which is the amount you will be able to withdraw monthly for 20 years.

Years	Annual Interest Rate			
	7%	8%	9%	10%
5	$5,050	$4,932	$4,817	$4,706
10	8,613	8,242	7,894	7,567
15	11,125	10,464	9,860	9,306
20	12,898	11,955	11,114	10,362
25	14,149	12,956	11,916	11,005
30	15,030	13,628	12,428	11,395

Numbers to Know: Inflation Rates over the Years

For the years 1984 through 1994, the average rate of inflation was 3.6 percent, as measured by the Consumer Price Index. For the 30-year period from 1963 through 1992, the average rate was 5.3 percent.

Individual Retirement Arrangements (IRAs)

Anyone who has income from employment and is under age 70½ may contribute $2,000 a year to an IRA. If your spouse earns less than $250 a year, that amount increases to $2,250. You will not have to pay tax on interest or dividends earned until you begin to withdraw the funds at retirement, when you will probably be in a lower tax bracket.

You may have more than one IRA, but your total contribution for the year cannot exceed the maximum for that year.

The amount you contribute may also be tax deductible. You can deduct the $2,000, or some portion of that, from your income taxes if (1) neither you nor your spouse participates in an employer-sponsored retirement plan, such as a 401(k), or (2) your adjusted gross income is under a certain amount (see the table below). Note, however, that changes in the law are being considered as of late 1995.

Allowable Tax Deduction	Adjusted Gross Income (line 14, IRS Form 1040)		
	Single or Head of Household	Married Filing Jointly or Qualifying Widow/ Widower	Married Filing Separately
$ 0	$35,000+	$50,000+	$10,000+
200	34,000	49,000	9,000
400	33,000	48,000	8,000
600	32,000	47,000	7,000
800	31,000	46,000	6,000
1,000	30,000	45,000	5,000
1,200	29,000	44,000	4,000
1,400	28,000	43,000	3,000
1,600	27,000	42,000	2,000
1,800	26,000	41,000	1,000
2,000	25,000 or less	40,000 or less	(a)

Note: IRA deductions are prorated for income that falls beneath the amounts shown.
a. Limited to 100 percent of earnings.

If you overcontribute to your IRA, the excess amount may be subject to a 6 percent excise tax. You can avoid the tax if you withdraw the excess before the tax return's due date or if you apply the excess to your next year's IRA contribution.

You can roll over, or move, each of your IRAs once every 12 months (not once every calendar year). If you move an IRA more frequently than that, you will be taxed as if you had received a distribution.

You cannot borrow from your IRA. If you withdraw for any reason, the withdrawal will be subject to a 10 percent penalty and income tax. The only exception to this rule is when you move your IRA from one account to another; in this case, you have 60 days in which to make the move. You need to designate to the original holder that the lump sum is to be rolled over; otherwise, taxes may be withheld and only the net amount rolled over. You will be taxed on that portion not rolled over unless you make up the difference with your own funds.

You can start drawing money out of your IRA without penalty at age 59$\frac{1}{2}$. At that point, you will begin paying tax on the amount taken out. You must stop contributing and start withdrawing by age 70$\frac{1}{2}$ according to a government schedule.

IRAs may be opened as late as the April 15 tax-filing deadline for the previous year's tax return. The earlier in the year you make your contribution, however, the faster your account will grow because of the effects of compound interest.

401(k) Retirement Plans

These company-sponsored retirement plans are named after the section of the IRS tax code that established them in 1981. Employees who participate in these plans (78 percent of all eligible workers do, according to a 1994 survey) can contribute up to a certain amount of their income, tax deferred, each year; the maximum amount for 1996 was $9,500 (or 15 percent of wages, whichever is less). Most companies (84 percent, according to one survey) offer to match the employee's contribution by a certain percentage amount, and they may also distribute profit-sharing monies into employees' 401(k) accounts.

Most companies require a 1-year waiting period before allowing an employee to participate in a 401(k).

You can begin to withdraw from your plan at age 59½. As with an IRA, if you take money out sooner, you may have to pay a 10 percent penalty and income tax on the amount if it is withdrawn for reasons other than those permitted. Unlike with an IRA, you can take out a loan, usually up to an amount that is one-half of your investment and for a term no longer than 5 years, against your 401(k) at an interest rate that is set by your company's plan. Money can also be withdrawn without penalty for a variety of reasons, including hardship, emergency medical expenses, college tuition, or purchasing a home; each plan may set its own rules.

If you leave your company, you have 60 days to move your 401(k) money to another company's plan or to an IRA.

Your contributions are not included when calculating your taxable income; in other words, if your annual salary is $30,000 and you contribute $2,000 to your 401(k), your income for federal tax purposes is only $28,000.

403(b) Retirement Plans

These plans are intended for employees of tax-exempt and nonprofit organizations—schools, hospitals, and so on. They are similar to 401(k) plans except that maximum contribution rates may be higher—$9,500 or 16.66 percent of gross income (whichever is less) in 1995.

Company Pension Plans

A 1989 law required companies to vest employees after 5 years of employment or to use a sliding scale that vests employees 20 percent after 3 years and up to 100 percent after 7 years. Vesting means that the money that *the company has contributed* to your account is yours after that period; if you leave the company before vesting occurs, that money belongs to the company, not to you.

The amount of your pension is usually determined by multiplying the number of years you have participated in the plan by a number from 1.4 to 1.7 percent of your average pay over the last 3 or 5 years of employment.

Most companies reduce the amount of your pension check by a percentage of your Social Security payments. A typical amount is 50 percent. If you elect to take a survivor's annuity, in which your spouse would continue to receive a monthly check upon your death, your check will be cut by another 10 percent or so; the younger your spouse, the greater the percentage.

If you retire before age 65, you can expect to take a reduction in your pension check. A typical cut is 33 percent for workers who retire at age 60 and 50 percent for those who retire at age 55.

Keogh Plans

If you are self-employed, you can invest, tax deferred, up to $30,000 or 20 percent of your net income (whichever is less) in a Keogh plan; the amounts are $22,500 and 13.0435 percent for profit-sharing plans. (These are 1994 limits; they usually change yearly.) As with an IRA, there is a 10 percent penalty for withdrawals before age 59$^{1}/_{2}$. Income tax must also be paid on the withdrawn amounts, but the amounts may be eligible for 5-year forward averaging.

A defined-benefit Keogh plan lets people age 55 and over save up to a certain limit (it was $118,000 or 100 percent of average net income during the 3 highest earning years in 1994); these savings are also tax deferred.

A Keogh must be established by December 31 of the tax year, but contributions do not have to be made until the following April 15.

Simplified Employee Pensions (SEPs)

SEPs are similar to IRAs, except that the employer makes the contributions and the employee establishes the plan. Contributions of up to 15 percent of the employee's net earnings to a maximum of $22,500 (in 1995) may be made annually.

Employers using SEPs must make contributions for all employees who are over age 21, who have worked for the company for at least 3 of the immediately preceding 5 years, and who received at least $300 in compensation from the employer for the year. Individuals who work alone can also set up a SEP; they can deduct up to 13.04 percent of self-employment earnings from taxable income, with a maximum contribution of $22,500 a year.

To establish a SEP that allows contributions to be made through salary deductions, an employer must have 25 or fewer employees, and at least 50 percent of them must choose to participate.

SEPs can be established as late as April 15 for the previous tax year.

Annuities

Annuities are purchased through insurance companies. There are no limits on the amount of your investment, and interest, dividends, and appreciation are tax deferred until withdrawal. Fixed annuities pay a variable rate of interest, and your principal is guaranteed. Various fees and sales charges may apply. Variable annuities are based on stock, bond, or money market performances.

As with other types of retirement plans, if you withdraw before age 59$\frac{1}{2}$, you must pay a 10 percent penalty and income tax on the amount.

Social Security

Social Security eligibility is based on the number of credits earned while working. You earn a credit for each quarter of a year during which you have at least a certain amount in earnings (see the table below). When you have 40 credits (or slightly fewer if you were born before 1929), you are eligible for retirement benefits.

To get a record of your Social Security contributions, including the number of credits earned to date, and a description of your estimated benefits at retirement, call (800) 772-1213 and request Form SSA-7004, Request for Statement of Earnings.

Earnings Needed in a Calendar Quarter to Receive a Social Security Credit

Calendar Year	Maximum Yearly Taxable Amount	Earnings Needed to Earn Quarter of Coverage
1951	$ 3,600	$ 50
1952	3,600	50
1953	3,600	50
1954	3,600	50
1955	4,200	50
1956	4,200	50
1957	4,200	50
1958	4,200	50
1959	4,800	50
1960	4,800	50
1961	4,800	50
1962	4,800	50
1963	4,800	50
1964	4,800	50
1965	4,800	50
1966	6,600	50
1967	6,600	50
1968	7,800	50
1969	7,800	50
1970	7,800	50
1971	7,800	50
1972	9,000	50
1973	10,800	50
1974	13,200	50
1975	14,100	50
1976	15,300	50
1977	16,500	50
1978	17,700	250
1979	22,900	260
1980	25,900	290
1981	29,700	310
1982	32,400	340
1983	35,700	370
1984	37,800	390
1985	39,600	410
1986	42,000	440

Earnings Needed in a Calendar Quarter to Receive a Social Security Credit (continued)

Calendar Year	Maximum Yearly Taxable Amount	Earnings Needed to Earn Quarter of Coverage
1987	43,800	460
1988	45,000	470
1989	48,000	500
1990	51,300	520
1991	53,400	540
1992	55,500	570
1993	57,600	590
1994	60,600	620
1995	61,200	630
1996+	(a)	(a)

a. To be determined.

Monthly Social Security Benefits at Age 65

Age in 1995	Who Receives Benefits	Present Annual Earnings				
		$12,000	$20,000	$30,000	$44,000	$61,200+
65	You	$540	$739	$985	$1,123	$1,199
	Spouse or child	270	369	492	561	599
64	You	544	744	992	1,135	1,217
	Spouse or child	272	372	496	567	608
63	You	557	760	1,015	1,167	1,256
	Spouse or child	278	380	507	583	628
62	You	547	747	997	1,151	1,244
	Spouse or child	273	373	498	575	622
61	You	547	748	998	1,156	1,255
	Spouse or child	273	374	499	578	627
55	You	533*	729*	974*	1,149*	1,280*
	Spouse** or child	275	377	504	594	662
50	You	517*	709*	948*	1,128*	1,283*
	Spouse** or child	277	380	508	605	691
45	You	520*	714*	955*	1,136*	1,318*
	Spouse** or child	279	383	512	609	709
40	You	517*	710*	951*	1,127*	1,319*
	Spouse** or child	280	385	516	612	716

Monthly Social Security Benefits at Age 65 (continued)

Age in 1995	Who Receives Benefits	Present Annual Earnings				
		$12,000	$20,000	$30,000	$44,000	$61,200+
35	You	488*	672*	901*	1,064*	1,249*
	Spouse** or child	282	388	521	615	722
30	You	491*	676*	907*	1,068*	1,255*
	Spouse** or child	283	390	523	616	724

Note: Figures for spouse apply to a spouse who has not worked enough quarters to collect Social Security on his or her own record.

*These amounts are reduced for retirement at age 65 because persons this age will not qualify for full benefits until after age 65; see the table below.

** The amount shown is for the spouse at full retirement age who is caring for an eligible child (under age 16 or disabled); the benefit for a spouse younger than full retirement age who is not caring for an eligible child would be reduced for early retirement.

Ages to Qualify for Full Social Security Retirement Benefits

If you were born in 1937 or earlier, you can collect full benefits on your 65th birthday. Persons born after 1937 will have to be somewhat older. The Social Security Administration reduces benefits by a certain percentage for each month you retire before age 65. For example, if you retire at age 62, your benefits will be permanently reduced by 20 percent; at 63, 13.3 percent; at 63½, 10 percent; and at 64, 6.66 percent.

Year of Birth	Full Retirement Age
1937 or earlier	65
1938	65 and 2 months
1939	65 and 4 months
1940	65 and 6 months
1941	65 and 8 months
1942	65 and 10 months
1943–1954	66
1955	66 and 2 months
1956	66 and 4 months
1957	66 and 6 months
1958	66 and 8 months
1959	66 and 10 months
1960 and after	67

Increases in Benefits If You Work Past Full Retirement Age

If you work beyond age 65 and continue to contribute, you can increase your benefits by the following percentages. Note that you may have to pay federal income tax (and perhaps state tax) on up to 85 percent of your Social Security income if (1) your adjusted gross income, (2) your tax-exempt interest, and (3) one-half of your Social Security benefit equal more than $44,000 (married couple) or $34,000 (single). These figures were effective in 1994 and could rise in the future.

Year of Birth	Yearly Increase (%)
1927–1928	4.0
1929–1930	4.5
1931–1932	5.0
1933–1934	5.5
1935–1936	6.0
1937–1938	6.5
1939–1940	7.0
1941–1942	7.5
1943 or later	8.0

If You Continue to Work: Social Security Earnings Limits and Benefit Reductions

You can continue to work while collecting Social Security, but certain limits apply; those shown are for 1995 and will change in the future.

Age	Limit on Earnings	Benefit Reduction
70 and up	None	None
65–69	$11,280	$1 for every $3 earned over the limit
Under 65	$8,160	$1 for every $2 earned over the limit

Who Funds Social Security? Social Security is funded entirely by a tax on current workers. If you were employed in 1996, your first $62,700 in income was subject to a 6.2 percent Social Security tax; all of your income was subject to the 1.45 percent Medicare tax. Your employer also paid the same amounts for you. If you were self-employed in 1996, you paid your share and the employer's share, or a full 12.4 percent, plus the Medicare tax; however, one-half of that tax was deductible as a business expense on your federal income taxes. The income limit usually rises each year as the national average wage rises. In 1994, the earnings limit for Medicare tax was eliminated; that tax now applies to all earned income.

Income Taxes and Social Security Earnings

Income*	Benefits May Be Taxed Up to—
Individual return	
Less than $25,000	Tax free
$25,000–$34,000	50%
Over $34,000	85%
Joint return	
Less than $32,000	Tax free
$32,000–$44,000	50%
Over $44,000	85%

*Includes adjusted gross income, tax-free interest, and one-half of Social Security benefits for you and your spouse.

Disability Provisions

If you are disabled physically or mentally and can prove the disability has prevented or will prevent you from working (defined as earning more than $500 a month for at least 1 year), you may be able to collect benefits. You are also eligible if your doctor states that your condition will result in death. Payments start in the 6th month of disability and end at age 65, when your regular Social Security benefits begin. Certain family members may also qualify to receive benefits. After 24 months of disability, you will become eligible for Medicare.

Survivors' Benefits

Social Security will pay benefits to your survivors, even if you have not accumulated the required number of work credits, if they meet these conditions:

- a surviving spouse at least age 60
- a disabled surviving spouse at least age 50
- a surviving spouse caring for a child under age 16
- an unmarried child under age 18 (or age 19 if a full-time secondary school student)
- an unmarried child age 18 or older who has a disability that began before age 22
- a parent who depended on you for at least half of his or her income
- an ex-spouse at least age 60 (or age 50 and disabled) who was married to you for at least 10 years before the divorce
- an ex-spouse of any age if he or she cares for a child eligible for benefits on your record

Medicare

You are eligible for Medicare at age 65, whether or not you are retired. To avoid paying a higher premium, you should enroll 3 months before your 65th birthday; if you don't, you will have to wait until the next open enrollment period, which is January 1 to March 1 each year. You will begin to be covered by Medicare on July 1 of the year you enroll.

Life Insurance: How Much Do You Need?

The purpose of life insurance is to provide financial support for your dependents in the event of your death. The amount you carry should be enough to make up for the lost income to the family for as long as necessary. Here's a simple worksheet to help you calculate the face value of the policy you need.

Total annual living costs, including car, home loans, etc. $ _____ (A)

Dependents' annual income
   Salary   _____
   Investments   _____
   Social Security   _____
   Pensions/retirement plans   _____
   Total $ _____ (B)

Subtract B from A $ _____ (C)

Divide C by average expected interest rate on investments   _____ ÷ _____ %

Face value of policy needed $ _____

Estate Planning

You may leave your heirs up to $600,000 free of federal estate taxes; legislation to raise this limit is being proposed at the time of this writing. If you are married, your spouse most likely will not have to pay estate taxes on any amount you leave to him or her.

To get down to that limit, some people take advantage of a tax law that allows annual tax-free gifts of $10,000 per person. Another way is to set up a trust, which is not subject to taxes.

The Unified Tax Credit is the amount of taxes that would be due on assets of $600,000: $192,800. Each person gets this credit, but it is cumulative and includes any taxable gifts you've made since 1976. For example, if you gave your son a car worth $20,000, at your death your Unified Tax Credit would be reduced by that amount and therefore reduce the amount you could leave tax-free to your heirs. The exception to this law is the nontaxable gift exemption of $10,000 described above.

Federal and State Taxes

Numbers to Know: Federal Income Taxes

You must file a tax return if your gross income is at a certain minimum level. For the 1995 tax year, the requirements are—

Filing Status	Annual Income (all sources)
Single	$ 6,400
Head of household	8,250
Married filing jointly	11,550
Married filing separately	2,500
Qualifying widow/widower	9,050

If you or your spouse is over age 65 or if you can be claimed as a dependent on someone else's return, different amounts apply. The standard deductions for 1995 for those filing nonitemized returns are—

Filing Status	Annual Income (all sources)
Single	$ 3,900
Head of household	5,750
Married filing jointly	6,550
Married filing separately	3,275
Qualifying widow/widower	6,550

The personal exemption for 1995 is $2,500, which is subject to reduction for high-income taxpayers.

If your child (defined as someone you claim as a dependent on your tax return who is under age 19 or is a full-time student under age 24) has interest, dividend, or capital gains income of at least $600, you must file a return for that child. If the child earns more than $600 from a job and has *any* dividend or interest (unearned) income, you must also file a return; use Form 8814. If the child has no unearned income and has a job that pays more than $3,600 in the tax year, you do not need to file a return unless you need to claim a refund on taxes withheld.

Federal Income Tax Brackets, 1995

Filing Status	Tax Bracket
Single	
Up to $23,350	15%
$23,351–$56,550	28%
$56,551–$117,950	31%
$117,951–$256,500	36%
Over $256,500	39.6%
Married filing jointly or qualifying widow/widower	
Up to $39,000	15%
$39,001–$94,250	28%
$94,251–$143,600	31%
$143,601–$256,500	36%
Over $256,500	39.6%
Married filing separately	
Up to $19,500	15%
$19,501–$47,125	28%
$47,126–$71,800	31%
$71,801–$128,250	36%
Over $128,250	39.6%
Head of household	
Up to $31,250	15%
$31,251–$80,750	28%
$80,751–$130,800	31%
$130,801–$256,500	36%
Over $256,500	39.6%

What's a 1099?	Like a W-2 form, which reports wage earnings, the 1099 forms reports various kinds of earnings to the IRS:

1099-R	Retirement income or distributions
1099-DIV	Dividend earnings
1099-INT	Taxable interest
1099B	Capital gains
1099M	Miscellaneous earnings

Estimated Federal Income Taxes

If you think you will owe, in addition to the amounts withheld from your paycheck or any credits due, at least $500 in taxes, you must make estimated quarterly tax payments. The IRS expects you to pay, over the course of the year, at least 90 percent of the taxes you expect you will owe that tax year, or 100 percent of the taxes you paid in the previous year. If your adjusted gross income for the previous tax year was more than $150,000, you must pay 110 percent of your prior year's taxes in the current year.

If you fail to pay your taxes, the penalty is $\frac{1}{2}$ of 1 percent a month for each month or portion of a month you are late.

How Long Tax Records Should Be Kept

Record	Keep This Long
Income, expenses	At least 3 years; 7 if possible
Property, investments	Until you sell
Real estate	At least for 7 years after you sell; keep records of primary residences until sale of final home
Tax returns	At least 6 years; forever if possible

The IRS can audit returns filed up to 3 years ago, unless you underreported your income by more than 25 percent, in which case the IRS has 6 years to audit. If the IRS suspects fraud, there is no time limit on audits.

Federal Income Tax Filing Dates

The filing date is April 15 unless that day falls on a weekend, in which case the filing date is the next workday. You may request a 4-month automatic extension by filing Form 4868, but only if you have paid an amount equal to all of the previous year's taxes or 90 percent of the current year's taxes by April 15. By filing Form 2688, you may request an additional 2-month extension, but the IRS may reject your request.

Internal Revenue Service (IRS) Address and Phone Numbers

The IRS headquarters is at—

1111 Constitution Avenue, NW
Washington, DC 20224
Phone: (202) 622-5000

Each state has a tax help phone number; check your local directory or a recent IRS publication. The general number is (800) 829-1040.

To request tax forms or check on the status of your tax refund, call (800) 829-4477.

The Problem Resolution Program has a toll-free number: (800) 829-1040.

Phone help for people with impaired hearing is available at (800) 829-4059 (answered by TDD equipment only).

The Tax Bite In each 8-hour workday, the average American worked 2 hours and 46 minutes to pay all federal, state, and local taxes in 1995. In 1945, the figure was 1 hour and 59 minutes.

Per Capita Federal Tax Burden by State

1995 Rank	State	1994	1995*	1996*
	United States	$4,728	$4,996	$5,206
1	Connecticut	7,300	7,769	8,154
2	New Jersey	6,482	6,889	7,218
3	New York	5,817	6,185	6,494
4	Massachusetts	5,754	6,113	6,408
5	Delaware	5,665	5,969	6,212
6	Alaska	5,585	5,797	5,932
7	Maryland	5,478	5,777	6,011
8	Illinois	5,390	5,739	6,030
9	New Hampshire	5,169	5,441	5,645
10	Nevada	5,262	5,401	5,459
11	Washington	5,123	5,374	5,557
12	Hawaii	5,104	5,370	5,575
13	Rhode Island	4,977	5,279	5,531
14	Minnesota	4,930	5,220	5,452
15	Colorado	4,931	5,182	5,369
16	Pennsylvania	4,859	5,173	5,435
17	California	4,905	5,130	5,295
18	Virginia	4,830	5,090	5,287
19	Michigan	4,765	5,072	5,326
20	Florida	4,771	4,974	5,110
21	Wyoming	4,524	4,844	5,099
22	Ohio	4,543	4,836	5,080
23	Wisconsin	4,505	4,783	5,007
24	Kansas	4,495	4,773	4,998
25	Missouri	4,314	4,585	4,804
26	Oregon	4,343	4,585	4,772
27	Nebraska	4,302	4,583	4,814
28	Indiana	4,240	4,518	4,742
29	Texas	4,287	4,501	4,658
30	Vermont	4,246	4,493	4,687
31	Georgia	4,203	4,414	4,567
32	Tennessee	4,144	4,391	4,587
33	Iowa	4,008	4,287	4,520

Per Capita Federal Tax Burden by State (continued)

1995 Rank	State	1994	1995*	1996*
34	North Carolina	$4,001	$4,220	$4,388
35	South Dakota	3,926	4,181	4,386
36	Maine	3,864	4,094	4,277
37	Montana	3,823	4,060	4,249
38	North Dakota	3,762	4,026	4,245
39	Idaho	3,782	3,986	4,143
40	Arizona	3,832	3,981	4,076
41	Alabama	3,691	3,922	4,107
42	Oklahoma	3,677	3,908	4,095
43	Louisiana	3,597	3,848	4,059
44	Kentucky	3,603	3,836	4,026
45	South Carolina	3,587	3,788	3,944
46	Arkansas	3,525	3,751	3,936
47	New Mexico	3,477	3,654	3,784
48	Utah	3,408	3,574	3,695
49	West Virginia	3,309	3,547	3,749
50	Mississippi	2,980	3,170	3,323
—	District of Columbia	6,896	7,396	7,837

Source: Tax Foundation calculations based on President Clinton's proposed FY 1996 budget.
*Estimated.

State Taxes and Rates (scheduled in 1995)

State	General Sales and Use Tax (%)	Gasoline Tax (¢/gallon)	Cigarette Tax (¢/20-pack)	Spirits Tax ($/gallon)	Table Wine Tax ($/gallon)	Beer Tax ($/gallon)
Alabama	4	18	16.5	56%[a]	0.18/ 1.64[b]	0.53
Alaska	0	8	29	5.60	0.85	0.3
Arizona	5	18	58	3.00	0.84	0.16
Arkansas	4.5	18.5	31.5	2.50	0.75	0.2
California	6	18	37	3.30	0.20	0.2
Colorado	3	22	20	3.04	0.62/ 0.42[b]	0.08
Connecticut	6	33	50	4.50	0.60	0.194
Delaware	0	23	24	5.46	0.97	0.156
District of Columbia	5.75	20	65	1.50	0.30	0.09
Florida	6	4[c]	33.9	6.50	2.25	0.48
Georgia	4	7.5	12	6.06	2.02	0.32
Hawaii	4	16	60	5.81	1.32	0.9
Idaho	5	22	28	0%[a]	0.45	0.15
Illinois	6.25	19	44	2.00	0.23	0.07
Indiana	5	15	15.5	2.68	0.47	0.115
Iowa	5	20	36	0%[a]	1.75	0.19
Kansas	4.9	18	24	2.5	0.30	0.18
Kentucky	6	15	3	1.92	0.50	0.08
Louisiana	4	20	20	3.33	0.152	0.32
Maine	6	19	37	14%[a]	0.60	0.33
Maryland	5	23.5	36	1.50	0.40	0.09
Massachusetts	5	21	51	4.05	0.55	0.106
Michigan	6	15	75	13.85%[a]	0.682	0.2
Minnesota	6	20	48	5.03	0.30	0.077
Mississippi	7	18	18	2.50%[a]	0.35	0.4268
Missouri	4.225	15	17	2.00	0.36	0.06
Montana	0	27	18	26%[a]	1.364	0.139
Nebraska	5	24[d]	34	3.00	0.75	0.23
Nevada	6.5	22.5	35	2.05	0.40	0.09
New Hampshire	0	18	25	0.30[a]	0	0.3
New Jersey	6	10.5	40	4.4	0.70	0.12

State Taxes and Rates (scheduled in 1995) (continued)

State	General Sales and Use Tax (%)	Gasoline Tax (¢/gallon)	Cigarette Tax (¢/20-pack)	Spirits Tax ($/gallon)	Table Wine Tax ($/gallon)	Beer Tax ($/gallon)
New Mexico	5	17	21	8.08	2.27	0.41
New York	4	8	36	8.586	0.1893	0.21
North Carolina	4	22[d]	5	3.5%[a]	1.06	0.534
North Dakota	5	18	44	4.05	0.50	0.16
Ohio	5	22	24	3.38[a]	0.3	0.18
Oklahoma	4.5	17	23	7.42	0.96	0.4
Oregon	0	24	28	0%[a]	0.67	0.084
Pennsylvania	6	22.35	31	1.25	0.032	0.08
Rhode Island	7	28	56	3.75	0.3/ 0.6[b]	0.097
South Carolina	5	16	7	2.72	1.28	0.77
South Dakota	4	18	23	3.93	0.93	0.27
Tennessee	6	22.4	13	4.00	1.10	0.11
Texas	6.25	21	41	2.40	0.204	0.194
Utah	4.875	19.5	26.5	13%[a]	13%	0.355
Vermont	4	16	20	25%[a]	0.55	0.265
Virginia	3.5	17.5	2.5	20%[a]	2.02	0.283
Washington	6.5	23	81.5	42.6%[a]	1.023	0.161
West Virginia	6	20.5	17	0%[a]	1.33	0.177
Wisconsin	5	23.4[d]	38	0.95	0.6	0.065
Wyoming	4	9	12	0.91[a]	0.28	0.025

Source: Compiled by Tax Foundation from survey of state revenue offices.

a. Control states. Rates represent tax over and above state store markup. Often taxes are built into the markup. For example, Oregon has a zero tax rate, but a 99 percent markup. The average markup for stores in control states is about 48 percent. Private outlets have an average markup of about 25 percent.

b. Rates represent native wine/nonnative wine.

c. Florida's gas tax rates vary by county.

d. Nebraska's gas tax rate is indexed and changes quarterly. North Carolina's gas tax rate is indexed and changes every 6 months. Wisconsin's gas tax rate is indexed annually.

Currency

What the Numbers Mean

Various numbers appear on U.S. paper currency; see the next page for an explanation of what they mean.

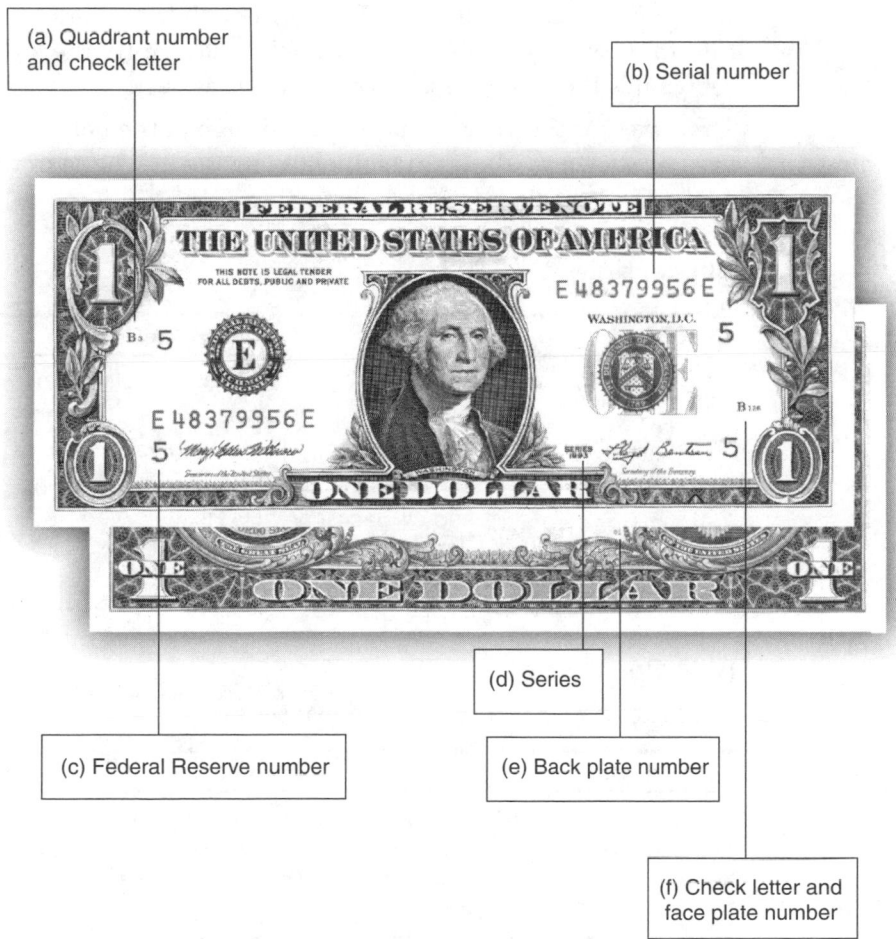

(a) Quadrant number and check letter

(b) Serial number

(c) Federal Reserve number

(d) Series

(e) Back plate number

(f) Check letter and face plate number

The quadrant number and check letter, check letter and face plate number, and back plate number (a, f, e, respectively) are designations used to identify the plate that printed a specific note and to identify the placement of the note on the printing plate.

The Federal Reserve number is shown at (c); 1 is Boston, 2 is New York, 3 is Philadelphia, 4 is Cleveland, 5 is Richmond, 6 is Atlanta, 7 is Chicago, 8 is St. Louis, 9 is Minneapolis, 10 is Kansas City, 11 is Dallas, and 12 is San Francisco.

The serial number (b) is specific to each note printed. The letter before and after the number represents the print run; the letter "O" is never used to avoid confusion with zero. If a star appears instead of a letter after the number, it means one of two things: (1) if the original note was mutilated during manufacturing and a replacement note was printed, necessarily out of sequence with the others in the series, the star shows that the note was a substitute; or (2) the star designates the one-millionth note in a series.

The series, or the year the particular design of the note was first used, is shown at (d). If only minor changes were made in the design, a letter is placed after the year, starting with A.

Bills in Print

Denomination	Portrait/Back Design
$1	George Washington/Great Seal of the United States
$2	Thomas Jefferson/Declaration of Independence
$5	Abraham Lincoln/Lincoln Memorial
$10	Alexander Hamilton/U.S. Treasury Building
$20	Andrew Jackson/White House
$50	Ulysses S. Grant/U.S. Capitol
$100	Benjamin Franklin/Independence Hall

Bills Out of Print

These notes are no longer being printed:

Denomination	Portrait
$500	McKinley
$1,000	Cleveland
$5,000	Madison
$10,000	Chase

Bill Facts   The Bureau of Engraving and Printing replaces the 1-dollar bill most frequently. More than 4 billion are in circulation, each with a life expectancy of 18 months.

Banks will replace a destroyed bill if more than half the original remains.

If you want to reproduce U.S. paper currency for a publication or in an advertisement, you may do so as long as you print it in black and white (not color) and the bill is less than 3/4 or greater than 1 1/2 times the size, in linear dimension, of any part of the bill. The negatives and plates used for printing must be destroyed. You may reproduce bills in color for motion picture films, microfilms, videotapes, and slides; no prints can be made from these images unless they conform with the color and size restrictions mentioned above.

Using U.S. Currency as Measuring Tools

U.S. coins can be used as rough measures of weight:

Dime	=	2 1/2 grams
Penny	=	3 1/8 grams
Nickel	=	5 grams
Quarter	=	6 1/4 grams
Half-dollar	=	12 1/2 grams

Four quarters and a dime weigh almost 1 ounce.

U.S. paper currency is 6 1/8 inches long and slightly more than 2 1/2 inches deep. Here are the approximate diameters of coins:

Dime	=	Slightly less than 3/4 inch
Penny	=	Slightly less than 3/4 inch
Nickel	=	7/8 inch
Quarter	=	Slightly less than 1 inch
Half-dollar	=	1 1/8 inch

Compensation: How Much Does That Job Pay?

The table that follows provides information on 281 occupations in the United States, with figures on total estimated work force in each occupation, lowest or starting salaries, and average or median salaries.

The Bureau of Labor Statistics of the U.S. Department of Labor offers these general observations on employment trends for the next 10 years:

- The civilian labor force is expected to grow to 151 million by 2005, from 127 million in 1992.

- White non-Hispanics will make up 73 percent of the labor force by 2005, down from 78 percent in 1992.

- Blacks, Hispanics, Asians, and other groups will make up about 35 percent of all labor force entrants between 1992 and 2005.

- Women, who were 42 percent of the labor force in 1979, will make up 48 percent in 2005.

- The median age of the labor force will be 40.5 years in 2005; it was 37.2 in 1992.

- Workers between ages 45 and 54 will make up 24 percent of the labor force in 2005, compared to 18 percent in 1992.

- Service-producing industries (transportation, communications, retail and wholesale trade, health, government, real estate, finance, insurance, etc.) will account for 24.5 million of the 26.4 million new jobs expected to be created between 1992 and 2005.

Salaries and Wages for 281 U.S. Occupations

- Salaries are for 1992 or 1993.

- Salaries and wages will of course vary by geographical location, with jobs in large cities paying more; by presence of supervisory or management duties; by educational level achieved; and by number of years on the job.

- To calculate approximate yearly salary when an hourly rate is given, multiply the hourly rate by 2,080.

Occupation	Number of Jobs (1992)	Lowest or Starting Pay ($)	Average or Median Pay ($)
*Accountants/auditors	939,000	28,000	76,000
*Actors/directors/producers	129,000		
Movie/TV actors		1,685/wk	—
Broadway stage actors		950/wk	—
Broadway stage directors		—	36,750/5 wks
*Actuaries	15,000	31,800	46,000
Adjusters/investigators/collectors	1,185,000	—	20,000
Administrative services managers	226,000		
Facilities managers		—	48,000
Office managers		—	40,000
Aircraft mechanics/engine specialists	131,000	17,700	32,500
*Aircraft pilots	85,000		
Airline		—	80,000
Non-airline		—	62,000
Air traffic controllers	23,000	22,700	53,800
*Animal caretakers (except farm)	103,000	—	13,000
Apparel workers	986,000	213/wk	258/wk
Architects	96,000	24,500	36,700
Archivists/curators	19,000	18,300	46,600

Salaries and Wages for 281 U.S. Occupations (continued)

Occupation	Number of Jobs (1992)	Lowest or Starting Pay ($)	Average or Median Pay ($)
Armed forces (Army designations)			
Lt. Colonel	—	—	2,661/mo
Major	—	—	2,217/mo
1st Lieutenant	—	—	1,797/mo
Sergeant	—	—	1,079/mo
Private	—	—	913/mo
Recruit	—	—	814/mo
*Automotive body repairers	202,000	227/wk	401/wk
Automotive mechanics	739,000	230/wk	408/wk
Bank tellers	525,000	10,500	14,800
*Barbers/cosmetologists	746,000	—	20,000–30,000
Blue-collar worker supervisors	1,757,000	323/wk	590/wk
Boilermakers	26,000	12.50/hr	553/wk
Bricklayers/stonemasons	139,000	260/wk	480/wk
Broadcast technicians	35,000	—	22,725
Budget analysts	67,000	24,000	39,700
Butchers & meat/poultry/fish cutters	349,000	—	310/wk
Carpenters	990,000	255/wk	425/wk
Carpet installers	62,000	185/wk	375/wk
Cashiers	2,747,000	—	11,388
*Counter/rental clerks	242,000	—	11,604
*Chefs/cooks/kitchen workers	3,092,000		
Executive chefs		—	40,000+
Cooks		5.50/hr	6.67/hr
Fast-food preparers		4.25/hr	4.68/hr
Child-care workers	—	—	154/wk
*Chiropractors	46,000	21,000	70,000
Clerical supervisors/managers	1,267,000	16,200	28,000

Salaries and Wages for 281 U.S. Occupations (continued)

Occupation	Number of Jobs (1992)	Lowest or Starting Pay ($)	Average or Median Pay ($)
Commercial/industrial electronic equipment repairers	68,000	—	484/wk
Communications equipment repairers	108,000	—	484/wk
Computer & peripheral equipment operators	296,000	13,400	21,100
*Computer programmers	555,000	19,700	35,600
*Computer repairers	83,000	—	619/wk
*Computer scientists/systems analysts	666,000	13,300	42,100
Concrete masons/terrazzo workers	100,000	—	15–37/hr
*Construction/building inspectors	66,000	21,700	31,200
*Construction contractors/managers	180,000	32,000	35,000–110,000
*Corrections officers	282,000	18,600	23,000
*Cost estimators	163,000	17,000	75,000
*Counselors	154,000		
Educational/vocational		—	30,000
School		—	40,400
Credit clerks/authorizers	218,000	18,000	20,800
Dancers/choreographers	18,000		
Dancers		587/wk	—
Choreographers		970/wk	3,000/wk
*Dental assistants	183,000	284/wk	332/wk
*Dental hygienists	108,000	—	32,000
Dental laboratory technicians	48,000	—	13.30/hr
Dentists	183,000	—	90,000
Designers	302,000		
Industrial		27,900	44,500
Interior		25,000	38,000
Diesel mechanics	263,000	11.60/hr	14.10/hr

Salaries and Wages for 281 U.S. Occupations (continued)

Occupation	Number of Jobs (1992)	Lowest or Starting Pay ($)	Average or Median Pay ($)
Dietitians/nutritionists	50,000	28,500	36,000
Drafters	314,000	15,900	27,400
*Drywall workers/lathers	121,000	235/wk	420/wk
Economists/marketing research analysts	51,000	25,200	65,000
Education administrators	351,000		
Principals (high school)		—	63,000
Academic deans (business school)		—	73,700
Admissions officers/registrars		—	47,500
Electricians	518,000	321/wk	550/wk
Electronic home entertainment equipment repairers	39,000	—	—
Elevator installers/repairers	22,000	—	740/wk
Employment interviewers	79,000	17,000	25,000
*Engineering/science/data processing managers	337,000	35,000	50,000–100,000
Engineering technicians	695,000	18,900	28,810
Engineers (all)	1,354,000	34,000	—
Senior level		—	87,000
Mid-level		—	52,500
Aerospace	66,000	—	—
Chemical	52,000	—	—
Civil	173,000	—	—
Electrical/electronics	370,000	—	—
Industrial	119,000	—	—
Mechanical	227,000	—	—
*Metallurgical/ceramic/materials	19,000	—	—
Mining	3,600	—	—
Nuclear	17,000	—	—
Petroleum	14,000	—	—
Farm equipment mechanics	47,000	—	355/wk

Salaries and Wages for 281 U.S. Occupations (continued)

Occupation	Number of Jobs (1992)	Lowest or Starting Pay ($)	Average or Median Pay ($)
Farm operators/managers	1,218,000	—	—
Farm managers		185/wk	382/wk
Financial managers	701,000	20,000	39,700
Firefighters	305,000	18,824	33,072
Fishers/hunters/trappers	60,000	—	—
*Flight attendants	93,000	13,000	20,000
Food/beverage service workers	4,365,000	—	—
Food inspectors/testers/graders	625,000	209/wk	381/wk
Forestry/logging workers	131,000	159/wk	296/wk
Funeral directors	27,000	—	—
*Gardeners/groundskeepers	884,000	9,100	14,300
*General maintenance mechanics	1,145,000	—	9.37/hr
*Glaziers	39,000	—	24.75/hr
Graphic/fine artists	273,000	14,600	23,000
*Guards	803,000	4.25/hr	6.00/hr
Handlers/equipment cleaners/helpers/laborers	4,451,000	180/wk	300/wk
*Health service managers	302,000	59,900	—
Hospital CEOs		77,000	140,900
Nursing home administrators		36,500	44,100
Small group practice administrators		—	46,600
*Heating/air conditioning/refrigeration technicians	212,000	280/wk	474/wk
Home appliance/power tool repairers	74,000	257/wk	467/wk
*Homemaker/home health aides	475,000	6.31/hr	8.28/hr
Hotel managers/assistants	99,000		
General managers		—	59,100
Assistant managers		—	32,500
*Human services workers	189,000	12,000–20,000	15,000–25,000

Salaries and Wages for 281 U.S. Occupations (continued)

Occupation	Number of Jobs (1992)	Lowest or Starting Pay ($)	Average or Median Pay ($)
Industrial machinery repairers	477,000	296/wk	498/wk
Industrial production managers	203,000	—	60,000
*Information clerks	1,333,000	11,900	18,600
*Insulation workers	57,000	279/wk	446/wk
Insurance agents/brokers	415,000	15,400	30,100
Janitors/cleaners/cleaning supervisors	3,018,000	9,152	14,384
Jewelers	30,000	—	28,000
Landscape architects	19,000	20,400	41,900
Lawyers/judges	716,000		
*Lawyers		36,600	134,000
Judges			
Circuit Court		—	141,700
Federal District Court		—	133,600
Librarians	141,000	25,900	37,900
Library technicians	71,000	—	23,900
Life scientists			
Agricultural scientists	29,000	20,189	45,000
*Biological/medical scientists	117,000	21,850	34,500
Foresters/conservation scientists	35,000	18,340	42,440
Line installers/cable splicers	273,000	350/wk	648/wk
*Loan officers/counselors	172,000		
Loan officers		25,000	45,000
Loan counselors		15,000	35,000
Machinists/tool programmers	359,000	275/wk	492/wk
Mail clerks/messengers	271,000	—	15,600
*Management analysts/consultants (salaried)	208,000	26,500	40,300
*Marketing/advertising/public relations managers	432,000	21,000	41,000

Salaries and Wages for 281 U.S. Occupations (continued)

Occupation	Number of Jobs (1992)	Lowest or Starting Pay ($)	Average or Median Pay ($)
Material moving equipment operators	983,000	253/wk	432/wk
Mathematicians	16,000	28,400	53,000
*Medical assistants	181,000	15,059	18,334
Medical technologists/technicians			
Cardiovascular technologists/technicians	31,000	15,223	28,756
Clinical lab technologists/technicians	268,000	14,664	26,312
*EEG technologists	6,300	19,695	23,369
*Emergency medical technicians	114,000	20,092	28,079
*Licensed practical nurses	659,000	15,392	21,476
*Medical record technicians	76,000	20,000	23,000
*Nuclear medicine technologists	12,000	26,402	32,843
Ophthalmic laboratory technicians	19,000	15,040	21,700
*Radiologic technologists	162,000	19,708	28,236
*Surgical technologists	44,000	18,087	21,741
Metalworking/plastics-working machine operators	1,378,000	236/wk	413/wk
Millwrights	73,000	335/wk	596/wk
Mobile heavy equipment mechanics	96,000	318/wk	516/wk
Motorcycle/boat/small-engine mechanics	46,000	263/wk	435/wk
Musical instrument repairers/tuners	12,000	—	20,000
Musicians	236,000		
Orchestra (major)		1,000/wk	—
Movies/TV		226/3 hrs	—
Newscasters/announcers	56,000		
Radio		13,000	17,000
TV		28,000	41,000
Nursing aides/psychiatric aides	1,389,000	9,500	13,800
*Occupational therapists	40,000	—	35,625
Office clerks	2,688,000	12,700	18,500

Salaries and Wages for 281 U.S. Occupations (continued)

Occupation	Number of Jobs (1992)	Lowest or Starting Pay ($)	Average or Median Pay ($)
*Office machine repairers	60,000	—	476/wk
*Operations research analysts	45,000	30,000	50,000
*Opticians (dispensing)	63,000	20,971	26,000
Optometrists	31,000	45,500	75,000
*Painters/paperhangers	440,000	202/wk	376/wk
Painting/coating machine operators	151,000	—	373/wk
*Paralegals	95,000	18,000	28,300
*Personnel/training/labor relations managers/specialists	474,000		
Managers		22,900	37,000
Specialists		17,000	32,000
*Pharmacists	163,000	26,100	45,000
Photographers/camera operators	118,000	12,300	21,200
Photographic process workers	63,000	210/wk	330/wk
*Physical therapists	90,000	17,784	35,464
Physical scientists			
Chemists	92,000	24,000	51,537
Geologists/geophysicists	48,000	25,704	51,800
Meteorologists	6,100	18,340	48,266
Physicists/astronomers	21,000	30,000	61,596
*Physician assistants	58,000	32,466	41,038
*Physicians (M.D.s/D.O.s)	556,000	28,618	139,000
Plasterers	32,000	—	15–33/hr
Plumbers/pipefitters	351,000	—	18.05/hr
*Podiatrists	15,000	35,578	100,287
Police officers/detectives/special agents	700,000	18,400	32,000
Postal clerks/mail carriers	361,000	23,737	32,832
Precision assemblers	334,000	201/wk	318/wk

Salaries and Wages for 281 U.S. Occupations (continued)

Occupation	Number of Jobs (1992)	Lowest or Starting Pay ($)	Average or Median Pay ($)
*Preschool workers	941,000	7,280	13,520
Printing workers			
Prepress workers	167,000		
Lithographers/photoengravers		—	518/wk
Typesetters/compositors		—	402/wk
Scanner operators		—	21.86/hr
Strippers		—	17.57/hr
Printing press operators	241,000	215/wk	420/wk
Bindery workers	76,000	200/wk	350/wk
Private household workers	869,000	—	179/wk
*Property/real estate managers	243,000	14,600	21,800
*Psychologists	143,000		
Counseling		—	48,000
Clinical		—	53,000
Research		—	50,000
School		—	55,000
Industrial		—	76,000
Full professors		—	55,000
Public relations specialists	98,000	21,000	32,000
Purchasers/buyers	624,000	13,959	33,067
Real estate agents/brokers/ appraisers	397,000	—	26,364
Record clerks	3,573,000		
Order		—	22,200
Payroll		—	21,000
Bookkeeping		—	19,100
Personnel		—	20,300
*Recreation workers	204,000	7,700	14,900
*Recreational therapists	30,000	—	25,557
*Registered nurses	1,835,000	21,944	34,424

Salaries and Wages for 281 U.S. Occupations (continued)

Occupation	Number of Jobs (1992)	Lowest or Starting Pay ($)	Average or Median Pay ($)
Religious workers			
Protestant ministers	290,000	—	27,000
Rabbis	3,900	—	38,000–60,000
Roman Catholic priests	53,000	—	9,000
Reporters/correspondents	58,000	21,312	34,000
*Respiratory therapists	74,000	21,528	32,084
*Restaurant/food service managers	496,000	20,200	27,900
Retail managers	1,070,000		
Managers		—	18,400
Assistant managers		—	13,100
Retail sales workers	4,086,000		
Cars/boats		—	25,000
Clothes		—	13,260
Furniture		—	18,400
Roofers	127,000	230/wk	315/wk
Roustabouts	33,000	—	11.90/hr
Sales representatives			
Manufacturers'/wholesale	1,613,000	16,400	32,000
*Securities/financial services	200,000	28,000	40,300
*Services	488,000	—	30,000
Science technicians	244,000	14,600	25,300
Secretaries	3,324,000	16,400	26,700
*Sheet-metal workers	91,000	—	27.62/hr
Shoe/leather workers/repairers	22,000	—	—
*Social scientists/urban planners	258,000	19,000	36,700
*Social workers	484,000	—	30,000
*Speech/language pathologists/ audiologists	73,000	29,050	36,036
Stationary engineers	31,000	302/wk	618/wk
Statisticians	16,000	—	51,893
Stenographers/court reporters	115,000	—	21,320

Salaries and Wages for 281 U.S. Occupations (continued)

Occupation	Number of Jobs (1992)	Lowest or Starting Pay ($)	Average or Median Pay ($)
Structural/reinforcing ironworkers	66,000	—	27/hr
Surveyors	99,000	13,400	26,800
*Teacher aides	885,000	—	8.31/hr
Teachers			
*Adult education	540,000	—	26,900
College/university	812,000	27,700	46,300
*Kindergarten/elementary/ secondary	3,255,000	—	34,000–36,000
Telephone installers/repairers	40,000	—	626/wk
Telephone operators	314,000	—	20,020
Textile machinery operators	284,000	—	353/wk
Tilesetters	30,000	—	10–28/hr
Tool & die makers	138,000	409/wk	642/wk
Transportation workers			
Bus drivers	562,000	206/wk	400/wk
Rail workers	116,000		
Through-freight engineers		—	59,600
Conductors		—	40,400
Commuter rail operators		—	18.46/hr
Taxi drivers/chauffeurs	120,000	187/wk	313/wk
Truck drivers	2,270,000		
Heavy straight trucks		—	11.91/hr
Light trucks		—	8.51/hr
Medium trucks		—	13.50/hr
Tractor-trailers		—	12.94/hr
Water transportation workers	54,000	350/wk	611/wk
*Travel agents	115,000	12,248	21,000
Typists/data entry keyers/word processors	1,238,000	13,400	20,000
Underwriters	100,000		
Commercial lines		28,000	40,600
Personal lines		25,000	40,400

Salaries and Wages for 281 U.S. Occupations (continued)

Occupation	Number of Jobs (1992)	Lowest or Starting Pay ($)	Average or Median Pay ($)
Upholsterers	60,000	200/wk	350/wk
Urban/regional planners	28,000	27,800	42,000
Vending machine servicers/repairers	20,000	5.00/hr	7.63/hr
*Veterinarians	44,000	27,858	63,069
Waiters/waitresses	1,800,000	180/wk (with tips)	220/wk (with tips)
Bartenders	382,000	200/wk (with tips)	250/wk (with tips)
Counter/fast-food attendants	1,600,000	170/wk	220/wk
Dining room/cafeteria attendants	441,000	175/wk	210/wk
Hosts/hostesses	223,000	—	—
Water/wastewater treatment plant operators	86,000	15,700	26,200
Welders/cutters/welding machine operators	403,000	278/wk	440/wk
Woodworkers	341,000	244/wk	385/wk
Writers/editors	283,000	20,000	40,000

Source: *Occupational Outlook Handbook, 1994–95*, Bureau of Labor Statistics, U.S. Department of Labor, 1994.
*Indicates that job opportunities in this field are expected to grow faster or much faster than average through 2005.

Fastest Growing Occupations
1992 to 2005

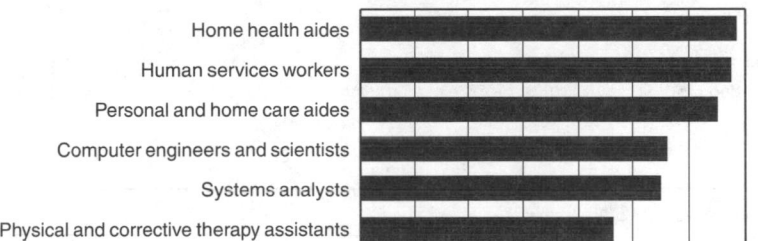

Source: U.S. Bureau of the Census.

Employed Workers with Alternative Work Arrangements by Occupation and Industry (percent distribution)

Characteristic	Workers with Alternative Arrangements				Workers with Traditional Arrangements
	Independent Contractors	On-call Workers and Day Laborers	Temporary Help Agency Workers	Workers Provided by Contract Firms	
Occupation					
Total workers, 16 years and over (in thousands)	8,309	2,078	1,181	652	111,052
Total (%)	100.0	100.0	100.0	100.0	100.0
Executive, administrative, and managerial	18.6	2.9	6.5	5.7	13.6
Professional specialty	16.3	20.9	8.3	25.6	14.7
Technicians and related support	1.1	1.5	3.7	6.9	3.4
Sales occupations	18.8	6.0	2.6	3.2	11.7
Administrative support, including clerical	3.8	9.5	30.1	4.8	16.0
Service occupations	10.6	19.7	9.0	27.8	13.6
Precision production, craft, and repair	19.2	14.3	5.6	14.6	10.1
Operators, fabricators, and laborers	6.5	20.5	33.2	10.4	14.6
Farming, forestry, and fishing	5.1	4.7	1.0	0.9	2.4
Industry					
Total workers, 16 years and over (in thousands)	8,309	2,078	1,181	652	111,052
Total (%)	100.0	100.0	100.0	100.0	100.0
Agriculture	5.0	4.4	0.4	0.3	2.4
Mining	0.2	0.5	0.2	2.4	0.6
Construction	21.2	15.2	2.8	8.4	4.4
Manufacturing	5.0	5.9	33.4	17.6	17.9
Transportation and public utilities	5.0	8.7	7.6	13.4	7.2
Wholesale and retail trade	13.2	13.8	8.1	6.0	21.4
Finance, insurance, and real estate	9.6	1.8	7.5	6.9	6.4
Services	40.6	46.0	38.7	32.3	34.4
Public administration	0.3	3.3	1.2	12.6	5.4

Source: U.S. Bureau of Labor Statistics.

Note: Data are for February 1995. Workers with traditional arrangements are those who do not fall into any of the "alternative arrangements" categories. Detail may not sum to totals due to rounding.

Federal Government General Schedule Pay Scale (1996)*

Level	Within-Grade Step Increases									
	1	2	3	4	5	6	7	8	9	10
GS-1	$12,384	$12,797	$13,208	$13,619	$14,032	$14,274	$14,679	$15,089	$15,107	$15,489
GS-2	13,923	14,255	14,717	15,107	15,274	15,723	16,172	16,621	17,070	17,519
GS-3	15,193	15,699	16,205	16,711	17,217	17,723	18,229	18,735	19,241	19,747
GS-4	17,055	17,624	18,193	18,762	19,331	19,900	20,469	21,038	21,607	22,176
GS-5	19,081	19,717	20,353	20,989	21,625	22,261	22,897	23,533	24,169	24,805
GS-6	21,269	21,978	22,687	23,396	24,105	24,814	25,523	26,232	26,941	27,650
GS-7	23,634	24,422	25,210	25,998	26,786	27,574	28,362	29,150	29,938	30,726
GS-8	26,175	27,048	27,921	28,794	29,667	30,540	31,413	32,286	33,159	34,032
GS-9	28,912	29,876	30,840	31,804	32,768	33,732	34,696	35,660	36,624	37,588
GS-10	31,839	32,900	33,961	35,022	36,083	37,144	38,205	39,266	40,327	41,388
GS-11	34,981	36,147	37,313	38,479	39,645	40,811	41,977	43,143	44,309	45,475
GS-12	41,926	43,324	44,722	46,120	47,518	48,916	50,314	51,712	53,110	54,508
GS-13	49,856	51,518	53,180	54,842	56,504	58,166	59,828	61,490	63,152	64,814
GS-14	58,915	60,879	62,843	64,807	66,771	68,735	70,699	72,663	74,627	76,591
GS-15	69,300	71,610	73,920	76,230	78,540	80,850	83,160	85,470	87,780	90,090

Note: Pay rates in 1995 for Senior Level (SL) and Scientific & Professional (ST) positions ranged from $81,529 to $115,700. Senior Executive Service (SES) pay rates ranged from $92,900 to $115,700. This table does not include locality pay adjustments.

*Proposed.

2
Food and Nutrition

Nutrition

Recommended Daily Dietary Allowances: Calories, Protein, and Vitamins

The National Academy of Sciences has established the following daily dietary allowances as a guide to maintaining good nutrition:

Category	Age	Weight (lbs.)	Calories (per day)	Protein (g)	Vitamin A*	Vitamin C (mg)	Thiamine (mg)	Ribo-flavin (mg)	Niacin (mg)
Infants	0–6 mos.	13	680	13	420	35	0.3	0.4	6
	6–12 mos.	20	954	18	400	35	0.5	0.6	8
Children	1–3	29	1,300	23	400	45	0.7	0.8	9
	4–6	44	1,700	30	500	45	0.9	1.0	11
	7–10	62	2,400	34	700	45	1.2	1.4	16
Boys/men	11–14	99	2,700	45	1,000	50	1.4	1.6	18
	15–18	145	2,800	56	1,000	60	1.4	1.7	18
	19–22	154	2,900	56	1,000	60	1.5	1.7	19
	23–50	154	2,700	56	1,000	60	1.4	1.6	18
	51–74	154	2,400	56	1,000	60	1.2	1.4	16
	75+	154	2,050	56	1,000	60	1.2	1.4	16
Girls/women	11–14	101	2,200	46	800	50	1.1	1.3	15
	15–18	120	2,100	46	800	60	1.1	1.3	14
	19–22	120	2,100	44	800	60	1.1	1.3	14
	23–50	120	2,000	44	800	60	1.0	1.2	13
	51–74	120	1,800	44	800	60	1.0	1.2	13
	75+	120	1,600	44	800	60	1.0	1.2	13
Pregnant women			add 300	add 30	add 200	add 20	add 0.4	add 0.3	add 20
Breastfeeding women			add 500	add 20	add 400	add 40	add 0.5	add 0.5	add 40

Source: U.S. Department of Agriculture, *Nutritive Value of Foods*, 1991.

*Retinol equivalent. To determine international units, multiply by 10 for fruits and vegetables and by 3.3 for animal source foods.

Recommended Daily Intake: Fiber, Iron, Calcium, Sodium, and Cholesterol

These are the generally recommended daily amounts for adults:

Fiber	25–35 g
Iron	15 mg
Calcium	800 mg
Sodium	2,400 mg or less
Cholesterol	300 mg or less

Numbers to Know: Sodium

It is estimated that 75 percent of the sodium we consume is added during the food preparation process, 10 to 15 percent is added at the table (table salt contains 40 percent sodium), and 10 percent occurs naturally in food. The U.S.-government-recommended daily intake of sodium, less than 2,400 milligrams, amounts to about 1 teaspoon of salt. Examples of foods with high sodium content are canned baked beans (606 mg in 1 cup), plain skim milk yogurt (174 mg in 1 cup), English muffins (293 mg in 1 medium), Parmesan cheese (528 mg in 1 tablespoon), American process cheese (406 mg in 1 ounce), and tomato sauce (1,498 mg in 1 cup).

Recommended Daily Intake: Carbohydrates and Fats

The need for carbohydrates and fats varies by individual. A low-fat diet is thought to prevent or stem the onset of many illnesses, including cancer, heart disease, and diabetes. The current measure is that no more than 30 percent of total calories consumed per day should come from fat; many nutritionists advise a number closer to 20 percent. One quick way to determine how many grams of fat you should consume per day is to—

■ Divide your ideal body weight (in pounds) by 2 (for the 30 percent diet) **or**

■ Divide your ideal body weight by 3 (for the 20 percent diet)

In other words, if your ideal body weight is 160 pounds, you should take in no more than 80 grams of fat per day if you follow the 30 percent limit; if you follow the 20 percent limit, you should consume no more than 53 grams of fat per day. (See the height and weight table on page 215.)

Another way to arrive at your daily fat limit is to consider the total calories consumed daily. To determine the maximum grams of fat to be consumed daily, multiply the number of calories you consume by 0.30 (or 0.20 for the 20 percent diet), then divide the result by 9 (which is the number of calories in 1 gram of fat). The next table gives some examples for the 30 percent recommendation.

Daily Fat Limits

Calories per Day	30% of Calories	Grams of Fat
1,200	360	40
1,500	450	50
1,800	540	60
2,000	600	66
2,200	660	73
2,500	750	83
2,800	840	93

Calorie-wise, Not All Food Is Equal

Fat has more than twice the calories per gram—9—than do carbohydrates and proteins, each of which has 4.

Fat Levels in Ground Beef and Ground Turkey

Here are calorie and fat breakdowns for ground beef and turkey with these labels (for a 3-ounce burger after cooking):

Type	Calories	Fat Grams
73% lean ground beef	248	18
80% lean ground chuck	228	15
93% lean ground beef	169	8
93% lean ground turkey	160	8

The Food Pyramid: Number of Recommended Servings or Amounts per Day

Food Group	1,600 Calories per Day	2,200 Calories per Day	2,800 Calories per Day
Bread	6	9	11
Vegetable	3	4	5
Fruit	2	3	4
Milk*	2–3	2–3	2–3
Meat (oz.)	5	6	7
Total fat (g)	53	73	93
Total added sugars (teaspoons)	6	12	18

*Teenagers, young adults, and pregnant and breastfeeding women need 3 servings.

What Is a Serving?

Food Group Item	Serving Size
Bread	1 slice
Bagel, English muffin	1/2
Tortilla	1
Crackers (small plain)	3–4
Pancake (4-in. diameter)	1
Cereal, rice, pasta (cooked)	1/2 cup
Cereal (dry)	1 oz.
Vegetable (raw, leafy)	1 cup
Vegetable (cooked, chopped raw)	1/2 cup
Vegetable juice	3/4 cup
Fruit (medium-size fresh)	1
Fruit (chopped, cooked, canned)	1/2 cup
Fruit juice	3/4 cup
Milk, yogurt	1 cup
Natural cheese	1 1/2 oz.
Process cheese	2 oz.
Lean meat, fish, poultry (cooked) (1/2 cup cooked dry beans, 1 egg, or 2 tablespoons peanut butter equal 1 oz. of lean meat)	2–3 oz.

Rule of Thumb: Losing Weight by Limiting Fat

Although calories still count, limiting fat intake can lead to weight loss. Women should consume no more than 30 to 35 grams of fat per day, and men no more than 45 to 50 grams per day if they want to lose weight.

What the FDA Says Label Claims Must Mean

Claim	Definition for Standard Serving Size
Calories	
Calorie free	Fewer than 5 calories
Low calorie	40 or fewer calories
Light/lite	1/3 fewer calories *or* 50% less fat than a compared food; if more than half the calories are from fat, fat must be reduced by 50% or more
Reduced calories	At least 25% fewer calories than a compared food
Sugar	
Sugar free	Less than 0.5 g
No added sugar	No sugar added during processing
Reduced sugar	At least 25% less sugar than a compared food
Fat	
Fat free	Less than 0.5 g of fat
Saturated fat free	Less than 0.5 g of saturated fat; level of trans-fatty acids does not exceed 1% of total fat
Low in saturated fat	1 g or less and not more than 15% of calories from saturated fatty acids
Low fat*	3 g or less of fat
Lower, reduced, or less fat	25% less fat than a compared food
Cholesterol	
Cholesterol free*	Less than 2 mg cholesterol and 2 g or less of saturated fat
Low cholesterol	20 mg or less of cholesterol and 2 g or less of saturated fat
Reduced cholesterol	At least 75% less than a compared food
Sodium	
Sodium free*	Less than 5 mg sodium
Light in sodium*	50% less sodium than a compared food
Very low sodium*	35 mg or less of sodium
Low sodium*	140 mg or less of sodium
No salt added	No salt added during processing of a food that usually contains added salt
Fiber	
High fiber	5 g or more of fiber
Good source of fiber	2.5–4.9 g of fiber
More or added fiber	At least 2.5 g more than a compared food

*Also per 50 g for products with small serving sizes.

Added Sugars Present in Selected Foods

Many packaged foods contain sugars added during the preparation process. Here are a few examples:

Food	Teaspoons
Bread	
Bread (1 slice)	0
Muffin (1 medium)	1
Cookies (2 medium)	1
Danish (1 medium)	1
Doughnut (1 medium)	2
Pound cake (1 oz.)	2
Angel food cake (1 slice)	5
Cake with frosting (1 slice)	6
Fruit pie/2-crust (1 slice)	6
Fruit	
Canned/light syrup (1/2 cup)	2
Canned/heavy syrup (1/2 cup)	4
Dairy	
Chocolate milk (2%) (1 cup)	3
Low-fat yogurt, flavored (8 oz.)	5
Low-fat yogurt, fruit (8 oz.)	7
Ice cream, ice milk, frozen yogurt (1/2 cup)	3
Chocolate milk shake	9
Other	
Chocolate bar (3 oz.)	3
Fruit sorbet (1/2 cup)	3
Gelatin dessert (1/2 cup)	4
Sherbet (1/2 cup)	5
Cola (12 fl. oz.)	9
Fruit drink (12 fl. oz.)	12

| **How Much Sugar in a Gram?** | Nutrition labels on food products list the amount of sugar present in terms of grams. To translate grams of sugar to teaspoons, multiply grams by 4; for example, 2 grams of sugar equals 8 teaspoons. |

Percentages of Fat in Vegetable Oils

Monounsaturated fats are preferred by nutritionists, who recommend that no more than 10 percent of total daily calories come from saturated and polyunsaturated fats. Canola and olive oils are both high in monounsaturated fats. Note that percentages do not add up to 100 percent because oils may contain other fat-like substances.

Oil	Polyunsaturated (%)	Monounsaturated (%)	Saturated (%)
Canola	32	62	6
Safflower	75	12	9
Sunflower	66	20	10
Corn	59	24	13
Olive	9	72	14
Peanut	32	46	17
Sesame	40	40	18
Cottonseed	52	18	26
Palm kernel	2	10	80
Coconut	2	6	87

Percentages of Fat in Milk

Consider the following when you're feeling virtuous about drinking 2 percent milk (also keep in mind that heavy cream is 40 percent fat):

Type of Milk (1 cup)	Fat (g)	Calories	% Calories from Fat
Whole (3.3%)	8.5	150	51
Low-fat (2%)	4.7	120	35
Low-fat (1%)	2.5	100	22
Skim	0.4	85	4

Nutritive Value of Foods: Calories, Protein, Fat, Cholesterol, Carbohydrates, and Sodium

The numbers given are approximations only and will vary based on size, brand, and cooking methods used.

Food	Amt.	Calories	Protein (g)	Fat (g)	Chol. (mg)	Carb. (g)	Sodium (mg)
Dairy Products							
Butter	1 tbsp.	100	Trace	11	31	Trace	116
Cheese, natural							
Blue	1 oz.	100	6	8	21	1	396
Cheddar							
Cut pieces	1 oz.	115	7	9	30	Trace	176
Shredded	1 cup	455	28	37	119	1	701
Cottage cheese							
Large curd (4% fat)	1 cup	235	28	10	34	6	911
Low-fat (2%)	1 cup	205	31	4	19	8	918
Cream	1 oz.	100	2	10	31	1	84
Feta	1 oz.	75	4	6	25	1	316
Mozzarella, part skim milk	1 oz.	80	8	5	15	1	150
Parmesan, grated	1 tbsp.	25	2	2	4	Trace	93
Provolone	1 oz.	100	7	8	20	1	248
Ricotta, part skim milk	1 cup	340	28	19	76	13	307
Swiss	1 oz.	105	8	8	26	1	74
Pasteurized process cheese							
American	1 oz.	105	6	9	27	Trace	406
Swiss	1 oz.	95	7	7	24	1	388
Pasteurized process cheese food							
American	1 oz.	95	6	7	18	2	337
Cream, sour	1 tbsp.	25	Trace	3	5	1	6
Cream, sweet							
Half-and-half	1 cup	315	7	28	89	10	98
Light	1 cup	470	6	46	159	9	95
Whipping, heavy	1 cup	820	5	88	326	7	89
Whipped topping (pressurized)	1 tbsp.	10	Trace	1	2	Trace	4
Milk (fresh)							
Whole (3.3% fat)	1 cup	150	8	8	33	11	120
Low-fat (2%)	1 cup	120	8	5	18	12	122
Low-fat (1%)	1 cup	100	8	3	10	12	123
Non-fat (skim)	1 cup	85	8	Trace	4	12	126
Buttermilk	1 cup	100	8	2	9	12	257
Milk (canned)							
Condensed, sweetened	1 cup	980	24	27	104	166	389
Evaporated							
Whole	1 cup	340	17	19	74	25	267
Skim	1 cup	200	19	1	9	29	293

Nutritive Value of Foods (continued)

Food	Amt.	Calories	Protein (g)	Fat (g)	Chol. (mg)	Carb. (g)	Sodium (mg)
Dairy Products (continued)							
Ice cream, vanilla							
Regular	1 cup	270	5	14	59	32	116
Soft serve	1 cup	375	7	23	153	38	153
Ice milk, vanilla	1 cup	185	5	6	18	29	105
Sherbet	1 cup	270	2	4	14	59	88
Yogurt, low-fat							
Fruit-flavored	8 oz.	230	10	2	10	43	133
Plain	8 oz.	145	12	4	14	16	159
Eggs							
Fried in margarine	1 egg	90	6	7	211	1	162
Hard-cooked	1 egg	75	6	5	213	1	62
Poached	1 egg	75	6	5	212	1	140
Scrambled	1 egg	100	7	7	215	1	171
Fats and Oils							
Fats, cooking	1 cup	1,810	0	205	0	0	0
Lard	1 cup	1,850	0	205	195	0	0
Margarine							
Hard	1 tbsp.	100	Trace	11	0	Trace	132
Soft	1 tbsp.	100	Trace	11	0	Trace	151
Spread							
Hard	1 tbsp.	75	Trace	9	0	0	139
Soft	1 tbsp.	75	Trace	9	0	0	139
Oils, salad or cooking							
Corn	1 tbsp.	125	0	14	0	0	0
Olive	1 tbsp.	125	0	14	0	0	0
Peanut	1 tbsp.	125	0	14	0	0	0
Safflower	1 tbsp.	125	0	14	0	0	0
Soybean oil	1 tbsp.	125	0	14	0	0	0
Soybean/cottonseed	1 tbsp.	125	0	14	0	0	0
Sunflower	1 tbsp.	125	0	14	0	0	0
Salad dressings							
Blue cheese	1 tbsp.	75	1	8	3	1	164
French							
Regular	1 tbsp.	85	Trace	9	0	1	188
Low-calorie	1 tbsp.	25	Trace	2	0	2	306
Italian							
Regular	1 tbsp.	80	Trace	9	0	1	162
Low-calorie	1 tbsp.	5	Trace	Trace	0	2	136
Mayonnaise							
Regular	1 tbsp.	100	Trace	11	8	Trace	80
Imitation	1 tbsp.	35	Trace	3	4	2	75

Nutritive Value of Foods (continued)

Food	Amt.	Calories	Protein (g)	Fat (g)	Chol. (mg)	Carb. (g)	Sodium (mg)
Fats and Oils (continued)							
Thousand Island							
Regular	1 tbsp.	60	Trace	6	4	2	112
Low-calorie	1 tbsp.	25	Trace	2	2	2	150
Fish and Shellfish							
Crabmeat, canned	1 cup	135	23	3	135	1	1,350
Fish sticks, frozen	1 stick	70	6	3	26	4	53
Flounder or sole, baked							
with butter	3 oz.	120	16	6	68	Trace	145
with margarine	3 oz.	120	16	6	55	Trace	151
without added fat	3 oz.	80	17	1	59	Trace	101
Halibut, broiled with butter	3 oz.	140	20	6	62	Trace	103
Ocean perch, breaded, fried	1 fillet	185	16	11	66	7	138
Oysters							
Raw	1 cup	160	20	4	120	8	175
Breaded, fried	1 oyster	90	5	5	35	5	70
Salmon							
Canned	3 oz.	120	17	5	34	0	443
Baked (red)	3 oz.	140	21	5	60	0	55
Smoked	3 oz.	150	18	8	51	0	1,700
Shrimp, fried	3 oz.	200	16	10	168	11	304
Trout, broiled with butter	3 oz.	175	21	9	71	Trace	122
Tuna, canned							
Oil pack	3 oz.	165	24	7	55	0	303
Water pack	3 oz.	135	30	1	48	0	468
Fruits and Fruit Juices							
Apples, unpeeled	1	80	Trace	Trace	0	21	Trace
Apple juice	1 cup	115	Trace	Trace	0	29	7
Applesauce, canned							
Sweetened	1 cup	195	Trace	Trace	0	51	8
Unsweetened	1 cup	105	Trace	Trace	0	28	5
Apricots							
Raw	3	50	1	Trace	0	12	1
Canned, heavy syrup	1 cup	215	1	Trace	0	55	10
Dried, uncooked	1 cup	310	5	1	0	80	13
Avocados, raw							
California	1	305	4	30	0	12	21
Florida	1	340	5	27	0	27	15
Bananas, raw	1	105	1	1	0	27	1
Blackberries, raw	1 cup	75	1	1	0	18	Trace
Blueberries, raw	1 cup	80	1	1	0	20	9
Cherries, sweet raw	10	50	1	1	0	11	Trace

Nutritive Value of Foods (continued)

Food	Amt.	Calories	Protein (g)	Fat (g)	Chol. (mg)	Carb. (g)	Sodium (mg)
Fruits and Fruit Juices (continued)							
Cranberry juice, cocktail	1 cup	145	Trace	Trace	0	38	10
Cranberry sauce	1 cup	420	1	Trace	0	108	80
Dates, whole	10	230	2	Trace	0	61	2
Fruit cocktail							
Heavy syrup	1 cup	185	1	Trace	0	48	15
Juice pack	1 cup	115	1	Trace	0	29	10
Grapefruit, raw	1/2	40	1	Trace	0	10	Trace
Grapefruit juice, canned, unsweetened	1 cup	95	1	Trace	0	22	2
Grapes, Thompson seedless	10	35	Trace	Trace	0	9	1
Grape juice, frozen, diluted	1 cup	125	Trace	Trace	0	32	5
Mangoes, raw	1	135	1	1	0	35	4
Melons, raw							
Cantaloupe	1/2	95	2	1	0	22	24
Honeydew	1/10	45	1	Trace	0	12	13
Nectarines, raw	1	65	1	1	0	16	Trace
Oranges, raw	1	60	1	Trace	0	15	Trace
Orange juice, frozen, diluted	1 cup	110	2	Trace	0	27	2
Peaches, raw	1	35	1	Trace	0	10	Trace
Canned							
Heavy syrup	1 cup	190	1	Trace	0	51	15
Juice pack	1 cup	110	2	Trace	0	29	10
Dried, uncooked	1 cup	380	6	1	0	98	11
Pears, raw							
Bartlett	1	100	1	1	0	25	Trace
Bosc	1	85	1	1	0	21	Trace
D'Anjou	1	120	1	1	0	30	Trace
Canned							
Heavy syrup	1 cup	190	1	Trace	0	49	13
Juice pack	1 cup	125	1	Trace	0	32	10
Pineapple, raw	1 cup	75	1	1	0	19	2
Canned							
Heavy syrup, crushed	1 cup	200	1	Trace	0	52	3
Heavy syrup, slices	1 slice	45	Trace	Trace	0	12	1
Juice pack, chunks	1 cup	150	1	Trace	0	39	3
Juice pack, slices	1 slice	35	Trace	Trace	0	9	1
Pineapple juice, unsweetened, canned	1 cup	140	1	Trace	0	34	3
Plantains, raw	1	220	2	1	0	57	7
Plums, raw	1	35	1	Trace	0	9	Trace
Prunes, dried, uncooked	5 large	115	1	Trace	0	31	2
Raisins, seedless	1 cup	435	5	1	0	115	17
Raspberries, raw	1 cup	60	1	1	0	14	Trace
Rhubarb, cooked, added sugar	1 cup	280	1	Trace	0	75	2
Strawberries, raw	1 cup	45	1	1	0	10	1

Nutritive Value of Foods (continued)

Food	Amt.	Calories	Protein (g)	Fat (g)	Chol. (mg)	Carb. (g)	Sodium (mg)
Fruits and Fruit Juices (continued)							
Tangerines, raw	1	35	1	Trace	0	9	1
Watermelon, raw	4" x 8"	155	3	2	0	35	10
Diced	1 cup	50	1	1	0	11	3
Grain Products							
Bagels, plain	1	200	7	2	0	38	245
Biscuits, small							
From mix	1	95	2	3	Trace	14	262
From refrigerated dough	1	65	1	2	1	10	249
Breadcrumbs, dry	1 cup	390	13	5	5	73	736
Breads							
Boston brown	1 slice	95	2	1	3	21	113
Cracked-wheat	1 slice	65	2	1	0	12	106
French or Vienna	1 slice	100	3	1	0	18	203
Italian	1 slice	85	3	Trace	0	17	176
Mixed grain	1 slice	65	2	1	0	12	106
Oatmeal	1 slice	65	2	1	0	12	124
Pita	1	165	6	1	0	33	339
Pumpernickel	1 slice	80	3	1	0	16	177
Raisin	1 slice	65	2	1	0	13	92
Rye	1 slice	65	2	1	0	12	175
Wheat	1 slice	65	2	1	0	12	138
White	1 slice	65	2	1	0	12	129
Whole wheat	1 slice	70	3	1	0	13	180
Bread stuffing mix, dry	1 cup	500	9	31	0	50	1,254
Breakfast cereals							
Hot type							
Corn (hominy) grits	1 cup	145	3	Trace	0	31	0
Cream of Wheat	1 cup	140	4	Trace	0	29	5
Malt-O-Meal	1 cup	120	4	Trace	0	26	2
Oatmeal							
Regular	1 cup	145	6	2	0	25	2
Instant, plain	1 packet	105	4	2	0	18	285
Instant, flavored	1 packet	160	5	2	0	31	254
Ready-to-eat							
All-Bran	1 oz.	70	4	1	0	21	320
Cap'n Crunch	1 oz.	120	1	3	0	23	213
Cheerios	1 oz.	110	4	2	0	24	351
Corn Flakes							
Kellogg's	1 oz.	110	2	Trace	0	24	351
Toasties	1 oz.	110	2	Trace	0	24	297
40% Bran Flakes							
Kellogg's	1 oz.	90	4	1	0	22	264
Post	1 oz.	90	3	Trace	0	22	260

Nutritive Value of Foods (continued)

Food	Amt.	Calories	Protein (g)	Fat (g)	Chol. (mg)	Carb. (g)	Sodium (mg)
Grain Products (continued)							
Froot Loops	1 oz.	110	2	1	0	25	145
Golden Grahams	1 oz.	110	2	1	Trace	24	346
Grape Nuts	1 oz.	100	3	Trace	0	23	197
Honey Nut Cheerios	1 oz.	105	3	1	0	23	257
Lucky Charms	1 oz.	110	3	1	0	23	201
Nature Valley Granola	1 oz.	125	3	5	0	19	58
100% Natural Cereal	1 oz.	135	3	6	Trace	18	12
Product 19	1 oz.	110	3	Trace	0	24	325
Raisin Bran							
Kellogg's	1 oz.	90	3	1	0	21	207
Post	1 oz.	85	3	1	0	21	185
Rice Krispies	1 oz.	110	2	Trace	0	25	340
Shredded Wheat	1 oz.	100	3	1	0	23	3
Special K	1 oz.	110	6	Trace	Trace	21	265
Sugar Frosted Flakes	1 oz.	110	1	Trace	0	26	230
Sugar Smacks	1 oz.	105	2	1	0	25	75
Super Sugar Crisp	1 oz.	105	2	Trace	0	26	25
Total	1 oz.	100	3	1	0	22	352
Trix	1 oz.	110	2	Trace	0	25	181
Wheaties	1 oz.	100	3	Trace	0	23	354
Cakes, from cake mixes							
Angel-food	1 piece	125	3	Trace	0	29	269
Coffee cake, crumb	1 piece	230	5	7	47	38	310
Devil's food w/ chocolate frosting	1 piece	235	3	8	37	40	181
Gingerbread	1 piece	175	2	4	1	32	192
Yellow w/ chocolate frosting	1 piece	235	3	8	36	40	157
Cakes, from home recipes							
Carrot, with cream cheese frosting	1 piece	385	4	21	74	48	279
Fruitcake	1 piece	165	2	7	20	25	67
Pound	1 slice	120	2	5	32	15	96
Cakes, commercial							
Pound	1 slice	110	2	5	64	15	108
Cheesecake	1 piece	280	5	18	170	26	204
Cookies							
Brownies with nuts	1	95	1	6	18	11	51
Chocolate chip, commercial	4	180	2	9	5	28	140
From recipe	4	185	2	11	18	26	82
From refrigerated dough	4	225	2	11	22	32	173
Fig bars	4	210	2	4	27	42	180
Oatmeal with raisins	4	245	3	10	2	36	148
Peanut butter	4	245	4	14	22	28	142
Sandwich type	4	195	2	8	0	29	189
Shortbread, commercial	4	155	2	8	27	20	123

Nutritive Value of Foods (continued)

Food	Amt.	Calories	Protein (g)	Fat (g)	Chol. (mg)	Carb. (g)	Sodium (mg)
Grain Products (continued)							
Sugar, refrigerated dough	4	235	2	12	29	31	261
Vanilla wafers	10	185	2	7	25	29	150
Corn chips	1 oz.	155	2	9	0	16	233
Cornmeal, whole ground	1 cup	435	11	5	0	90	1
Crackers							
Cheese, plain	10	50	1	3	6	6	112
Sandwich type	1	40	1	2	1	5	90
Graham, plain	2	60	1	1	0	11	86
Melba toast	1	20	1	Trace	0	4	44
Rye wafers	2	55	1	1	0	10	115
Saltines	4	50	1	1	4	9	165
Snack type	1	15	Trace	1	0	2	30
Wheat, thin	4	35	1	1	0	5	69
Whole-wheat wafers	2	35	1	2	0	5	59
Croissants	1	235	5	12	13	27	452
Danish pastry, plain	1	220	4	12	49	26	218
Danish pastry, fruit	1	235	4	13	56	28	233
Doughnuts, cake	1	210	3	12	20	24	192
Doughnuts, yeast	1	235	4	13	21	26	222
English muffins, plain	1	140	5	1	0	27	378
French toast	1	155	6	7	112	17	257
Macaroni, cold	1 cup	115	4	Trace	0	24	1
Muffins, commercial mix							
Blueberry	1	140	3	5	45	22	225
Bran	1	140	3	4	28	24	385
Corn	1	145	3	6	42	22	291
Noodles, cooked	1 cup	200	7	2	50	37	3
Pancakes, 4"							
Buckwheat	1	55	2	2	20	6	125
Plain	1	60	2	2	16	9	115
Pies (baked)							
Apple	1 piece	405	3	18	0	60	476
Blueberry	1 piece	380	4	17	0	55	423
Cherry	1 piece	410	4	18	0	61	480
Creme	1 piece	455	3	23	8	59	369
Custard	1 piece	330	9	17	169	36	436
Lemon	1 piece	335	5	14	143	53	395
Peach	1 piece	405	4	17	0	60	423
Pecan	1 piece	575	7	32	95	71	305
Pumpkin	1 piece	320	6	17	109	37	325
Pies (fried)							
Apple	1 pie	255	2	14	14	31	326
Cherry	1 pie	250	2	14	13	32	371

Nutritive Value of Foods (continued)

Food	Amt.	Calories	Protein (g)	Fat (g)	Chol. (mg)	Carb. (g)	Sodium (mg)
Grain Products (continued)							
Popcorn							
Air-popped, unsalted	1 cup	30	1	Trace	0	6	Trace
Popped in vegetable oil	1 cup	55	1	3	0	6	86
Sugar-syrup coated	1 cup	135	2	1	0	30	Trace
Pretzels							
Stick	10	10	Trace	Trace	0	2	48
Twisted, dutch	1	65	2	1	0	13	258
Twisted, thin	10	240	6	2	0	48	966
Rice							
Brown, cooked	1 cup	230	5	1	0	50	0
White, enriched, cooked	1 cup	225	4	Trace	0	50	0
Instant, cooked	1 cup	180	4	0	0	40	0
Rolls, commercial							
Dinner	1	85	2	2	Trace	14	155
Frankfurter and hamburger	1	115	3	2	Trace	20	241
Hard	1	155	5	2	Trace	30	313
Hoagie	1	400	11	8	Trace	72	683
Spaghetti, cooked	1 cup	190	7	1	0	39	1
Toaster pastries	1	210	2	6	0	38	248
Waffles							
From home recipe	1	245	7	13	102	26	445
From mix	1	205	7	8	59	27	515
Wheat flours							
All-purpose, unsifted	1 cup	455	13	1	0	95	3
Cake	1 cup	350	7	1	0	76	2
Self-rising	1 cup	440	12	1	0	93	1,349
Whole wheat	1 cup	400	16	2	0	85	4
Legumes, Nuts, and Seeds							
Almonds, shelled							
Slivered, packed	1 cup	795	27	70	0	28	15
Whole	1 oz.	165	6	15	0	6	3
Beans							
Dry, cooked, drained							
Black	1 cup	225	15	1	0	41	1
Great Northern	1 cup	210	14	1	0	38	13
Lima	1 cup	260	16	1	0	49	4
Pea (navy)	1 cup	225	15	1	0	40	13
Pinto	1 cup	265	15	1	0	49	3
Canned, red kidney	1 cup	230	15	1	0	42	968
Black-eyed peas, dry, cooked	1 cup	190	13	1	0	35	20
Brazil nuts, shelled	1 oz.	185	4	19	0	4	1

Nutritive Value of Foods (continued)

Food	Amt.	Calories	Protein (g)	Fat (g)	Chol. (mg)	Carb. (g)	Sodium (mg)
Legumes, Nuts, and Seeds (continued)							
Cashew nuts, salted							
Dry roasted	1 oz.	165	4	13	0	9	181
Roasted in oil	1 oz.	165	5	14	0	8	177
Chestnuts, European, roasted, shelled	1 cup	350	5	3	0	76	3
Coconut, dried, sweetened, shredded	1 cup	470	3	33	0	44	244
Filberts (hazelnuts), chopped	1 oz.	180	4	18	0	4	1
Lentils, dry, cooked	1 cup	215	16	1	0	38	26
Macadamia nuts, roasted in oil, salted	1 oz.	205	2	22	0	4	74
Mixed nuts, with peanuts, salted							
Dry roasted	1 oz.	170	5	15	0	7	190
Roasted in oil	1 oz.	175	5	16	0	6	185
Peanuts, roasted in oil, salted	1 oz.	165	8	14	0	5	122
Peanut butter	1 tbsp.	95	5	8	0	3	75
Peas, split, dry, cooked	1 cup	230	16	1	0	42	26
Pecans, halves	1 oz.	190	2	19	0	5	Trace
Pine nuts	1 oz.	160	3	17	0	5	20
Pistachio nuts, dried, shelled	1 oz.	165	6	14	0	7	2
Refried beans, canned	1 cup	295	18	3	0	51	1,228
Soybeans, dry, cooked, drained	1 cup	235	20	10	0	19	4
Soy products							
Miso	1 cup	470	29	13	0	65	8,142
Tofu, 2½" x 2¾" x 1"	1 piece	85	9	5	0	3	8
Sunflower seeds, dry, hulled	1 oz.	160	6	14	0	5	1
Walnuts, black, chopped	1 oz.	4	170	7	0	3	Trace
Meat and Meat Products							
Beef roast, cooked							
Chuck blade, lean only	2.2 oz.	170	19	9	66	0	44
Bottom round, lean only	2.8 oz.	175	25	8	75	0	40
Rib, lean only	2.2 oz.	150	17	9	49	0	45
Eye of round, lean only	2.6 oz.	135	22	5	52	0	46
Ground beef, broiled, patty							
Lean	3 oz.	230	21	16	74	0	65
Regular	3 oz.	245	20	18	76	0	70
Liver, fried, slice	3 oz.	185	23	7	410	7	90
Steak, sirloin, broiled, lean only	2.5 oz.	150	22	6	64	0	48
Beef, canned, corned	3 oz.	185	22	10	80	0	802
Beef, dried, chipped	2.5 oz.	145	24	4	46	0	3,053
Lamb, cooked							
Chops, loin, broiled	2.3 oz.	140	19	6	60	0	54
Leg, roasted, lean only	2.6 oz.	140	20	6	65	0	50
Rib, roasted, lean only	2 oz.	130	15	7	50	0	46

Nutritive Value of Foods (continued)

Food	Amt.	Calories	Protein (g)	Fat (g)	Chol. (mg)	Carb. (g)	Sodium (mg)
Meat and Meat Products (continued)							
Pork, cured, cooked							
Bacon, regular	3 slices	110	6	9	16	Trace	303
Canadian style	2 slices	85	11	4	27	1	711
Ham, light cure, lean only	2.4 oz.	105	17	4	37	0	902
Ham, canned	3 oz.	140	18	7	35	Trace	908
Luncheon meat							
Chopped ham	2 slices	95	7	7	21	0	576
Cooked ham, regular	2 slices	105	10	6	32	2	751
Pork, fresh, cooked							
Chop, broiled, lean only	2.5 oz.	165	23	8	71	0	56
Chop, pan fried, lean only	2.4 oz.	180	19	11	72	0	57
Sausages							
Bologna	2 slices	180	7	16	31	2	581
Braunschweiger	2 slices	205	8	18	89	2	652
Brown and serve	1 link	50	2	5	9	Trace	105
Frankfurter	1	145	5	13	23	1	504
Pork link	1	50	3	4	11	Trace	168
Salami	2 slices	145	8	11	37	1	607
Vienna sausage	1	45	2	4	8	Trace	152
Veal, medium fat, cooked							
Cutlet	3 oz.	185	23	9	109	0	56
Rib	3 oz.	230	23	14	109	0	57
Mixed Dishes							
Chili con carne with beans, canned	1 cup	340	19	16	28	31	1,354
Macaroni and cheese							
Canned	1 cup	230	9	10	24	26	730
From home recipe	1 cup	430	17	22	44	40	1,086
Poultry and Poultry Products							
Chicken							
Fried, with skin							
Batter-dipped, breast	4.9 oz.	365	35	18	119	13	385
Batter-dipped, drumstick	2.5 oz.	195	16	11	62	6	194
Flour-coated, breast	3.5 oz.	220	31	9	87	2	74
Flour-coated, drumstick	1.7 oz.	120	13	7	44	1	44
Roasted							
Breast	3.0 oz.	140	27	3	73	0	64
Drumstick	1.6 oz.	75	12	2	41	0	42
Duck, roasted	1/2 duck	445	52	25	197	0	144
Turkey, roasted							
Dark meat, piece 2½" x 1⅝" x ¼"	4 pieces	160	24	6	72	0	67
Light meat, piece 4" x 2" x ¼"	2 pieces	135	25	3	59	0	54

Nutritive Value of Foods (continued)

Food	Amt.	Calories	Protein (g)	Fat (g)	Chol. (mg)	Carb. (g)	Sodium (mg)
Soups, Sauces, and Gravies							
Soups							
Canned, condensed; made with milk							
Clam chowder, New England	1 cup	165	9	7	22	17	992
Cream of chicken	1 cup	190	7	11	27	15	1,047
Cream of mushroom	1 cup	205	6	14	20	15	1,076
Tomato	1 cup	160	6	6	17	22	932
Canned, condensed; made with water							
Bean with bacon	1 cup	170	8	6	3	23	951
Beef broth, bouillon, consommé	1 cup	15	3	1	Trace	Trace	782
Beef noodle	1 cup	85	5	3	5	9	952
Chicken noodle	1 cup	75	4	2	7	9	1,106
Chicken rice	1 cup	60	4	2	7	7	815
Clam chowder, Manhattan	1 cup	80	4	2	2	12	1,808
Cream of chicken	1 cup	115	3	7	10	9	986
Cream of mushroom	1 cup	130	2	9	2	9	1,032
Minestrone	1 cup	80	4	3	2	11	911
Pea, green	1 cup	165	9	3	0	27	988
Tomato	1 cup	85	2	2	0	17	871
Vegetable beef	1 cup	80	6	2	5	10	956
Vegetarian	1 cup	70	2	2	0	12	822
Sauces							
From dry mix							
Cheese, prepared with milk	1 cup	305	16	17	53	23	1,565
Hollandaise, prepared with water	1 cup	240	5	20	52	14	1,564
White sauce, prepared with milk	1 cup	240	10	13	34	21	797
From home recipe							
White sauce	1 cup	395	10	30	32	24	888
Ready to serve							
Barbecue	1 tbsp.	10	Trace	Trace	0	2	130
Soy	1 tbsp.	10	2	0	0	2	1,029
Gravies							
Canned							
Beef	1 cup	125	9	5	7	11	1,305
Chicken	1 cup	190	5	14	5	13	1,373
Mushroom	1 cup	120	3	6	0	13	1,357
From dry mix							
Brown	1 cup	80	3	2	2	14	1,147
Chicken	1 cup	85	3	2	3	14	1,134

Nutritive Value of Foods (continued)

Food	Amt.	Calories	Protein (g)	Fat (g)	Chol. (mg)	Carb. (g)	Sodium (mg)
Sugars and Sweets							
Candy							
Caramels	1 oz.	115	1	3	1	22	64
Chocolate							
Milk, plain	1 oz.	145	2	9	6	16	23
Milk, with almonds	1 oz.	150	3	10	5	15	23
Milk, with peanuts	1 oz.	155	4	11	5	13	19
Milk, with rice cereal	1 oz.	140	2	7	6	18	46
Semisweet, small pieces	1 cup	860	7	61	0	97	24
Sweet (dark)	1 oz.	150	1	10	0	16	5
Fondant, uncoated (mints, candy corn, other)	1 oz.	105	Trace	0	0	27	57
Fudge, chocolate, plain	1 oz.	115	1	3	1	21	54
Gumdrops	1 oz.	100	Trace	Trace	0	25	10
Hard	1 oz.	110	0	0	0	28	7
Jellybeans	1 oz.	105	Trace	Trace	0	26	7
Marshmallows	1 oz.	90	1	0	0	23	25
Custard, baked	1 cup	305	14	15	278	29	209
Gelatin dessert, from mix	1/2 cup	70	2	0	0	17	55
Honey	1 cup	1,030	1	0	0	279	17
	1 tbsp.	65	Trace	0	0	17	1
Jams and preserves	1 tbsp.	55	Trace	Trace	0	14	2
	1 packet	40	Trace	Trace	0	10	2
Jellies	1 tbsp.	50	Trace	Trace	0	13	5
	1 packet	40	Trace	Trace	0	10	4
Popsicle, 3 oz.	1	70	0	0	0	18	11
Puddings							
Canned							
Chocolate	5 oz. can	205	3	11	1	30	285
Tapioca	5 oz. can	160	3	5	Trace	28	252
Vanilla	5 oz. can	220	2	10	1	33	305
Dry mix, prepared w/ whole milk							
Chocolate							
Instant	1/2 cup	155	4	4	14	27	440
Regular (cooked)	1/2 cup	150	4	4	15	25	167
Rice	1/2 cup	155	4	4	15	27	140
Tapioca	1/2 cup	145	4	4	15	25	152
Vanilla							
Instant	1/2 cup	150	4	4	15	27	375
Regular (cooked)	1/2 cup	145	4	4	15	25	178

Nutritive Value of Foods (continued)

Food	Amt.	Calories	Protein (g)	Fat (g)	Chol. (mg)	Carb. (g)	Sodium (mg)
Sugars and Sweets (continued)							
Sugars							
Brown, packed	1 cup	820	0	0	0	212	97
White, granulated	1 cup	770	0	0	0	199	5
	1 tbsp.	45	0	0	0	12	Trace
	1 packet	25	0	0	0	6	Trace
White, powdered, sifted, spooned	1 cup	385	0	0	0	100	2
Syrups							
Chocolate-flavored syrup or topping							
Thin type	2 tbsp.	85	1	Trace	0	22	36
Fudge type	2 tbsp.	125	2	5	0	21	42
Molasses, cane, blackstrap	2 tbsp.	85	0	0	0	22	38
Table syrup (corn, maple)	2 tbsp.	122	0	0	0	32	19
Vegetables and Vegetable Products							
Artichokes, cooked	1	55	3	Trace	0	12	79
Asparagus, cooked or raw	1 cup	45	5	1	0	8	7
Bamboo shoots, canned	1 cup	25	2	1	0	4	9
Beans, lima, cooked	1 cup	170	10	1	0	32	90
Beans, green							
Cooked from raw	1 cup	45	2	Trace	0	10	4
Cooked from frozen	1 cup	35	2	Trace	0	8	18
Canned	1 cup	25	2	Trace	0	6	339
Bean sprouts, cooked	1 cup	25	3	Trace	0	5	12
Beets							
Cooked	1 cup	55	2	Trace	0	11	83
Canned	1 cup	55	2	Trace	0	12	466
Broccoli							
Raw	1 spear	40	4	1	0	8	41
Cooked from raw	1 cup	45	5	Trace	0	9	17
Cooked from frozen	1 cup	50	6	Trace	0	10	44
Brussels sprouts, cooked							
From raw	1 cup	60	4	1	0	13	33
From frozen	1 cup	65	6	1	0	13	36
Cabbage							
Raw	1 cup	15	1	Trace	0	4	13
Cooked	1 cup	30	1	Trace	0	7	29
Cabbage, Chinese, cooked	1 cup	20	3	Trace	0	3	58
Cabbage, red, raw	1 cup	20	1	Trace	0	4	8
Cabbage, savoy, raw	1 cup	20	1	Trace	0	4	20

Nutritive Value of Foods (continued)

Food	Amt.	Calories	Protein (g)	Fat (g)	Chol. (mg)	Carb. (g)	Sodium (mg)
Vegetables and Vegetable Products (continued)							
Carrots							
Raw, whole or strips	1	30	1	Trace	0	7	25
Raw, grated	1 cup	45	1	Trace	0	11	39
Cooked							
From raw	1 cup	70	2	Trace	0	16	103
From frozen	1 cup	55	2	Trace	0	12	86
Cauliflower							
Raw	1 cup	25	2	Trace	0	5	15
Cooked							
From raw	1 cup	30	2	Trace	0	6	8
From frozen	1 cup	35	3	Trace	0	7	32
Celery, pascal	1 stalk	5	Trace	Trace	0	1	35
Collards, cooked							
From raw	1 cup	25	2	Trace	0	5	36
From frozen	1 cup	60	5	1	0	12	85
Corn, sweet, cooked							
From raw	1 ear	85	3	1	0	19	13
From frozen	1 ear	60	2	Trace	0	14	3
Kernels	1 cup	135	5	Trace	0	34	8
Canned							
Cream style	1 cup	185	4	1	0	46	730
Whole kernel	1 cup	165	5	1	0	41	571
Cucumber, with peel	6 slices	5	Trace	Trace	0	1	1
Eggplant, cooked	1 cup	25	1	Trace	0	6	3
Endive, curly	1 cup	10	1	Trace	0	2	11
Kale, cooked, from raw	1 cup	40	2	1	0	7	30
Kohlrabi, cooked	1 cup	50	3	Trace	0	11	35
Lettuce							
Boston, leaves	1 outer or 2 inner leaves	Trace	Trace	Trace	0	Trace	1
Iceberg, pieces	1 cup	5	1	Trace	0	1	5
Romaine	1 cup	10	1	Trace	0	2	5
Mushrooms							
Raw	1 cup	20	1	Trace	0	3	3
Cooked	1 cup	40	3	1	0	8	3
Canned	1 cup	35	3	Trace	0	8	663
Mustard greens	1 cup	20	3	Trace	0	3	22
Okra, cooked	8 pods	25	2	Trace	0	6	4
Onions							
Raw, chopped	1 cup	55	2	Trace	0	12	3
Cooked	1 cup	60	2	Trace	0	13	17

Nutritive Value of Foods (continued)

Food	Amt.	Calories	Protein (g)	Fat (g)	Chol. (mg)	Carb. (g)	Sodium (mg)
Vegetables and Vegetable Products (continued)							
Onions, spring	6	10	1	Trace	0	2	1
Onion rings, breaded, pan-fried	2 rings	80	1	5	0	8	75
Parsley, raw	10 sprigs	5	Trace	Trace	0	1	4
Parsnips, cooked	1 cup	125	2	Trace	0	30	16
Peas, green, cooked							
Canned	1 cup	115	8	1	0	21	372
Frozen	1 cup	125	8	Trace	0	23	139
Peppers							
Hot, raw	1	20	1	Trace	0	4	3
Sweet, raw	1	20	1	Trace	0	4	2
Potatoes, cooked							
Baked w/ skin	1	220	5	Trace	0	51	16
Boiled	1	120	3	Trace	0	27	5
French fried							
Oven heated	10 fries	110	2	4	0	17	16
Fried in vegetable oil	10 fries	160	2	8	0	20	108
Potato chips	10 chips	105	1	7	0	10	94
Potato products, prepared							
Au gratin							
From dry mix	1 cup	230	6	10	12	31	1,076
From home recipe	1 cup	325	12	19	56	28	1,061
Hashed browns, from frozen	1 cup	340	5	18	0	44	53
Mashed							
Milk added	1 cup	160	4	1	4	37	636
Milk and margarine added	1 cup	225	4	9	4	35	620
From dehydrated flakes w/o milk	1 cup	235	4	12	29	32	697
Potato salad, made with mayonnaise	1 cup	360	7	21	170	28	1,323
Scalloped							
From dry mix	1 cup	230	5	11	27	31	835
From home recipe	1 cup	210	7	9	29	26	821
Pumpkin, cooked							
From raw	1 cup	50	2	Trace	0	12	2
Canned	1 cup	85	3	1	0	20	12
Radishes, raw	4	5	Trace	Trace	0	1	4
Sauerkraut, canned	1 cup	45	2	Trace	0	10	1,560
Spinach							
Raw	1 cup	10	2	Trace	0	2	43
Cooked							
From raw	1 cup	40	5	Trace	0	7	126
From frozen	1 cup	55	6	Trace	0	10	163
Canned	1 cup	50	6	1	0	7	683

Nutritive Value of Foods (continued)

Food	Amt.	Calories	Protein (g)	Fat (g)	Chol. (mg)	Carb. (g)	Sodium (mg)
Vegetables and Vegetable Products (continued)							
Squash, cooked							
Summer	1 cup	35	2	1	0	8	2
Winter	1 cup	80	2	1	0	18	2
Sweet potatoes							
Cooked							
Baked in skin, peeled	1 potato	115	2	Trace	0	28	11
Boiled, w/o skin	1 potato	160	2	Trace	0	37	20
Candied	1 piece	145	1	3	8	29	74
Solid pack (mashed)	1 cup	260	5	1	0	59	191
Tomatoes							
Raw	1	25	1	Trace	0	5	10
Canned	1 cup	50	2	1	0	10	391
Tomato juice, canned	1 cup	40	2	Trace	0	10	881
Tomato products, canned							
Paste	1 cup	220	10	2	0	49	170
Purée	1 cup	105	4	Trace	0	25	50
Sauce	1 cup	75	3	Trace	0	18	1,482
Turnips, cooked, diced	1 cup	30	1	Trace	0	8	78
Turnip greens, cooked							
From raw	1 cup	30	2	Trace	0	6	42
From frozen	1 cup	50	5	1	0	8	25
Vegetable juice cocktail, canned	1 cup	45	2	Trace	0	11	883
Vegetables, mixed, cooked							
Canned	1 cup	75	4	Trace	0	15	243
Frozen	1 cup	105	5	Trace	0	24	64
Water chestnuts, canned	1 cup	70	1	Trace	0	17	11
Miscellaneous Items							
Catsup	1 tbsp.	15	Trace	Trace	0	69	2,845
Mustard, prepared	1 tsp.	5	Trace	Trace	0	Trace	63
Olives, canned							
Green	4 medium	15	Trace	2	0	Trace	312
Ripe	3 small	15	Trace	2	0	Trace	68
Pickles							
Dill, 3³/₄" long, 1¹/₄" diam.	1	5	Trace	Trace	0	1	928
Slices	2	10	Trace	Trace	0	3	101
Sweet, gherkin, small	1	20	Trace	Trace	0	5	107
Relish	1 tbsp.	20	Trace	Trace	0	5	107

Nutritive Value of Foods (continued)

Food	Amt.	Calories	Protein (g)	Fat (g)	Chol. (mg)	Carb. (g)	Sodium (mg)
Fast Foods							
Hamburgers							
Burger King Whopper	1	630	27	38	90	44	880
Hamburger	1	260	14	10	30	28	500
Cheeseburger	1	300	16	14	45	28	660
McDonald's Big Mac	1	500	25	26	100	42	890
Hamburger	1	255	12	9	39	30	490
Cheeseburger	1	305	15	13	50	30	725
Wendy's Bacon Cheeseburger Jr.	1	440	22	25	65	33	870
Hamburger (single)	1	440	26	23	75	36	850
Jr. Cheeseburger	1	320	18	13	45	34	760
Chicken sandwiches							
Burger King Chicken Sandwich	1	700	27	42	60	54	1,440
McDonald's McChicken	1	415	19	9	37	39	490
Wendy's grilled chicken	1	290	24	7	60	35	670
Wendy's breaded chicken	1	450	26	20	60	44	740
Pizza							
Domino's cheese & ham	1 slice	208	11	5.5	NA	29	402
Little Caesar's cheese & pepperoni	1 slice	201	11	8	23	22	430
Pizza Hut							
Thin crust cheese & pepperoni	1 slice	230	12	11.5	26	19.6	678
Pan cheese & pepperoni	1 slice	280	8.3	18.1	25	26.4	618
Fried chicken							
Burger King Chicken Tenders	6	236	16	13	38	14	541
Kentucky Fried							
Original (breast)	1	199	16	12	NA	7	558
Extra crispy (breast)	1	286	17	18	NA	14	564
McDonald's Chicken McNuggets	6	270	20	15	55	17	580
Wendy's chicken nuggets	6	280	14	20	50	12	600
French fries							
Burger King (medium)	1 order	372	5	20	0	43	238
McDonald's (medium)	1 order	320	4	17	0	36	150
Wendy's (medium)	1 order	360	5	17	0	50	220
Miscellaneous							
Hardee's ham biscuit w/ egg	1	458	19	26	NA	37	1,584
McDonald's Egg McMuffin	1	280	18	11	235	28	710
McDonald's Filet-O-Fish	1	370	14	18	50	38	730
Wendy's bacon & cheese baked potato	1	500	13	25	NA	54	430
Wendy's taco salad	1	640	34	30	80	70	960

Notes: NA = not available.

Cooking

Kitchen Conversion Chart: U.S. to Metric

If You Know—	Multiply by—	To Get—
ounces	28.35	grams (g)
pounds	0.45	kilograms (kg)
teaspoons	5	milliliters (ml)
tablespoons	15	milliliters
fluid ounces	30	milliliters
cups	0.24	liters (L)
pints	0.47	liters
quarts	0.95	liters
gallons	3.8	liters

Kitchen Conversion Chart: Metric to U.S.

If You Know—	Multiply by—	To Get—
grams	0.035	ounces
kilograms	2.2	pounds
milliliters	5	teaspoons
milliliters	15	tablespoons
milliliters	0.034	fluid ounces
liters	4	cups
liters	2.1	pints
liters	1.06	quarts
liters	0.26	gallons

British and U.S Measurements Compared

British	U.S.
1 fluid ounce (8 fl. drams)	0.961 fluid ounce
2¹/₂ teaspoons	1 tablespoon
2 tablespoons	3 tablespoons
1 gill (5 fl. oz.)	4 fluid ounces
1 cup (10 fl. oz.)	1 cup (8 fl. oz.)
1 Imperial pint (20 fl. oz.)	1 pint (16 fl. oz.)
1 quart (40 fl. oz.)	1 quart (32 fl. oz.)
1 Imperial gallon (160 fl. oz.)	1 gallon (128 fl. oz.)

A Bushel and a Peck: Miscellaneous Equivalent Measures

This—	Equals This—
Dry	
pinch	¹/₈ teaspoon
8 quarts	1 peck
4 pecks	1 bushel
1,000 grams (1 kilogram)	2 pounds, 3¹/₄ ounces
1 ounce	28.35 grams
1 pound	16 ounces; 453.59 grams; 7,000 grains
Liquid	
1 dash	6 drops or ¹/₈ teaspoon
1 pony	1 ounce
1 finger	1 ounce
1 jigger	1¹/₂ ounces
1 deciliter	6 tablespoons + 2 teaspoons
¹/₄ liter	1 cup + 2³/₄ teaspoons
¹/₂ liter	1 pint + 4¹/₂ teaspoons
1 liter	1 quart + scant ¹/₄ cup
4 liters	1 gallon + 1 scant cup
10 liters	2¹/₂ gallons + 2¹/₄ cups

Cooking Measurement Equivalents, U.S. and Metric

Teaspoon	Tablespoon	Fluid Ounce	Cup	Liquid Pint	Liquid Quart	Gallon	Metric Amount
$1/4$–$1/5$	—	—	—	—	—	—	1 ml
$1/2$	—	—	—	—	—	—	2.5 ml
1	$1/3$	—	—	—	—	—	5 ml
$1 1/2$	$1/2$	—	—	—	—	—	7.5 ml
3	1	$1/2$	$1/16$	—	—	—	15 ml
6	2	1	$1/8$	—	—	—	30 ml
12	4	2	$1/4$	—	—	—	60 ml
16	$5 1/3$	$2 2/3$	$1/3$	—	—	—	78.4 ml
24	8	4	$1/2$	—	—	—	117.6 ml
30	10	5	$5/8$	—	—	—	147 ml
32	$10 2/3$	$5 1/3$	$2/3$	—	—	—	156.8 ml
36	12	6	$3/4$	—	—	—	176.4 ml
42	14	7	$7/8$	—	—	—	205.8 ml
48	16	8	1	$1/2$	$1/4$	—	236.6 ml
—	—	16	2	1	$1/2$	—	473.2 ml
—	—	32	4	2	1	$1/4$	0.95 L
—	—	64	8	4	2	$1/2$	1.9 L
—	—	128	16	8	4	1	3.785 L

Ounces and Grams: Approximate Equivalents

U.S.	Metric
1/4 ounce	7 grams
1/2 ounce	14 grams
1 ounce	28 grams
1 1/4 ounces	35 grams
1 1/2 ounces	40 grams
1 2/3 ounces	45 grams
2 ounces	55 grams
2 1/2 ounces	70 grams
4 ounces	112 grams
5 ounces	140 grams
8 ounces	228 grams
10 ounces	280 grams
15 ounces	425 grams
16 ounces (1 pound)	454 grams

Metric	U.S.
1 gram	0.035 ounce
50 grams	1.75 ounces
100 grams	3.5 ounces
250 grams	8.75 ounces
500 grams	1.1 pounds
1 kilogram	2.2 pounds

Some Uncommon Measures

Certain food items have their own units of measurement, some of them dating back hundreds of years.

- A *hand* is a small bunch of bananas that grows off the larger stem; the individual bananas are called *fingers*.
- A *cran* is 45 gallons of fresh herring.
- A *firkin* is a British term for 56 pounds of butter.
- A *frail* is 50 pounds of raisins.
- A *clove* is 8 or 10 pounds of cheese.

Cups and Liters: Approximate Equivalents

U.S.	Metric
$^1/_4$ cup	60 milliliters
$^1/_3$ cup	80 milliliters
$^1/_2$ cup	120 milliliters
$^2/_3$ cup	160 milliliters
1 cup	230 milliliters
1$^1/_4$ cups	300 milliliters
1$^1/_2$ cups	360 milliliters
1 $^2/_3$ cups	400 milliliters
2 cups	460 milliliters
2$^1/_2$ cups	600 milliliters
3 cups	700 milliliters
4 cups (1 quart)	0.95 liter
4 quarts (1 gallon)	3.8 liters

Metric	U.S.
50 milliliters	0.21 cup
100 milliliters	0.42 cup
150 milliliters	0.63 cup
200 milliliters	0.84 cup
250 milliliters	1.06 cups
500 milliliters	1.06 pints
1 liter	1.06 quarts

Oven Temperature: Approximate Equivalents

Degrees Fahrenheit	Degrees Celsius	British Gas Mark
250	120	$1/2$
275	140	1
300	150	2
325	160	3
350	180	4
375	190	5
400	200	6
425	220	7
450	230	8
475	250	9
500	260	
525	270	

Standard Baking Pan Sizes

Pan Type	Pan Size	Approximate Volume
Cake		
Round	8 x $1^{1}/_{2}$ inches	4 cups
	9 x 2 inches	6 cups
Rectangular	13 x 9 x 2 inches	15 cups
Square	8 x 8 x 2 inches	8 cups
	9 x 9 x $1^{1}/_{2}$ inches	8 cups
	9 x 9 x 2 inches	10 cups
Tube	9 x 3 inches	12 cups
	10 x 4 inches	18 cups
Loaf	$8^{1}/_{2}$ x $4^{1}/_{2}$ x $2^{1}/_{2}$ inches	6 cups
	9 x 5 x 3 inches	8 cups
Pie	8 x $1^{1}/_{4}$ inches	3 cups level
	9 x $1^{1}/_{2}$ inches	4 cups level
	9 x 2 inches (deep dish)	6 cups level
Tart or quiche	4 x $1^{1}/_{4}$ inches	$1/2$ cup
	8 x 1 inch	$1^{1}/_{2}$ cups
	9 x $1^{3}/_{8}$ inches	4 cups
Springform	8 x 3 inches	12 cups
	9 x 3 inches	16 cups

Emergency Recipe Substitutions

This Ingredient Amount—	Can Be Replaced with This—
1 teaspoon baking powder	$1/2$ teaspoon baking soda + $1/2$ teaspoon cream of tartar
1 tablespoon baking powder	$1^1/2$ teaspoons baking soda + $1^1/2$ teaspoons cream of tartar
1 cup beef stock	1 cup water + 2 teaspoons soy or tamari sauce
1 cup buttermilk	1 cup plain yogurt *or* $1/2$ cup plain yogurt + $1/2$ cup skim milk *or* 1 tablespoon vinegar plus milk to equal 1 cup *or* 1 tablespoon lemon juice + milk to equal 1 cup
1 cup cake flour	1 cup minus 2 tablespoons sifted all-purpose flour
1 cup self-rising flour	1 cup all-purpose flour + 1 teaspoon baking powder + $1/2$ teaspoon salt
1 ounce (square) unsweetened chocolate	3 tablespoons unsweetened cocoa powder + 1 tablespoon butter (or shortening or cooking oil)
1 ounce (square) semisweet chocolate	3 tablespoons semisweet chocolate pieces *or* 1 square unsweetened chocolate + 1 tablespoon sugar
4 ounces (squares) sweet chocolate	$1/4$ cup unsweetened cocoa powder + $1/3$ cup sugar + 1 tablespoon shortening
1 tablespoon cornstarch	2 tablespoons flour
1 cup corn syrup	1 cup granulated sugar + $1/4$ cup water *or* 1 cup honey
1 cup heavy (whipping) cream	$1/3$ cup butter + whole milk to equal 1 cup
1 whole egg	2 egg yolks
1 whole egg	2 egg whites
2 large eggs	3 small eggs
1 small clove garlic	$1/8$ teaspoon garlic powder
1 tablespoon fresh ginger	$1/8$ teaspoon ground ginger
1 tablespoon fresh herbs	1 teaspoon dried herbs
1 cup honey	1 cup granulated sugar + $1/4$ cup water
1 tablespoon lemon juice	1 tablespoon distilled white vinegar
1 tablespoon lemongrass	$1/2$ teaspoon finely shredded lemon peel
$1/4$ cup marsala wine	$1/4$ cup dry white wine + 1 teaspoon brandy
1 cup molasses	$3/4$ cup granulated sugar
1 cup whole milk	$1/2$ cup evaporated milk + $1/2$ cup water

Emergency Recipe Substitutions (continued)

This Ingredient Amount—	Can Be Replaced with This—
1 pound fresh mushrooms	12 ounces canned mushrooms, drained *or* 3 ounces dried, rehydrated
1 teaspoon dry mustard	1 tablespoon prepared mustard
1 tablespoon fresh parsley	1 teaspoon dried parsley
1 small onion, chopped	1 teaspoon onion powder *or* 1 tablespoon dried minced onion
$1/8$ teaspoon cayenne pepper	3 to 4 drops liquid hot pepper
1 teaspoon poultry seasoning	$3/4$ teaspoon dried sage + $1/4$ teaspoon ground nutmeg + $1/8$ teaspoon ground allspice + dash ground cloves or ginger
3 tablespoons shallots	2 tablespoons chopped onion + 1 tablespoon minced garlic
1 cup sour cream	3 tablespoons butter + buttermilk or yogurt to equal 1 cup *or* 1 cup plain yogurt
1 cup granulated sugar	$1/2$ to $2/3$ cup honey or pure maple syrup (for baking)
1 cup brown sugar	1 cup granulated sugar + $1/4$ cup molasses
3 cups tomato juice	$1 1/2$ cups tomato sauce + $1 1/2$ cups water
2 cups tomato sauce	$3/4$ cup tomato paste + 1 cup water
1 tablespoon tapioca	$1 1/2$ tablespoons all-purpose flour
1 cup yogurt	1 cup buttermilk

Recipe Ingredient Equivalents

This Amount—	Is Equal to—
1 pound of apples (about 3 medium)	3 cups sliced
8 slices cooked bacon	$1/2$ cup crumbled
1 pound bananas (about 3 medium)	$1 1/3$ cups mashed
1 pint berries	$1 3/4$ to 2 cups
1 quart berries	3 to 4 cups
1-pound loaf of bread	13 to 18 slices
$1 1/2$ slices bread with crust	$1/2$ cup bread crumbs
1 pound butter or margarine	2 cups
$1/4$ pound butter or margarine	1 stick or $1/2$ cup
1-pound head of cabbage	$4 1/2$ cups shredded; 2 to 3 cups cooked

Recipe Ingredient Equivalents (continued)

This Amount—	Is Equal to—
1 pound carrots	3 cups shredded
2 stalks celery	$3/4$ to 1 cup sliced
$1/4$ pound cheese	1 cup shredded
$1/4$ pound Parmesan cheese	1 cup grated
1 large chicken breast	2 cups cooked
19 chocolate wafer cookies	1 cup crumbs
22 vanilla wafer cookies	1 cup finely crushed
4 medium ears of corn	1 cup kernels
1 pound cornmeal	3 cups
8 ounces cottage cheese	1 cup
1 pound crab in shell	$3/4$ to 1 cup flaked crabmeat
14 squares graham crackers	1 cup fine crumbs
22 to 28 saltine crackers	1 cup finely crushed
3 ounces cream cheese	6 tablespoons
1 ounce unsweetened chocolate	1 square
6-ounce package chocolate chips	1 cup
$3^1/2$-ounce can flaked coconut *or* 4-ounce can shredded coconut	$1^1/3$ cups
1 cup heavy cream ($1/2$ pint)	2 cups whipped cream
8 ounces sour cream	1 cup
1 pound pitted dates	3 cups chopped
5 large whole eggs	1 cup
8 to 10 egg whites (large eggs)	1 cup
12 to 14 egg yolks (large eggs)	1 cup
1 pound flour (all purpose)	$3^1/2$ cups unsifted; 4 cups sifted
1 pound flour (cake)	$4^3/4$ to 5 cups sifted
1 pound rye flour	$4^1/2$ cups
1 pound whole wheat flour	$3^1/3$ to $3^3/4$ cups
1 medium lemon	2 to 3 tablespoons juice; 1 tablespoon grated peel
1-pound head of lettuce	$6^1/4$ cups torn
1 medium lime	2 tablespoons juice
1 cup macaroni, dry	$2^1/4$ cups cooked
11 large marshmallows	1 cup

Recipe Ingredient Equivalents (continued)

This Amount—	Is Equal to—
10 miniature marshmallows	1 large marshmallow
5$\frac{1}{3}$- or 6-ounce can evaporated milk	$\frac{2}{3}$ cup
13- or 14-ounce can evaporated milk	1$\frac{2}{3}$ cups
14-ounce can sweetened condensed milk	1$\frac{1}{4}$ cups
3 cups (8 ounces) raw mushrooms	1 cup sliced cooked
$\frac{1}{4}$ pound (3 ounces) dried mushrooms	1 pound fresh
1 cup noodles, dry	1$\frac{3}{4}$ cup cooked
1 pound nuts, in shell	
Almonds	1 to 1$\frac{1}{2}$ cups nutmeats
Peanuts	2 to 2$\frac{1}{2}$ cups nutmeats
Pecans	2$\frac{1}{4}$ cups nutmeats
Walnuts	1$\frac{2}{3}$ to 2 cups nutmeats
1 pound nuts, shelled	
Almonds	3 cups
Cashews	3$\frac{1}{4}$ cups
Peanuts	3 to 4 cups
Pecans	3 to 4 cups
Walnuts	3 to 4 cups
1 cup regular or quick-cooking oats	1$\frac{3}{4}$ cups cooked
1 medium onion	$\frac{1}{2}$ to $\frac{2}{3}$ cup chopped
1 large onion	$\frac{3}{4}$ to 1 cup chopped
1 green onion	2 tablespoons sliced
1 medium orange	$\frac{1}{3}$ to $\frac{1}{2}$ cup juice; 2 tablespoons grated peel
1 cup small dried pasta	1$\frac{3}{4}$ cup cooked
1 pound peas in pod	1 cup shelled
2 medium peaches or pears	1 cup sliced
1 large green pepper	1 cup diced
1 pound potatoes (3 medium)	2$\frac{1}{4}$ cups diced; 1$\frac{3}{4}$ to 2 cups mashed
1 cup regular long-grain rice	3 cups cooked
1 cup wild rice	4 cups cooked
16 ounces salad oil	2 cups
1 pound raw shrimp in shell	$\frac{1}{2}$ pound shelled, cleaned, cooked
2 ounces dried spaghetti	1 cup cooked

Recipe Ingredient Equivalents (continued)

This Amount—	Is Equal to—
1 quart strawberries	4 cups sliced
1 pound granulated sugar	$2^{1}/_{4}$ cups
1 pound brown sugar	$2^{1}/_{4}$ cups packed
1 pound powdered sugar	$3^{1}/_{2}$ cups unsifted
16 ounces corn syrup	2 cups
1 pound fresh tomatoes	2 cups diced
16-ounce can tomatoes, drained	$1^{1}/_{4}$ cups
1 pound tea leaves	125 brewed cups

How Many Oysters in a Pint?

One pint of shucked oysters contains this many oysters:

> 20 extra large, or
> 20 to 26 large, or
> 26 to 38 medium, or
> 30 to 63 small

Container Sizes: How Much Do They Hold?

Package/Can Size	Amount
8-ounce can	1 cup
$10^{1}/_{2}$- to 12-ounce can (#1 tall)	$1^{1}/_{4}$ cups
12-ounce vacuum can	$1^{1}/_{2}$ cups
14- to 16-ounce (#300) can	$1^{3}/_{4}$ cups
16- to 17-ounce (#303) can	2 cups
20-ounce (#2) can	$2^{1}/_{2}$ cups
29-ounce (#$2^{1}/_{2}$) can	$3^{1}/_{2}$ cups
46-ounce can	$5^{3}/_{4}$ cups
$6^{1}/_{2}$-pound (#10) can	12 to 13 cups
10-ounce package frozen vegetables	$1^{1}/_{2}$ to 2 cups
16-ounce package frozen vegetables	3 to 4 cups
1-pound box brown sugar (packed)	$2^{1}/_{3}$ cups
1-pound box powdered sugar (unsifted)	$3^{1}/_{2}$ cups
5-pound bag sugar	$11^{1}/_{4}$ cups
5-pound bag all-purpose flour (unsifted)	$17^{1}/_{2}$ cups

Cooking Times: Meats

Type	Size/Amount	Cooking Method	Temperature	Cooking Time
Beef				
Flank steak	³/₄ inch thick	grilled	medium	12 to 14 min. (medium)
		broiled		12 to 14 min.
Steak (T-bone, tenderloin, rib eye, etc.)	1 inch thick	grilled	medium-hot	8 to 12 min. (rare)
		broiled		12 to 17 min. (medium)
				16 to 22 min. (well-done)
	1 inch thick	pan fried	medium	8 to 11 min. (rare)
				12 to 14 min. (medium)
				15 to 17 min. (well-done)
	1¹/₂ inches thick	grilled	medium-hot	14 to 18 min. (rare)
Ground patties	³/₄ inch thick	grilled	medium	14 to 18 min. (medium to well)
	³/₄ inch thick	broiled	broiler	14 to 18 min. (medium to well)
Boneless rolled rump roast	4 to 6 pounds	indirectly grilled	medium-low	1¹/₄ to 2¹/₂ hours (medium)
	4 to 6 pounds	roasted	325°	1¹/₂ to 3 hours (medium)
Rib roast	4 to 6 pounds	roasted	325°	1³/₄ to 3 hours (rare)
				2¹/₄ to 3¹/₄ hours (medium)
				2³/₄ to 4¹/₄ hours (well-done)
Eye round roast	2 to 3 pounds	roasted	325°	1¹/₄ to 1³/₄ hours (rare)
				1³/₄ to 2¹/₄ hours (medium)
				2¹/₄ to 2³/₄ hours (well-done)
Veal				
Chop	1 inch thick	grilled	medium	19 to 23 min. (medium to well)
	1 inch thick	broiled	broiler	12 to 15 min. (medium to well)
Cutlet	¹/₄ inch thick	pan fried	medium	4 to 6 min. (medium to well)
Loin roast	3 to 5 pounds	indirectly grilled	medium-low	1³/₄ to 3 hours (medium)
	3 to 5 pounds	roasted	325°	1³/₄ to 3 hours (medium to well)
Boneless rolled	3 to 5 pounds	roasted	325°	2³/₄ to 3¹/₄ hours (well-done)
Lamb				
Chop	1 inch thick	grilled	medium	10 to 14 min. (rare)
				14 to 16 min. (medium)
	1 inch thick	broiled	broiler	8 to 10 min. (rare)
				10 to 12 min. (medium)
	1 inch thick	pan fried	medium	7 to 9 min. (medium)
Boneless rolled leg roast	4 to 7 pounds	indirectly grilled	medium-slow	2¹/₄ to 3³/₄ hours (medium to well)
Leg (bone in)	5 to 7 pounds	roasted	325°	1³/₄ to 2¹/₄ hours (medium)

Cooking Times: Meats (continued)

Type	Size/Amount	Cooking Method	Temperature	Cooking Time
Pork				
Blade steak	½ inch thick	grilled	medium-hot	10 to 12 min. (well-done)
	½ inch thick	broiled	broiler	12 to 14 min. (well-done)
Chop	¾ inch thick	grilled	medium-hot	12 to 14 min. (medium to well)
	¾ inch thick	broiled	broiler	8 to 14 min. (medium to well)
	¾ inch thick	pan fried	medium	7 to 10 min. (well-done)
	1¼ to 1½ inches thick		medium-hot	30 to 40 min. (well-done)
Boneless top loin roast, single loin	2 to 4 pounds	indirectly grilled	medium-low	1 to 1¼ hours (medium)
	2 to 4 pounds	roasted	325°	1 to 1¼ hours (medium to well)
Spareribs (loin back)	2 to 4 pounds	indirectly grilled	medium	1¼ to 1½ hours (well-done)
Country-style ribs	2 to 4 pounds	indirectly grilled	medium	1½ to 2 hours (well-done)
Fresh bratwurst		grilled	medium-hot	12 to 14 min. (well-done)
Frankfurters		grilled, broiled	medium-hot	3 to 5 min.
Ham				
Fully cooked	4 to 6 pounds	indirectly grilled	medium-slow	1¼ to 2 hours
Boneless half	4 to 6 pounds	roasted	325°	1¼ to 2½ hours
Smoked picnic	5 to 8 pounds	indirectly grilled	medium-slow	2 to 3 hours
	5 to 8 pounds	roasted	325°	2 to 4 hours

Note: A meat thermometer is the best way to determine doneness: Rare is 140°F, medium is 160°F, and well-done is 170°F. The U.S. Department of Agriculture has revised its recommended cooking temperature for pork to 160°F, or until juices run clear.

Hints for Cooking a Roast

Ideally, a roast should be at room temperature before cooking. But because bacteria can grow if a roast, especially a large one, is left out for more than 1 hour, the next best thing is to set the oven temperature slightly lower than the required temperature and then cook the roast slightly longer.

If you put a frozen or partially thawed roast in the oven, increase cooking time by as much as 50 percent.

Once done, let the roast sit for about 15 to 25 minutes before carving so that the juices will have a chance to redistribute evenly throughout the meat.

Cooking Times: Fish and Shellfish

Type	Size/Amount	Cooking Method	Temperature	Cooking Time
Filets, steaks, cubes	1/2 to 1 inch thick	grilled	medium-hot	4 to 6 min. per 1/2 inch
		baked	450°	4 to 6 min. per 1/2 inch
		broiled	broiler	4 to 6 min. per 1/2 inch
		poached	simmer	4 to 6 min. per 1/2 inch
		microwaved	100% (high)	3 to 4 min. per 1/2 inch
	1/2 inch thick	pan fried	hot	3 to 4 min. per side
Dressed	1/2 to 1 1/2 pounds	grilled	medium-hot	7 to 9 min. per 1/2 pound
		baked	350°	6 to 9 min. per 1/2 lb.
	1 to 1 1/2 pounds	microwaved	100% (high)	10 to 12 min.
	8 to 12 ounces	pan fried	hot	5 to 8 min. per side
Clams	24 clams in shells	steamed	steam in Dutch oven	5 to 7 min.
Crab, blue	3 pounds	boiled	boil, then simmer in 8 quarts water	15 min.
Lobster, whole	1 to 1 1/2 pounds	boiled	boil then, simmer in 8 quarts water	20 min.
	1 to 1 1/2 pounds	broiled	broiler	5 to 7 min.
Lobster tails	6 ounces	grilled	medium-hot	6 to 10 min.
	8 ounces	grilled	medium-hot	12 to 15 min.
Sea scallops	12 to 15 per pound	grilled	medium-hot	5 to 8 min.
		pan fried	medium	1 1/2 to 3 min.
Shrimp, peeled	medium (20/pound)	grilled	medium-hot	6 to 8 min.
	jumbo (12–15/pound)	grilled	medium-hot	10 to 12 min.
	1 pound	boiled	boil, then simmer in 4 cups water	1 to 3 min.

Best Temperatures for Frying Food

The ideal frying temperature is about 375°F. Putting cold foods into the pan reduces this temperature, which results in greasiness. Try bringing the food to room temperature before frying (if you can do so safely); if this isn't possible, increase the starting temperature of the oil by 10 or 15 degrees. Also, don't crowd the pan—fry only a few pieces at a time to maintain the ideal temperature.

Cooking Times: Poultry

Type	Size/Amount	Cooking Method	Temperature	Cooking Time
Chicken, whole	2¹/₂ to 3 pounds	indirectly grilled	medium	1 to 1¹/₄ hours
	2¹/₂ to 3 pounds	roasted	375°	1 to 1¹/₄ hours
Chicken, broiler fryer, half	1¹/₄ to 1¹/₂ pounds	grilled	medium	40 to 50 min.
	1¹/₄ to 1¹/₂ pounds	broiled	broiler	30 min.
Chicken breast, boneless, skinless	4 to 5 ounces	grilled	medium hot	15 to 18 min.
	4 to 5 ounces	broiled	broiler	12 to 15 min.
Chicken breast halves, thighs, legs		grilled	medium	35 to 45 min.
		broiled	broiler	25 to 35 min.
Chicken kabobs		grilled	medium-hot	8 to 10 min.
		broiled	broiler	8 to 10 min.
Cornish hen, half	¹/₂ to ³/₄ pound	grilled	medium-hot	45 to 50 min.
	¹/₂ to ³/₄ pound	broiled	broiler	30 to 40 min.
Cornish hen, whole	1 to 1¹/₂ pounds	indirectly grilled	medium	1 to 1¹/₄ hours
	1 to 1¹/₂ pounds	roasted	375°	1 to 1¹/₄ hours
Duckling	3 to 5 pounds	roasted	375°	1³/₄ to 2¹/₄ hours
Turkey breast tenderloin	4 to 6 ounces	grilled	medium	12 to 15 min.
	4 to 6 ounces	broiled	broiler	8 to 10 min.
Turkey, unstuffed (stuffed, add 45 min.)	6 to 8 pounds	indirectly grilled	medium	1³/₄ to 2¹/₄ hours
	8 to 12 pounds		medium	2¹/₂ to 3¹/₂ hours
	12 to 16 pounds		medium	3 to 4 hours
	6 to 8 pounds	roasted	325°	3 to 3¹/₂ hours
	8 to 12 pounds		325°	3 to 4 hours
	12 to 16 pounds		325°	4 to 5 hours
	16 to 20 pounds		325°	4¹/₄ to 5 hours
	20 to 24 pounds		325°	5 to 6 hours
Turkey breast, whole	4 to 6 pounds	indirectly grilled	medium	1³/₄ to 2¹/₄ hours
	6 to 8 pounds		medium	2¹/₂ to 3¹/₂ hours
	4 to 6 pounds	roasted	325°	1¹/₂ to 2¹/₄ hours
	6 to 9 pounds		325°	2¹/₄ to 3¹/₄ hours
Turkey drumstick	¹/₂ to 1¹/₂ pounds	grilled	medium	³/₄ to 1¹/₄ hours
	1 to 1¹/₂ pounds	roasted	325°	1¹/₄ to 1³/₄ hours
Quail	4 to 6 ounces	roasted	375°	30 to 50 min.

<table>
<tr><td>Potatoes: America's Favorite Vegetable</td><td>On average, each American consumed 141 pounds of potatoes in 1994—the largest amount since 1935. Most of that year's increase went to make frozen french fries. Consumption of sweet potatoes also increased, to 4.6 pounds per person. All told, on average each American ate about 425 pounds of vegetables in 1994, including 1.9 pounds of garlic.</td></tr>
</table>

Cooking Times: Vegetables

Type	Size/Amount	Cooking Method	Temperature	Cooking Time
Artichokes	2 10-ounce	boiled		20 to 30 min.
		steamed		20 to 25 min.
		microwaved	high	7 to 9 min.
Asparagus	1 pound	boiled		4 to 8 min.
		steamed		4 to 8 min.
		microwaved	high	4 to 6 min.
Beans, green, yellow wax (whole)	³/₄ pound	boiled		20 to 25 min.
		steamed		18 to 22 min.
		microwaved	high	13 to 15 min.
Beets, whole	1 pound	boiled		40 to 50 min.
Beets, cubed, sliced	1 pound	boiled		20 min.
		microwaved	high	9 to 12 min.
Broccoli	³/₄ pound	boiled		8 to 12 min.
		steamed		8 to 12 min.
		microwaved	high	4 to 7 min.
Brussels sprouts	³/₄ pound	boiled		10 to 12 min.
		steamed		10 to 15 min.
		microwaved	high	4 to 6 min.
Cabbage	half 1-pound head	boiled		3 to 5 min.
		steamed		10 to 12 min.
		microwaved	high	9 to 12 min.
Carrots	1 pound	boiled		7 to 9 min.
		steamed		6 to 8 min.
		microwaved	high	7 to 10 min.
Cauliflower	1¹/₂-pound head	boiled		10 to 15 min.
		steamed		8 to 10 min.
		microwaved	high	7 to 10 min.
Celeriac	1 pound	boiled		5 to 6 min.
		steamed		5 min.
		microwaved	high	3 to 4 min.
Celery	5 stalks	boiled		6 to 9 min.
		steamed		7 to 10 min.
		microwaved	high	7 to 10 min.

Cooking Times: Vegetables (continued)

Type	Size/Amount	Cooking Method	Temperature	Cooking Time
Corn	2 cups	boiled		4 min.
		steamed		4 to 5 min.
		microwaved	high	4 to 5 min.
Corn on the cob	1 ear	boiled		5 to 7 min.
		microwaved	high	3 to 5 min.
Eggplant	1 medium	boiled		4 to 5 min.
		steamed		4 to 5 min.
		microwaved	high	5 to 8 min.
Fennel	2 heads	boiled		6 to 10 min.
		microwaved	high	4 to 6 min.
Greens	¾ pound	boiled		9 to 12 min.
Jerusalem artichokes	1 pound	boiled		8 to 10 min.
		steamed		10 to 12 min.
		microwaved	high	5 to 7 min.
Jicama	10 ounces	boiled		5 min.
		steamed		5 min.
		microwaved	high	5 min.
Kohlrabi	1 pound	boiled		6 to 8 min.
		steamed		8 min.
		microwaved	high	6 to 8 min.
Leeks	1½ pound	boiled		5 min.
		steamed		5 min.
		microwaved	high	4 to 5 min.
Mushrooms	1 pound	pan fried		5 min.
		steamed		10 to 12 min
		microwaved	high	5 min.
Okra	½ pound	boiled		8 to 15 min.
		microwaved	high	4 to 6 min.
Onions	1 large	pan fried		5 to 10 min.
		microwaved	high	3 to 4 min.
Parsnips	¾ pound	boiled		7 to 9 min.
		steamed		8 to 10 min.
		microwaved	high	4 to 6 min.
Pea pods	½ pound	boiled		2 to 4 min.
		steamed		2 to 4 min.
		microwaved	high	2 to 5 min.
Peas, green	2 pounds	boiled		10 to 12 min.
		steamed		12 to 15 min.
		microwaved	high	6 to 8 min.
Peppers	2 large	boiled		6 to 7 min.
		steamed		6 to 7 min.
		microwaved	high	5 to 8 min.
Potatoes	1 pound	boiled		20 to 25 min.
		steamed		20 min.
		microwaved	high	8 to 10 min.

179

Cooking Times: Vegetables (continued)

Type	Size/Amount	Cooking Method	Temperature	Cooking Time
Rutabagas	1 pound	boiled		18 to 20 min.
		steamed		18 to 20 min.
		microwaved	high	11 to 13 min.
Spinach	1 pound	boiled		3 to 5 min.
		steamed		3 to 5 min.
		microwaved	high	4 to 6 min.
Squash, acorn	1 pound	baked	350°	55 min.
		microwaved	high	6 to 9 min.
Squash, butternut	1½-pound	baked	350°	55 min.
		microwaved	high	9 to 12 min.
Squash, yellow and zucchini	¾ pound	boiled		3 to 5 min.
		steamed		4 to 6 min.
		microwaved	high	4 to 5 min.
Squash, spaghetti	2½- to 3-pound	baked	350°	30 to 40 min.
		microwaved	high	15 to 20 min.
Sweet potatoes	1 pound	boiled		25 to 35 min.
		microwaved	high	10 to 13 min.
Turnips	1 pound	boiled		10 to 12 min.
		steamed		10 to 15 min.
		microwaved	high	12 to 14 min.

How Big Is a Jumbo Egg?

Here is a breakdown of egg sizes based on their net weight per dozen:

Size	Ounces
Jumbo	30
Extra Large	27
Large	24
Medium	21
Small	18
Pee Wee	15

Cooking Times: Rice

Type	Dry Amount	Water	Cooking Time	Cooked Amount
Wild	1 cup	3 cups	1 hour+	4 cups
Brown	1 cup	2 cups	35 min.	3 cups
White long-grain	1 cup	2 cups	15 min.	3 cups
Quick	1½ cups	1½ cups	see package	3 cups

Cooking Times: Pasta

Type	Cooking Time (min.)
Fresh	
Fettucine	1½ to 2
Lasagna	2 to 3
Linguine	1½ to 2
Ravioli	7 to 9
Tagliatelle	1½ to 2
Tortellini	7 to 9
Packaged (dry)	
Bow ties (large)	10
Capellini	5 to 6
Cavatelli	12
Fettucine	8 to 10
Fusilli	15
Lasagna	10 to 12
Linguine	8 to 10
Macaroni, elbow	10
Manicotti	18
Noodles	6 to 8
Orzo	5 to 8
Rotini	8 to 10
Shells (large)	12 to 14
Spaghetti	10 to 12
Tortellini	15
Vermicelli	5 to 7
Ziti	14

Cooking Times: Beans

Rinse and soak dried beans overnight in covered pan (or place 1 pound beans in 8 cups cold water in Dutch oven, bring to boil, simmer for 2 minutes, remove from heat, and let stand for 1 hour). Drain and rinse, put beans in pan with 8 cups fresh water, bring to boil, then reduce heat and cook (covered and simmering) as specified below, stirring occasionally.

Type (1 lb.)	Cooking Time	Amount of Cooked Beans
Black	45 min. to $1\frac{1}{2}$ hours	6 cups
Garbanzo, chick pea	$1\frac{1}{2}$ to 2 hours	6 cups
Great Northern	1 to $1\frac{1}{2}$ hours	7 cups
Kidney	1 to $1\frac{1}{2}$ hours	$6\frac{1}{2}$ cups
Lentil	30 min.	6 cups
Lima (baby)	45 min. to 1 hour	$6\frac{1}{2}$ cups
Lima (large)	1 to $1\frac{1}{4}$ hours	6 cups
Navy	1 to $1\frac{1}{2}$ hours	7 cups
Pinto	$1\frac{1}{4}$ to $1\frac{3}{4}$ hours	$6\frac{1}{2}$ cups
Soybean	2 to 3 hours	7 cups
Split pea	$1\frac{1}{2}$ hours	5 cups

How Much to Buy to Serve a Crowd

Food	8 Servings	12 Servings	16 Servings
Meat			
Boneless cuts (ground meat, roasts, ham, brisket, veal, etc.)	2 to 3 pounds	3 to 4 pounds	5 to 6 pounds
Boneless poultry, fish	2 to 3 pounds	3 to 4 pounds	5 to 6 pounds
Pork ribs, other bony cuts	6 to 8 pounds	9 to 12 pounds	12 to 15 pounds
Lamb chops, T-bone steaks, pork chops, pork and beef roasts	3 to 5 pounds	5 to 7 pounds	7 to 9 pounds
Whole poultry with bones	8 to 10 pounds	12 to 14 pounds	16 to 20 pounds
Vegetables			
Vegetables that hold their shape (broccoli, carrots, cauliflower, squash)	2 pounds	3 pounds	4 pounds
Greens that shrink when cooked (kale, collards, spinach)	4 pounds	6 pounds	8 pounds
Peas to be shelled	4 pounds	6 pounds	8 pounds
Frozen vegetables (peas, corn, beans)	2 16-ounce bags	3 16-ounce bags	4 16-ounce bags
Potatoes, mashed or scalloped	2 to 3 pounds	4 to 5 pounds	6 to 7 pounds
Rice and Pasta			
Spaghetti	1 pound uncooked	2 pounds uncooked	3 pounds uncooked
Rice, white	2 cups uncooked	4 cups uncooked	6 cups uncooked
Salads			
Lettuce	1 to 2 heads	2 to 3 heads	4 to 5 heads
Potato or pasta	1 quart	2 quarts	3 quarts

How Much to Buy to Serve a Crowd (continued)

Food	8 Servings	12 Servings	16 Servings
Desserts			
Cake (13" x 9" sheet)	1	1	2
Ice cream, frozen yogurt	1 quart	2 quarts	3 quarts
Pie (9")	1	2	3
Beverages			
Coffee (ground)	$1/4$ pound	$1/2$ pound	$3/4$ to 1 pound
Milk	$1/2$ gallon	$3/4$ gallon	1 gallon
Soft drinks	1 2-liter bottle	2 2-liter bottles	3 2-liter bottles
Tea, hot	$1^{1}/2$ quarts	3 quarts	4 quarts
Tea, iced	2 to 3 quarts	3 to 4 quarts	4 to 5 quarts

Serving Sizes: Meat, Poultry, and Fish

Type	Allow per Person (oz. before cooking)
Meat	
Beef	
Rib roast, bone in	12 ounces
Rib roast, boneless	6 to 8 ounces
Pot roast, bone in	8 to 12 ounces
Pot roast, boneless	6 to 8 ounces
Stewing beef, bone in	8 to 12 ounces
Steaks, bone in	8 to 12 ounces
Ground beef	6 to 8 ounces
Lamb	
Roast leg or shoulder, bone in	12 to 16 ounces
Roast leg or shoulder, boneless	6 to 8 ounces
Rack or rib roast	1 rack = 2 servings
Crown roast	2 ribs
Chops, shoulder	1 chop
Chops, loin	1 or 2 chops
Chops, rib	2 or 3 chops

Serving Sizes: Meat, Poultry, and Fish (continued)

Type	Allow per Person (oz. before cooking)
Meat (continued)	
Pork	
Roast, bone in	12 to 16 ounces
Roast, boneless	6 to 8 ounces
Chops	1 or 2 chops
Spareribs	12 to 16 ounces
Poultry	
Chicken	
Whole	12 ounces
Parts	8 to 10 ounces
Turkey	
Under 12 pounds	$3/4$ pound
More than 12 pounds	$1/2$ pound to $3/4$ pound
Parts	$1/2$ pound
Duck	
Under 4 pounds	1 duck = 2 servings
More than 4 pounds	1 duck = 2 to 3 servings
Goose, whole	1 to $1\frac{1}{2}$ pounds
Cornish game hen, whole	$1/2$ to 1 bird
Squab, whole	1 squab
Pheasant, whole (2 to 3 pounds)	1 pheasant = 2 servings
Quail, whole (5 ounces each)	2 quail
Partridge, whole (12 to 14 ounces)	1 partridge
Fish	
Whole, dressed	8 ounces
Fillets, steaks	6 to 8 ounces
Shellfish	
Clams	
Steamers	20 clams
Hard shell	2 dozen = 4 to 6 appetizers
Crabmeat, cooked	4 to 5 ounces
Crabs, soft shell	2 to 3 crabs
Mussels	
In shell	3 quarts = 4 servings
Shucked	1 quart (undrained) = 4 servings

Serving Sizes: Meat, Poultry, and Fish (continued)

Type	Allow per Person (oz. before cooking)
Shellfish (continued)	
Oysters	
On half shell	6 to 12
Shucked	1 to 1½ quarts = 6 servings
Shrimp, in shell	6 to 8 ounces
Squid	8 ounces

Rule of Thumb: Food Storage

Food items with a large exposed surface area are more likely to serve as sites for air-borne bacteria—ground meats, for example. Also, the higher the ratio of unsaturated to saturated fat, the shorter the refrigerator and freezer life, which is why beef can be stored for longer periods than chicken. Foods containing salt have a shorter freezer life than unsalted foods because the salt lowers the freezing point.

Safe Food Storage Times: In the Refrigerator

Food	Storage Time
Fresh beef, lamb, pork, veal	
Roasts	3 to 5 days
Steaks, chops	2 to 5 days
Stew meat	2 days
Ground meat	2 days
Leftovers	2 days
Processed meats (opened package)	
Ham, whole and half	7 days
Bacon	5 to 7 days
Sausage	2 days
Frankfurters	4 to 5 days
Luncheon meats	3 days
Fresh fish	1 to 2 days
Fresh poultry	1 to 2 days

Safe Food Storage Times: In the Refrigerator (continued)

Food	Storage Time
Dairy products	
Butter, margarine	1 month
Buttermilk	1 to 2 weeks
Cheese (opened), chunks	1 to 2 months
Cheese, cottage or ricotta	5 to 7 days
Cheese, cream	2 weeks
Cheese, Parmesan (grated)	1 year
Cheese, slices	2 weeks
Eggs (in shell)	1 month
Eggs (hardcooked)	1 week
Milk	1 week
Half-and-half	7 to 10 days
Heavy cream	10 days
Sour cream	3 to 4 weeks
Yogurt	2 weeks
Fruits	
Apples	1 month
Apricots	5 days
Bananas (ripened)	2 to 3 days
Berries, cherries	3 days
Citrus (oranges, lemons, etc.)	2 weeks
Grapes	4 to 6 days
Juices	6 days
Kiwi fruit	3 to 6 weeks
Melons	1 week
Pears	3 to 5 days
Pineapples (ripened)	2 days
Plums	5 days
Vegetables	
Artichokes	3 to 4 days
Asparagus	2 to 3 days
Beans	2 to 5 days
Beets	1 to 2 weeks
Broccoli	3 to 5 days
Brussels sprouts	2 to 4 days
Cabbage	1 to 2 weeks
Carrots	1 to 2 weeks
Cauliflower	3 to 5 days

Safe Food Storage Times: In the Refrigerator (continued)

Food	Storage Time
Vegetables (continued)	
Celery	1 to 2 weeks
Corn (in husks)	1 to 2 days
Cucumbers	3 to 5 days
Eggplant	2 to 4 days
Fennel	5 to 7 days
Greens	3 to 5 days
Kohlrabi (green tops)	2 to 3 days
Kohlrabi (bulbs)	1 to 2 weeks
Mushrooms (in paper bag)	1 week
Okra	3 to 4 days
Onions*	1 month
Onions, green (leeks, scallions)	3 to 4 days
Parsnips	2 weeks
Peas, snow or sugar snap	1 to 2 days
Peas, sweet green (in pod)	2 to 4 days
Peppers	3 to 5 days
Potatoes*	2 months
Radishes	2 weeks
Rutabagas*	Several months
Squash, summer	3 to 5 days
Squash, winter*	Several months
Sweet potatoes*	Several months
Tomatoes (ripened)	1 week
Turnips	1 week
Condiments (after opening)	
Catsup (in glass)	3 months
Catsup (in plastic)	2 months
Coffee	6 to 8 weeks
Nuts (shelled)	6 months
Olives	1 month
Pickles	2 months
Salad dressings	3 months
Spaghetti sauce	5 to 7 days

Note: Times shown are for raw foods except as noted.

*Keep in cool, dry place; refrigeration not necessary.

A Test for Egg Freshness

If you're not sure how fresh an egg is, place the egg in a pan of cold water.

- If it lies on its side, it's fresh.
- If it tilts at an angle, it's about 1 week old.
- If it stands straight up, it's about 10 days old.
- If it floats, it's too old; discard it.

Safe Food Storage Times: In the Freezer

If frozen foods are properly wrapped and don't thaw out, they can be kept longer than the times shown below; their flavor and texture, however, will begin to decline after these times.

Food	Storage Time
Breads and desserts	
Breads	3 months
Cakes	3 to 5 months
Cookies	6 months
Pies and pastries	2 months
Dairy products*	
Butter	6 months
Cheese, cottage	4 to 6 months
Cheese, cream	4 months
Cheese, hard or semihard	6 months
Cheese, soft	4 months
Cream, heavy	3 to 6 months
Egg whites	6 months
Egg yolks	8 months
Ice cream	1 to 3 months
Milk	1 to 3 months
Sour cream	3 months
Fish and shellfish	
Crabmeat (Dungeness)	3 months
Crabmeat (King)	10 months
Cod	6 months
Flounder	6 months
Haddock	6 months
Halibut	6 months

Safe Food Storage Times: In the Freezer (continued)

Food	Storage Time
Fish and Shellfish (continued)	
Ocean perch	2 months
Salmon	2 months
Shrimp	1 year
Striped bass	3 months
Fruits	12 months
Meats	
Bacon	1 month
Beef	6 to 12 months
Cooked meat	3 months
Frankfurters	1 month
Ground meats	3 to 4 months
Ham	1 to 2 months
Lamb	6 to 9 months
Pork	3 to 6 months
Sausage	2 months
Veal	6 to 9 months
Poultry	
Chicken, whole	6 to 12 months
Chicken, pieces	6 to 9 months
Chicken, cooked	1 month
Duck, whole	6 months
Goose, whole	6 months
Turkey	6 to 12 months
Vegetables	
Commercially frozen	8 months
Home frozen	12 months

*Cheeses may become crumbly when thawed.

Safe Food Storage Times: In the Pantry

Food	Storage Time
Canned foods (discard bulging or leaking cans)	
High-acid (citrus juices, pickles, etc.)	1 year
Low-acid (corn, peas, stew, etc.)	2+ years
Packaged mixes	
Biscuit	9 months
Brownie	9 months
Cake	1 year
Cocoa	8 months
Frosting	8 months
Pancake	6 months
Pie crust	6 months
Potato, instant	18 months
Pudding	1 year
Sauce, soup, gravy	6 months
Staples	
Baking powder and soda	18 months
Beans and peas, dried	18 months
Bouillon cubes or powder	1 year
Cereal, ready-to-eat	Check package
Cereal, uncooked	1 year
Chocolate, cooking	1 year
Coconut	1 year
Coffee, unopened	2 years
Cornmeal	10 months
Cornstarch	18 months
Crackers	3 months
Flour	10 to 15 months
Fruit, dried	6 months
Gelatin	18 months
Herbs and spices	1 year
Honey	1 year
Milk, evaporated and condensed	1 year
Milk, non-fat dry	6 months
Molasses	2 years
Pasta, dry	1 to 2 years
Peanut butter	6 months
Rice, brown or wild	1 year
Rice, white	2 years

Safe Food Storage Times: In the Pantry (continued)

Food	Storage Time
Staples (continued)	
Shortening	8 months
Sugar	2 years
Tea bags	18 months
Vanilla	1 year
Vegetable oil	3 months
Vinegar (opened)	1 year
Worcestershire sauce	1 year

A Test for Baking Powder Freshness

To see if your baking powder is still fresh, add 1 teaspoon to ¼ cup hot water. If the mixture fizzes, the powder is good.

Poultry Thawing Times

Type	Weight (lbs.)	In Refrigerator	In Cold Water
Chicken, whole	3 to 4	12 to 16 hours	1 to 2 hours
Chicken, pieces	under 1	3 to 9 hours	1 hour
Duck	3 to 6	1 to 2 days	4 to 6 hours
Goose	6 to 12	1 to 2 days	4 to 6 hours
Pheasant	2 to 4	12 to 16 hours	1 to 2 hours
Turkey, whole	4 to 12	1 to 2 days	4 to 6 hours
	12 to 20	2 to 3 days	6 to 8 hours
	20 to 24	3 to 4 days	8 to 12 hours
Turkey, breast	5 to 11	1 to 2 days	4 to 6 hours
Turkey, boneless roast	3 to 10	1 to 2 days	1 to 3 hours

Poultry: Fresh, Hard-Chilled, and Frozen Defined

The U.S. Department of Agriculture has established 3 categories for labeling poultry products. A product can be labeled "fresh" if it was never chilled below 26°F (the temperature at which poultry, because of its salt content, freezes). The bird can be labeled "hard-chilled" if it was chilled between 0° and 26°F. The product is labeled "frozen" or "previously frozen" if it was chilled below 0°F.

How Long Foods Will Keep If the Power Fails

If you keep the door closed, your refrigerator will keep foods cold enough for about 6 hours; your freezer will keep foods frozen for between 36 and 48 hours if it is fully loaded and for about 24 hours if it is only half full. If the power outage lasts longer than those times, follow these guidelines:

- *If these refrigerated foods have been kept at above 40°F (normal refrigerator temperature) for more than 8 hours, throw them away:* milk, cream, soft cheese, mayonnaise, raw meat or poultry, luncheon meat, hot dogs.

- *After 1 day above 40°F,* throw out fruit juice.

- *After 5 to 7 days above 40°F,* throw out fresh eggs. You can keep fresh fruits and vegetables, hard cheese, and butter and margarine as long as they look and smell all right.

- *If frozen foods have thawed out but are still cold,* you can refreeze meat and poultry, hard cheese and butter, vegetables, and juices; but you should either cook and serve or discard casseroles, stews, and anything made with cream, milk, or eggs.

Food Hotlines and Other Useful Food and Nutrition Phone Numbers

- Meat and Poultry

U.S. Department of Agriculture Meat and Poultry Line
(800) 535-4555, weekdays 10 a.m. to 4 p.m. EST

Butterball Turkey Talk-Line
(800) 323-4848 (English and Spanish)
(800) TDD-3848 (hearing-impaired)
November and December only

Weber Grill-Line
(800) 474-5568, spring and summer months

**Food Hotlines and Other Useful Food and
Nutrition Phone Numbers (continued)**

■ Fruits and Vegetables

United Fresh Fruit and Vegetable Association
(703) 863-3410, weekdays 8:30 a.m. to 5:30 p.m. EST

■ General Nutrition

American Dietetic Association Consumer Nutrition Hotline
(800) 366-1655, weekdays 10 a.m. to 5 p.m. EST

American Heart Association
(800) 242-8721, weekdays 8:30 a.m. to 5 p.m. local time

American Diabetes Association
(800) 232-3472

**Interpreting
Dates on
Cans and
Packages**

Increasingly, food and manufacturers are placing "Sell by"
or "Best used by" dates on their products, a practice called
open dating. But many canned and bottled goods still have
mysterious codes stamped on them that are impossible to
translate without knowing the coding system the manufacturer
uses. Almost all of these codes specify the plant location and
the date when the item was produced; some even include the
shift, batch, or machine used.

Some codes use only the last digit of the year—say "5" for 1995;
others use a letter code for the month, and "A" may mean June,
not January. Others use the day of the year, with "1" being
"January 1" and "365" being "December 31." The bottom line is
that unless you know the manufacturer's system, you won't
have much luck in cracking the code. Calling the manufacturer
(many print toll-free phone numbers on their labels) may lead
you to someone who can tell you what the code means.

Beverages

Caffeine Content

Beverage	Serving Size (oz.)	Caffeine (mg)
Coffee		
Drip	5	110–150
Percolated	5	60–125
Instant	5	40–105
Decaffeinated	5	2–5
Tea		
5-minute steep	5	40–100
3-minute steep	5	20–50
Instant	8	31
Herbal	5	0
Hot Chocolate	5	2–10
Cola	12	45

Ideal Brewing Temperatures The ideal brewing temperature for tea and coffee is between 185°F and 205°F. Too low a temperature means that not enough flavor will be released; too high a temperature releases too much of the compound that causes bitterness.

Alcohol Content by Volume

Beverage	Average Alcohol Content (%)
Beer and Wine	
Beer, regular (12 fl. oz.)	4.5
Beer, light (12 fl. oz.)	4.0
Malt liquor (12 fl. oz.)	15.7
Wine, table (all types) (3.5 fl. oz.)	11.5
Wine, dessert (2 fl. oz.)	18.8
Wine, fortified (sherry, Madeira) (2 fl. oz.)	17.0
Mixed Drinks	
Bloody Mary (5 fl. oz.)	11.7
Bourbon and soda (4 fl. oz.)	16.1
Daiquiri (2 fl. oz.)	28.3
Gin and tonic (7 fl. oz.)	8.8
Manhattan (2 fl. oz.)	36.9
Martini (2.5 fl. oz.)	38.4
Piña colada (4.5 fl. oz.)	12.3
Screwdriver (7 fl. oz.)	8.2
Tom Collins (7.5 fl. oz.)	9.0
Whiskey sour (3 fl. oz.)	11.8
Liqueurs	
Coffee, 53 proof (1.5 fl. oz.)	26.5
Coffee, 63 proof (1.5 fl. oz.)	31.5
Crème de menthe (1.5 fl. oz.)	36.0

"Proof" Defined

The *proof* of an alcoholic beverage is defined differently in the United States, Great Britain, and Europe.

In the United States, the proof of a beverage is twice its alcohol content in percentage by volume at 68°F; in other words, a whiskey that is 80 proof contains 40 percent alcohol.

In Great Britain, the proof is based on the number of 57.1 percent alcohol by volume; a whiskey that is 100 proof contains 57.1 percent alcohol. To get the U.S. equivalent, multiply the British proof by 8 and divide the result by 7.

In Europe, proof is strictly the percentage of alcohol by volume; 40 percent proof equals 40 percent alcohol, or half the number for the U.S. proof.

Miscellaneous Measures for Spirits

Pony*	=	½ jigger (or 1 fluid ounce)
Shot	=	⅔ jigger (1 fluid ounce)
Jigger	=	1.5 shot (or fluid ounces)
Pint	=	16 shots (or fluid ounces)
	=	0.625 fifth
Fifth	=	25.6 shots (or fluid ounces)
	=	1.6 pints
	=	0.8 quart
	=	0.757 liter
Quart	=	32 shots (or fluid ounces)
	=	1.25 fifths
Magnum	=	2 quarts
	=	1.5 liters
½ gallon	=	64 shots (or fluid ounces)
	=	1.75 liters
Jeroboam (champagne, brandy only)	=	6.4 pints
	=	1.6 magnums
	=	0.8 gallon
	=	3 liters
	=	4 standard bottles
Rehoboam	=	6 standard bottles
Methuselah	=	8 standard bottles
Salmanazar	=	12 standard bottles
Balthazar	=	16 standard bottles
Nebuchadnezzar	=	20 standard bottles

* Definition varies among sources.

Standard Wine Bottle Sizes

Bottle	U.S. Oz.	Metric
Small (split)	6.3	187 milliliters
Medium (tenth) (half bottle)	12.7	375 milliliters
Regular (fifth) (standard bottle)	25.4	750 milliliters
Large (quart) (liter)	33.8	1 liter
Magnum	50.7	1.5 liters

3
Health

Health Statistics

Leading Causes of Death in the United States

Rank	Cause of Death	Number	% of Total Deaths
	All causes	2,268,000	100.0
1	Heart diseases	739,860	32.6
2	Cancer	530,870	23.4
3	Cerebrovascular diseases	149,740	6.6
4	Chronic obstructive pulmonary diseases/allied conditions	101,090	4.5
5	Accidents and adverse effects	88,630	3.9
	Motor vehicle accidents	40,880	1.8
	All other accidents and adverse effects	47,750	2.1
6	Pneumonia and influenza	81,730	3.6
7	Diabetes mellitus	55,110	2.4
8	Human immunodeficiency virus infection	38,500	1.7
9	Suicide	31,230	1.4
10	Homicide and legal intervention	25,470	1.1
11	Chronic liver disease and cirrhosis	24,730	1.1
12	Nephritis, nephrotic syndrome, and nephrosis	23,500	1.0
13	Septicemia	20,420	0.9
14	Athcrosclcrosis	17,090	0.8
15	Certain conditions originating in the perinatal period	15,820	0.7
	All other causes	324,160	14.3

Source: National Center for Health Statistics, *Monthly Vital Statistics Report*, October 1994.
Note: Data are provisional for 1994 and were estimated from a 10 percent sample of deaths.

Top 9 Underlying Causes of Death

The health-related factors shown below represent the root causes of death from disease in the United States. The leader, tobacco, causes cancer of the lung, esophagus, mouth and throat, pancreas, kidney, and bladder. In addition, it is a risk factor for heart disease, stroke, and high blood pressure and can cause pneumonia and lung disease. Diet and lack of exercise lead to cardiovascular diseases; cancer of the colon, breast, and prostate; and diabetes. Alcohol leads to 60 to 90 percent of deaths from cirrhosis, 40 to 50 percent of traffic accident deaths, and 3 to 5 percent of cancer deaths, in addition to several thousand accidental and violent deaths.

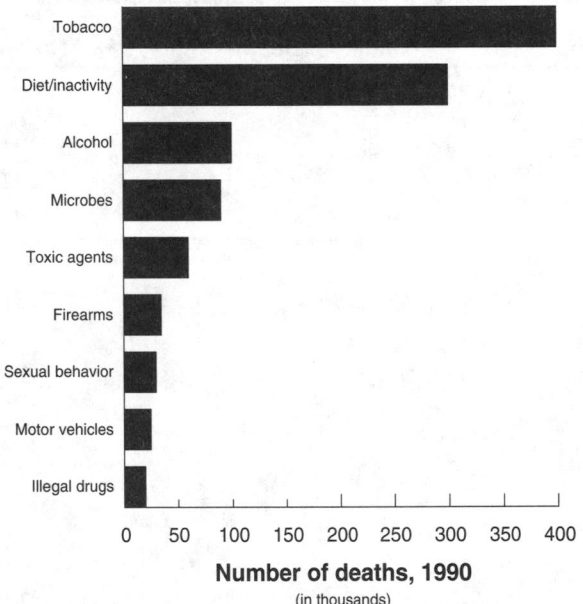

Number of deaths, 1990
(in thousands)

Source: National Center for Health Statistics, 1993.

Incidence of Diseases Reported in the United States, 1988 to 1993

Disease	1988	1989	1990	1991	1992	1993
AIDS	31,001	33,722	41,595	43,672	45,472	103,691
Amebiasis	2,860	3,217	3,328	2,989	2,942	2,970
Aseptic meningitis	7,234	10,274	11,852	14,526	12,223	12,848
Botulism[1]	84	89	92	114	91	97
Brucellosis	96	95	85	104	105	120
Chickenpox	192,900	185,400	173,100	147,100	158,400	134,700
Diphtheria	2	3	4	5	4	
Encephalitis:						
Primary infectious	882	981	1,341	1,021	774	919
Post infectious	121	88	105	82	129	170
Haemophilus influenzae	(2)	(2)	(2)	2,764	1,412	1,419
Hepatitis:						
B (serum)	23,200	23,400	21,100	18,000	16,100	13,400
A (infectious)	28,500	35,800	31,400	24,400	23,100	24,200
Unspecified	2,500	2,300	1,700	1,300	900	600
Non-A, non-B	2,600	2,500	2,600	3,600	6,000	4,800
Legionellosis	1,085	1,190	1,370	1,317	1,339	1,280
Leprosy (Hansen disease)	184	163	198	154	172	187
Leptospirosis	54	93	77	58	54	51
Lyme disease	(2)	(2)	(2)	9,465	9,895	8,257
Malaria	1,099	1,277	1,292	1,278	1,087	1,411
Measles	3,400	18,200	27,800	9,600	2,200	300
Meningococcal infections	2,964	2,727	2,451	2,130	2,134	2,637
Mumps	4,900	5,700	5,300	4,300	2,600	1,700
Pertussis (whooping cough)	3,500	4,200	4,600	2,700	4,100	6,600
Plague	15	4	2	11	13	10
Poliomyelitis, acute	9	5	7	8	4	3
Psittacosis	114	116	113	94	92	60
Rabies:						
Animal	4,651	4,724	4,826	6,910	8,589	9,377
Human	0	1	1	3	1	3
Rheumatic fever, acute[3]	158	144	108	127	75	112
Rubella (German measles)	200	400	1,100	1,400	200	200
Salmonellosis[4]	48,900	47,800	48,600	48,200	40,900	41,600
Shigellosis (bacillary dysentery)	30,600	25,000	27,100	23,500	23,900	32,200
Tetanus	53	53	64	57	45	48
Toxic shock syndrome	390	400	322	280	244	212
Trichinosis	45	30	129	62	41	16

Incidence of Diseases Reported in the United States, 1988 to 1993 (continued)

Disease	1988	1989	1990	1991	1992	1993
Tuberculosis[5]	22,400	23,500	25,700	26,300	26,700	25,300
Tularemia	201	152	152	193	159	132
Typhoid fever	436	460	552	501	414	440
Typhus fever:						
Flea-borne (endemic-murine)	54	41	50	43	28	25
Tick-borne (Rocky Mountain spotted fever)	609	623	651	628	502	456
Venereal diseases (civilian cases):						
Gonorrhea	720,000	733,000	690,000	620,000	501,000	439,000
Syphilis	103,000	111,000	134,000	129,000	113,000	101,000
Other	5,200	4,900	4,600	4,000	2,200	1,700

Source: *Statistical Abstract of the United States, 1995.*

Note: Figures should be interpreted with caution. Although reporting of some of these diseases is incomplete, the figures are of value in indicating trends of disease incidence. Includes cases imported from outside the United States.

1. Includes food-borne, infant, wound, and unspecified cases.
2. Disease was not notifiable.
3. Based on reports from states: 25 in 1987, 29 in 1988, 28 in 1989, 30 in 1990, 23 in 1991, and 26 in 1992 and 1993.
4. Excludes typhoid fever.
5. Newly reported active cases.

Major Causes of Death Worldwide

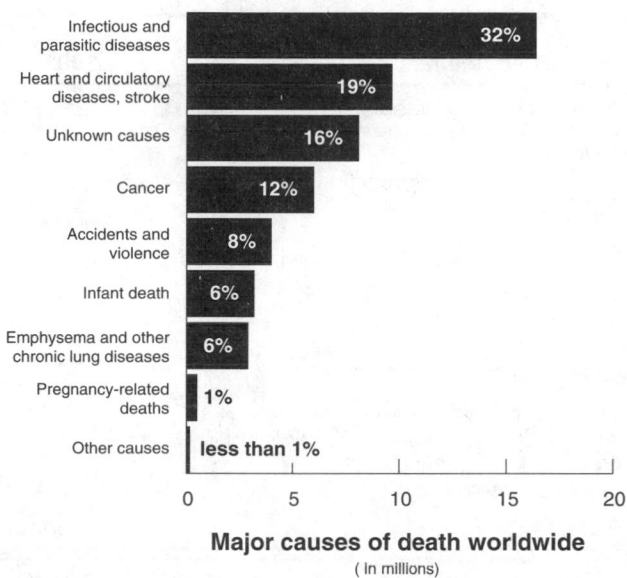

Source: World Health Report, 1995.

Estimated HIV Infections in Adults Worldwide

Region	HIV Infections
Sub-Saharan Africa	10.0 million
South and Southeast Asia	2.5 million
Latin America/Caribbean	2.0 million
North America	1.0 million
Western Europe	500 thousand
North Africa/Middle East	100 thousand
Eastern Europe/Central Asia	50 thousand
East Asia/Pacific	50 thousand
Australasia	25 thousand

Source: World Health Organization, 1994.

Number of Deaths from Accidents by Type in the United States

Type of Accident	1993	% Change from 1992
Motor vehicle accidents	42,000	+3
Falls	13,500	+5
Poisoning by solids and liquids	6,500	+12
Drowning	4,800	+7
Fires, burns, deaths associated with fires	4,000	0
Suffocation by ingested object	2,900	+4
Firearms	1,600	+7
Poisoning by gases, vapors	700	0
All other types (includes medical complications, machinery, water and air transport, mechanical suffocation, excessive cold)	14,000	+8
Total	90,000	+5

Source: National Safety Council, *Accident Facts*, 1994 edition. Used with permission.

Classification of Burns

Burns are classified by degrees. *First-degree* burns are usually caused by sun or steam; only the top layer of skin is affected, and these burns usually heal within a week.

Second-degree burns may be caused by scalding or touching hot metal. These burns affect the deep skin layer and cause blisters; they usually heal in 2 to 3 weeks.

Third-degree burns are caused by fire; the full skin layer is destroyed, and grafting is required.

A Sampling of Consumer-Product-Related Injuries Around the House

Product	Number of Injuries (1991)
Stairs, ramps, landings, floors	1,693,175
Bicycles, accessories	600,649
Cutlery, knives	448,525
Chairs, sofas, sofa beds	381,024
Tables, all types	332,775
Playground equipment	266,869
Glass doors, windows, panels	217,916
Desks, cabinets, shelves, racks	210,188
Cans, other containers	209,407
Ladders, stools	179,422
Toys	163,624
All-terrain vehicles, mopeds, minibikes	127,371
Home power tools, including saws	123,090
Workshop manual tools	118,799
Exercise equipment	86,210
Lawn mowers, all types	76,133
Glass bottles, jars	69,837
Skateboards	56,435
Cooking ranges, ovens, etc.	54,659
Chain saws	44,019
Trampolines	38,823
Cleaning agents (except soap)	38,397
Toboggans, sleds, snow disks, etc.	38,282
Nonpowder guns, BBs, pellets	32,562
Fireworks	11,390

Source: National Safety Council, *Accident Facts*, 1994 edition. Used with permission.
Note: Injury totals represent estimates of hospital emergency room cases nationwide associated with these products; only selected products are listed.

Health and the Human Body

Immunization Schedule

The following schedule is recommended by the U.S. Centers for Disease Control and Prevention (CDC) and the American College of Physicians:

Age	Vaccine
2 months	DTP,[1] polio, Hib,[2] hepatitis B
4 months	DTP, polio, Hib, hepatitis B
6 months	DTP, Hib, hepatitis B
12–15 months	Hib, chickenpox
15 months	Measles, mumps, rubella (MMR); DTP; polio
4–6 years	DTP, polio, MMR
13–16 years	Tetanus-diphtheria, chickenpox[3]

1. Diphtheria, tetanus, pertussis (whooping cough).
2. *Haemophilus influenzae* type b.
3. For children who have not contracted chickenpox by this age (initial CDC guidelines, 1995).

Childhood Diseases: Incubation Periods and Appearance of Fever

Disease	Incubation Period (days)	Fever Appears
Chickenpox	7–21	Slight
Rubella (German measles)	14–21	First 2 days
Measles	7–14	First 5 days
Mumps	14–28	First day
Pertussis (whooping cough)	7–14	First 7 days

When Teeth Appear: Eruption Times

Age	Teeth
Primary	
6–8 months	4 central incisors
7–9 months	4 lateral incisors
12–14 months	4 first molars
16–18 months	4 first canines (cuspids)
20–24 months	Second molars
Permanent	
6–7 years	First molars (6-year molars)
6–8 years	Central incisors
7–9 years	Lateral incisors
9–12 years	Canines
10–12 years	Second molars (bicuspids, 12-year molars)
16–25 years	Third molars (wisdom teeth)

Dental Checkups and X-Rays: How Often?

These recommendations are estimated ranges only; ranges are broad because individuals vary widely in their susceptibility to dental problems. Persons with healthy teeth and gums are probably safe following the longer periods shown, whereas persons whose general oral health needs closer attention should follow the shorter intervals.

Treatment	How Often
Basic checkup	6 months–1 year
Cleaning	3 months–1 year
X-rays	
Bitewing	Every 1–2 years
Full mouth	First at age 16; then every 5–8 years

Test Your Hearing by Phone

Hearing tests are available over the telephone. Call (800) 222-EARS for more information.

Physical Checkups and Tests: How Often?

These recommendations are intended as general guidelines only, and authorities may differ; the recent disagreement over mammogram screenings between the National Cancer Institute and the American Cancer Society is an example. Family health history also plays a role; consult your physician for a schedule that is right for you.

Procedure	How Often
Complete physical	
Ages 18–30	Every 5 years
Ages 30–40	Every 3 years
Ages 40–60	Every 2 years
Age 60 and over	Once a year
Vision exam	
Ages 4–50	Every 3 to 5 years
Age 50 and over	As necessary
Age 40	First glaucoma screening
Mammogram	
Ages 40–49	Every 2 years
Age 50 and over	Every year
Breast self-exam	Monthly
Pelvic exam/Pap smear	Every 1 to 2 years from time of first sexual activity or age 20
Pregnancy checkups	Start at 12th week, then monthly until last 2 months, when frequency increases
Colon cancer screening	
Ages 40–49	Every 2 years
Age 50 and over	Every year
Blood lipid measurements	First during teenage years; then every 5 years if within normal limits
Blood pressure	At least once every 5 years

Hearing: Decibel Levels of Common Sounds

Loudness is measured in decibels (dB). Constant exposure to levels above 90 decibels may lead to permanent hearing loss; a level of 120 decibels is painful to the human ear. The decibel scale is logarithmic, meaning that for every increase of 10 decibels the noise level increases by a factor of 10—a 20-decibel noise is 10 times as loud as a 10-decibel noise. But the human ear hears a 10-decibel increase as only twice as loud.

Sound	Decibel Level
Silence	0
Whisper	20
Quiet room	40
Light traffic*	55
Normal speaking	60
Noisy restaurant	70
Loud music	80
Heavy traffic*	90
Rock concert	120
Jet takeoff**	130

*From about 50 feet away.
**From about 100 feet away.

Hearing: Frequency Ranges of Various Animals

Animal	Frequency Range Heard (Hz)
Dog	15 to 50,000
Human	20 to 20,000
Cat	60 to 65,000
Dolphin	150 to 150,000
Bat	1,000 to 120,000

The Upper Limits of Hearing

According to the *Guinness Book of World Records, 1995*, there is a recorded case of a human voice being detectable at a distance of 10.5 miles across still water at night. The normal intelligible outdoor range of a human male voice in still air is 600 feet.

Temperature, Blood Pressure, Cholesterol, Triglycerides, Blood Sugar, and Respiration: Normal and Abnormal Rates and Levels in Adults

Condition	Reading
Temperature	*Normal:* 96.5°F–99°F; 98.6°F average; can vary by 1° or 2° during day (rectal 0.5°–1°F higher)
	Fever: Above 99°F
	Dangerous: 106°F and higher
Blood pressure	*Normal:* systolic (first number) below 140; diastolic (second number) below 85
	Abnormal: systolic more than 140; diastolic of 85–89 is high normal; 90–104 is mild hypertension; 105–114 is moderate hypertension; more than 115 is severe hypertension
Cholesterol	*Normal (in mg/dL):* total, 200 or less; LDL, 139 or less; HDL, 30–85 (women), 30–70 (men)
	Borderline: total, 200–239; LDL, 139–159
	High: total, 240 or more; LDL, 160 or more; HDL, 30 or less
Triglycerides	*Normal (in mg/dL):* 40–160 (men), 35–135 (women)
Blood sugar*	*Normal (in mg/dL):* to age 50, 70–140; age 50–60, 70–150; over age 60, 70–160
Respiration	*Normal:* 14–19 breaths per minute (men); 20–22 breaths per minute (women)

*Diabetes is usually indicated if a person's fasting blood sugar (i.e., when no food has been consumed for the previous 8 hours) is 140 mg/dL or more.

Cholesterol Ratio Another way to look at cholesterol is to consider the ratio of total cholesterol to HDL cholesterol (the "good" cholesterol). This number should not exceed 4.5. To determine the ratio, divide the HDL number into the total cholesterol number. For example,

220 (total cholesterol) ÷ 65 (HDL cholesterol) = 3.38

What Does 20/20 Mean?

This is a measure of distance vision: The first 20 refers to the distance between you and a line of printing on an eye chart; the second 20 means that someone with normal eyesight can read that line from 20 feet away. If you can read only the line of printing that a person with normal eyesight can see from 40 feet away, then you have 20/40 vision.

The corresponding number for perfect vision for reading or other close-up work is 14/14, with the 14 referring to the number of inches between your eyes and the object, for example, the pages of a book.

Contraception Effectiveness

Method	Number of pregnancies per 100 women in 1 year
Sterilization	Less than 1
Birth control pill	2.5
IUD	4
Condom with foam	10
Condom without foam	
Married women	14
Single women	11
Diaphragm	18
Sponge	18
Foams/jellies/suppositories	20
Withdrawal	20
Periodic abstinence	24
No method	60–80

Distribution of Blood Types in the United States

Type	% of Population
O	45
A	40
B	10
AB	5

Determining Ideal Pulse Rate

Before starting an exercise program, you should find the "target" heart rate (or number of beats per minute) that you need to achieve during exercise to get an effective workout without putting your heart under unnecessary strain. To do this, subtract your age from 220, then multiply the result by 70 percent (or less to start, if you are in poor shape). Here's an example for a 35-year-old:

$$220 - 35 = 185$$
$$185 \times .70 = 129.5$$

The easiest way to measure your heart rate is to take your pulse after exercise and count the beats for 10 seconds; then multiply that number by 6 to get beats per minute.

The average number of heartbeats per minute is 55 to 60 (while sleeping) and 70 to 85 (awake, at rest). The rate may go to 200 or more during heavy exercise.

Height and Weight Table

MEN					WOMEN			
Height (ft./in.)	Small Frame	Medium Frame	Large Frame		Height (ft./in.)	Small Frame	Medium Frame	Large Frame
5 2	128–134	131–141	138–150		4 10	102–111	109–121	118–131
5 3	130–136	133–143	140–153		4 11	103–113	111–123	120–134
5 4	132–138	135–145	142–156		5 0	104–115	113–126	122–137
5 5	134–140	137–148	144–160		5 1	106–118	115–129	125–140
5 6	136–142	139–151	146–164		5 2	108–121	118–132	128–143
5 7	138–145	142–154	149–168		5 3	111–124	121–135	131–147
5 8	140–148	145–157	152–172		5 4	114–127	124–138	134–151
5 9	142–151	148–160	155–176		5 5	117–130	127–141	137–155
5 10	144–154	151–163	158–180		5 6	120–133	130–144	140–159
5 11	146–157	154–166	161–184		5 7	123–136	133–147	143–163
6 0	149–160	157–170	164–188		5 8	126–139	136–150	146–167
6 1	152–164	160–174	168–192		5 9	129–142	139–153	149–170
6 2	155–168	164–178	172–197		5 10	132–145	142–156	152–173
6 3	158–172	167–182	176–202		5 11	135–148	145–159	155–176
6 4	162–176	171–187	181–207		6 0	138–151	148–162	158–179

Source: Statistical Bulletin, Metropolitan Life Insurance Company, 1983.

Determining Frame Size

To determine your frame size—small, medium, or large—try the following calculation:

12. Stretch one arm out in front of you, then bend the forearm up so that it forms a 45-degree angle.

13. Turn your wrist so that it faces toward your body. With your other hand, locate the two prominent bones on either side of your elbow.

14. Measure the distance between these bones and compare your result with the chart below, which shows a medium frame. Lower measurements than shown for your sex and height mean that you have a small frame; larger measurements mean that you have a large frame.

Height	Elbow Breadth
Men	
5'4"–5'7"	$2^1/_2$"–$2^7/_8$"
5'8"–5'11"	$2^7/_8$"–3"
6'–6'3"	$2^3/_4$"–$3^1/_8$"
6'4"	$2^7/_8$"–$3^1/_4$"
Women	
4'10"–5'3"	$2^1/_4$"–$2^1/_2$"
5'4"–5'11"	$2^3/_8$"–$2^5/_8$"
6'	$2^1/_2$"–$2^3/_4$"

Rule of Thumb: How to Lose (or Gain) Weight

First determine the daily calorie intake needed to maintain your current weight:

$$\text{Current weight} \times 15 = \text{daily calorie intake}$$

If you want to lose 1 pound a week, consume 500 fewer calories than the resulting daily calorie intake number. If you want to gain weight, increase your diet by 500 calories a day. Note that this is a general guideline only, and that nutrition experts advise against consuming less than 1,200 calories a day.

Note: Numbers shown are for a person weighing about 150 pounds. A heavier person will

Calories Burned During Various Activities

Activity	Calories Burned per Hour
Sitting quietly	50
Slow walking	140
Canoeing	180
Light gardening	200
Raking	220
Croquet	250
Window cleaning	260
Brisk walking	300
Golf (pulling bag, level course)	300
Softball	300
Stacking firewood	360
Heavy housework (floor scrubbing, etc.)	400
Bicycling (9 mph)	420
Horseback riding (trotting)	450
Tennis	460
Ice- or roller-skating	460
Lawn mowing (pushing, not riding)	470
Skiing, downhill	480
Skiing, cross county (moderate speed)	480
Splitting wood	480
Hiking	500
Shoveling	500
Swimming, slow	520
Aerobic dancing (intense)	540
Jogging	550
Basketball	560
Stair climbing	600
Swimming (fast)	660
Jumping rope (70 jumps per min.)	660
Bicycling, racing	690
Racquetball	725
Running (9-min. mile)	780
Squash	870
Scuba diving (very active)	1,000
Wood-chopping with ax (fast)	1,200

burn more calories than shown, and a lighter person will burn fewer.

The Stressful Events Rating Scale

Add up the numbers listed for events of your life that have happened within the past year. If you score more than 200, you have a 50 percent chance of becoming seriously ill; a score of 300 or more raises your chances to 80 percent.

Life Event	Score
1. Death of spouse	100
2. Divorce	73
3. Marital separation	65
4. Jail term	63
5. Death of close family member	63
6. Personal injury or illness	53
7. Marriage	50
8. Being fired	47
9. Marital reconciliation	45
10. Retirement	45
11. Change in health of family member	44
12. Pregnancy	40
13. Sex difficulties	39
14. Having a baby	39
15. Business readjustment	39
16. Change in financial state	38
17. Death of close friend	37
18. Change to different line of work	36
19. Change in number of arguments with spouse	35
20. Large mortgage in relation to income	31
21. Foreclosure of mortgage or loan	30
22. Change in responsibilities at work	29
23. Son or daughter leaving home	29
24. Trouble with in-laws	29
25. Outstanding personal achievement	28
26. Spouse begins or stops work	26
27. Begin or end school	26
28. Change in living conditions	25
29. Change in personal habits	24
30. Trouble with boss	23
31. Change in work hours or conditions	20
32. Change in residence	20
33. Change in schools	20

The Stressful Events Rating Scale (continued)

Life Event	Score
34. Change in church activities	19
35. Change in recreation	19
36. Change in social activities	18
37. Small mortgage in relation to income	17
38. Change in sleeping habits	16
39. Change in number of family get-togethers	15
40. Change in eating habits	13
41. Vacation	13
42. Christmas	12
43. Minor violations of the law	11

Source: Reprinted by permission of the publisher from Holmes and Rahe, "Social Readjustment Rating Scale," *Journal of Psychosomatic Research,* vol. 11. Copyright 1967 by Elsevier Science Inc.

Blood Alcohol Levels: The Danger Zones

The number known as the blood alcohol concentration (BAC) is used by many state and local police departments to determine impaired ability to drive. A 0.10 percent level of alcohol in the blood is the usual measurement of legally drunk, although many jurisdictions set lower concentrations.

Weight (lbs.)	Number of Drinks* over a 2-Hour Period									
100	1	2	3	4	5	6	7	8	9	10
120	1	2	3	4	5	6	7	8	9	10
140	1	2	3	4	5	6	7	8	9	10
160	1	2	3	4	5	6	7	8	9	10
180	1	2	3	4	5	6	7	8	9	10
200	1	2	3	4	5	6	7	8	9	10
220	1	2	3	4	5	6	7	8	9	10
240	1	2	3	4	5	6	7	8	9	10
	Be careful BAC to .05				**Driving impaired .05 – .09**			**DO NOT DRIVE .10 and up**		

Source: National Highway Traffic Safety Administration, U.S. Department of Transportation.
* A drink is defined as one 12-ounce beer, one 4-ounce glass of wine, or 1.5 ounces of 86-proof liquor.

**Rule of
Thumb:
Age by
Heartbeat**

According to one theory, regardless of their size most mammals live for about 800 million heartbeats. The smaller the animal, the faster its heart beats, so that squirrels die at a younger age, in terms of years, than do elephants.

Humans, the theory goes, are an exception to this rule: We live 3 times longer than we should, if our lifespan were determined by number of heartbeats.

The Human Body: Some Statistics

Here's the weight of body parts, in pounds, for a person weighing 152 pounds:

Part of Body	Weight (lbs.)
Head	$10^1/_2$
Neck and trunk	70
Arms	$16^1/_2$
Hands	$2^1/_2$
Legs	$47^1/_2$
Feet	5

The adult brain weighs from 38 to 60 ounces in men, and from 37 to 55 ounces in women.

The adult heart weighs about 10 ounces in men and about 8 ounces in women.

The skin of an average adult weighs close to 6 pounds and is considered to be a major organ of the body.

The human body contains about 75 trillion cells; each day, 2 billion cells die and are replaced. In slim people, body fluids make up about 70 percent of total body weight; in obese people, the proportion is 50 percent. There are 1.5 gallons of blood in men and 0.875 gallons in women.

Here are some other breakdowns (the percentages do not total 100 because fluids are present in muscles, fat, and bone):

Water	54%–60% of total body weight
Muscles	36%–42% of total body weight
Fat	18%–20% of total body weight
Bone	18% of total body weight

The adult human body has the following numbers of these components:

Bones	206
Muscles	639
Blood vessels (in miles)	60,000

Chemically, the human body contains these rough percentages of components:

Element	%	Element	%
Carbon	50.0	Sulfur	0.8
Oxygen	20.0	Sodium	0.4
Hydrogen	10.0	Chlorine	0.4
Nitrogen	8.5	Magnesium	0.1
Calcium	4.0	Iron	0.01
Phosphorus	2.5	Manganese	0.001
Potassium	1.0	Iodine	0.00005

Lung capacity is 4.5 to 9.5 quarts in adult men, and 3.3 to 5.7 quarts in adult women. If the air sacs, or alveoli, in the lungs were flattened out, they would cover between 600 and 1,000 square feet.

The Nervous System: Common Response Times

Action	Response Time (sec.)
Movement of eyes to focus on new spot	0.48
Walking one pace	0.60
Sitting down	1.32
Standing up	1.62

The human body responds to sound, touch, sight, cold, and warmth within about 0.1 to 0.02 second. The body's reaction times for smell and pain are slower, at about 0.3 and 0.7 second, respectively. Impulses can travel through the human nervous system at speeds as fast as 223 miles per hour.

Health Information Services

Phone Numbers for Federal Government Health Information Sources

Topic	Phone Number	Fax Number
Aging		
National Institute on Aging	(800) 222-2225	(301) 589-3014
AIDS		
National AIDS Clearinghouse	(800) 458-5231	(301) 738-6616
	(800) 342-AIDS (English hotline)	
	(800) 344-SIDA (Spanish hotline)	
National AIDS Clinical Trials Information Service	(800) 874-2572	(301) 738-6616
Alcohol and Drug Abuse		
National Clearinghouse for Alcohol and Drug Information	(800) 729-6686	(301) 468-6433
National Automated Drug and Alcohol Treatment Referral Hotline	(800) 662-4357	
Alzheimer's Disease and Related Disorders		
Alzheimer's Disease Education and Referral Center	(800) 438-4380	(301) 495-3334
Arthritis and Related Diseases		
National Arthritis and Musculoskeletal and Skin Diseases Information Clearinghouse	(301) 495-4484	(301) 495-3334

Phone Numbers for Federal Government
Health Information Sources (continued)

Topic	Phone Number	Fax Number
Asthma		
National Heart, Lung, and Blood Institute Information Center	(301) 251-1222	(301) 251-1223
Cancer		
Cancer Information Service	(800) 422-6237	—
Deafness		
National Institute on Deafness and Other Communication Disorders Clearinghouse	(800) 241-1044 (800) 241-1055 (TTD/TTY)	(301) 565-5112
Diabetes		
National Diabetes Information Clearinghouse	(301) 654-3327	—
Digestive Diseases		
National Digestive Diseases Information Clearinghouse	(301) 654-3810	(301) 770-5164
Heart Disease		
National Heart, Lung, and Blood Institute Information Center	(301) 251-1222	(301) 251-1223
High Blood Pressure		
National Heart, Lung, and Blood Institute Information Center	(301) 251-1222	(301) 251-1223
Kidney Disease		
National Kidney and Urologic Diseases Information Clearinghouse	(301) 654-4415	(301) 770-5164
Oral Health		
National Oral Health Information Clearinghouse	(301) 402-7364	—
Rare Diseases		
Office of Orphan Products and Rare Diseases	(800) 300-7469	(301) 443-4915

Phone Numbers for Federal Government
Health Information Sources (continued)

Topic	Phone Number	Fax Number
Smoking		
National Heart, Lung, and Blood Institute Information Center	(301) 251-1222	(301) 251-1223
Stroke and Neurological Disorders		
Information Office	(800) 352-9424	—

4

Travel

U.S. Travel

Road Distances Between Selected U.S. Cities (in miles)

The distances shown are approximate. Various factors can affect how distances are calculated, including whether city centers or city limit markers are used as measuring points; the route taken; and any recent highway improvements, which can decrease previously stated distances.

City	Atlanta	Birmingham	Boston	Buffalo	Chicago	Cleveland
Atlanta, GA	—	155	1,070	876	707	692
Birmingham, AL	155	—	1,194	947	657	734
Boston, MA	1,070	1,194	—	457	983	639
Buffalo, NY	876	947	457	—	536	192
Chicago, IL	707	657	983	536	—	344
Cleveland, OH	692	734	639	192	344	—
Dallas, TX	820	653	1,815	1,387	931	1,205
Denver, CO	1,431	1,318	1,991	1,561	1,050	1,369
Detroit, MI	726	754	702	252	279	175
El Paso, TX	1,438	1,278	2,358	1,928	1,439	1,746
Houston, TX	841	692	1,886	1,532	1,092	1,358
Indianapolis, IN	493	492	940	510	189	318
Kansas City, MO	823	703	1,427	997	503	815
Los Angeles, CA	2,254	2,078	3,036	2,606	2,112	2,424
Louisville, KY	382	378	996	571	305	379
Memphis, TN	371	249	1,345	965	546	773
Miami, FL	665	777	1,539	1,445	1,390	1,325
Minneapolis, MN	1,114	1,067	1,402	955	411	763
New Orleans, LA	517	347	1,541	1,294	947	1,102
New York, NY	863	983	213	436	840	514
Omaha, NE	986	907	1,458	1,011	493	819
Philadelphia, PA	771	894	304	383	758	432
Phoenix, AZ	1,888	1,680	2,664	2,234	1,729	2,052
Pittsburgh, PA	737	792	597	219	457	131
St. Louis, MO	553	508	1,179	749	293	567
Salt Lake City, UT	1,976	1,805	2,425	1,978	1,458	1,786
San Antonio, TX	1,022	860	2,108	1,630	1,245	1,488
San Francisco, CA	2,563	2,385	3,179	2,732	2,212	2,540
Seattle, WA	2,843	2,612	3,043	2,596	2,052	2,404
Washington, DC	640	751	440	386	695	369

Road Distances Between Selected U.S. Cities (in miles) (continued)

City	Dallas	Denver	Detroit	El Paso	Houston	Indianapolis
Atlanta, GA	820	1,431	726	1,438	841	493
Birmingham, AL	653	1,318	754	1,278	692	492
Boston, MA	1,815	1,991	702	2,358	1,886	940
Buffalo, NY	1,387	1,561	252	1,928	1,532	510
Chicago, IL	931	1,050	279	1,439	1,092	189
Cleveland, OH	1,205	1,369	175	1,746	1,358	318
Dallas, TX	—	801	1,167	625	242	877
Denver, CO	801	—	1,301	652	1,032	1,051
Detroit, MI	1,167	1,301	—	1,696	1,312	290
El Paso, TX	625	652	1,696	—	756	1,418
Houston, TX	242	1,032	1,312	756	—	1,022
Indianapolis, IN	877	1,051	290	1,418	1,022	—
Kansas City, MO	508	616	760	936	750	487
Los Angeles, CA	1,425	1,174	2,369	800	1,556	2,096
Louisville, KY	865	1,135	378	1,443	981	114
Memphis, TN	470	1,069	756	1,095	586	466
Miami, FL	1,332	2,094	1,409	1,957	1,237	1,225
Minneapolis, MN	969	867	698	1,353	1,211	600
New Orleans, LA	504	1,305	1,101	1,121	365	839
New York, NY	1,604	1,780	671	2,147	1,675	729
Omaha, NE	661	559	754	1,015	903	590
Philadelphia, PA	1,515	1,698	589	2,065	1,586	647
Phoenix, AZ	1,027	836	1,986	402	1,158	1,713
Pittsburgh, PA	1,237	1,411	288	1,778	1,395	360
St. Louis, MO	638	871	529	1,179	799	239
Salt Lake City, UT	1,239	512	1,721	877	1,465	1,545
San Antonio, TX	271	975	1,458	566	193	1,192
San Francisco, CA	1,765	1,266	2,475	1,202	1,958	2,299
Seattle, WA	2,122	1,373	2,339	1,760	2,348	2,241
Washington, DC	1,372	1,635	526	1,997	1,443	565

Road Distances Between Selected U.S. Cities (in miles) (continued)

City	Kansas City	Los Angeles	Louisville	Memphis	Miami	Minneapolis
Atlanta, GA	823	2,254	382	371	665	1,114
Birmingham, AL	703	2,078	378	249	777	1,067
Boston, MA	1,427	3,036	996	1,345	1,539	1,402
Buffalo, NY	997	2,606	571	965	1,445	955
Chicago, IL	503	2,112	305	546	1,390	411
Cleveland, OH	815	2,424	379	773	1,325	763
Dallas, TX	508	1,425	865	470	1,332	969
Denver, CO	616	1,174	1,135	1,069	2,094	867
Detroit, MI	760	2,369	378	756	1,409	698
El Paso, TX	936	800	1,443	1,095	1,957	1,353
Houston, TX	750	1,556	981	586	1,237	1,211
Indianapolis, IN	487	2,096	114	466	1,225	600
Kansas City, MO	—	1,609	519	454	1,479	466
Los Angeles, CA	1,609	—	2,128	1,847	2,757	2,041
Louisville, KY	519	2,128	—	396	1,111	716
Memphis, TN	454	1,847	396	—	1,025	854
Miami, FL	1,479	2,757	1,111	1,025	—	1,801
Minneapolis, MN	466	2,041	716	854	1,801	—
New Orleans, LA	839	1,921	725	401	892	1,255
New York, NY	1,216	2,825	785	1,134	1,328	1,259
Omaha, NE	204	1,733	704	658	1,683	373
Philadelphia, PA	1,134	2,743	703	1,045	1,239	1,177
Phoenix, AZ	1,226	398	1,749	1,464	2,359	1,644
Pittsburgh, PA	847	2,456	416	810	1,250	876
St. Louis, MO	255	1,864	264	295	1,241	559
Salt Lake City, UT	1,128	728	1,647	1,556	2,571	1,243
San Antonio, TX	788	1,388	1,059	692	1,435	1,188
San Francisco, CA	1,882	403	2,401	2,151	3,097	1,997
Seattle, WA	1,909	1,150	2,355	2,363	3,389	1,641
Washington, DC	1,071	2,680	601	902	1,101	1,114

Road Distances Between Selected U.S. Cities (in miles) (continued)

City	New Orleans	New York	Omaha	Philadelphia	Phoenix	Pittsburgh
Atlanta, GA	517	863	986	771	1,888	737
Birmingham, AL	347	983	907	894	1,680	792
Boston, MA	1,541	213	1,458	304	2,664	597
Buffalo, NY	1,294	436	1,011	383	2,234	219
Chicago, IL	947	840	493	758	1,729	457
Cleveland, OH	1,102	514	819	432	2,052	131
Dallas, TX	504	1,604	661	1,515	1,027	1,237
Denver, CO	1,305	1,780	559	1,698	836	1,411
Detroit, MI	1,101	671	754	589	1,986	288
El Paso, TX	1,121	2,147	1,015	2,065	402	1,778
Houston, TX	365	1,675	903	1,586	1,158	1,395
Indianapolis, IN	839	729	590	647	1,713	360
Kansas City, MO	839	1,216	204	1,134	1,226	847
Los Angeles, CA	1,921	2,825	1,733	2,743	398	2,456
Louisville, KY	725	785	704	703	1,749	416
Memphis, TN	401	1,134	658	1,045	1,464	810
Miami, FL	892	1,328	1,683	1,239	2,359	1,250
Minneapolis, MN	1,255	1,259	373	1,177	1,644	876
New Orleans, LA	—	1,330	1,043	1,241	1,523	1,118
New York, NY	1,330	—	1,315	93	2,442	386
Omaha, NE	1,043	1,315	—	1,233	1,305	932
Philadelphia, PA	1,241	93	1,233	—	2,360	304
Phoenix, AZ	1,523	2,442	1,305	2,360	—	2,073
Pittsburgh, PA	1,118	386	932	304	2,073	—
St. Louis, MO	696	968	459	886	1,485	599
Salt Lake City, UT	1,743	2,282	967	2,200	651	1,899
San Antonio, TX	574	1,889	928	1,795	1,009	1,528
San Francisco, CA	2,269	3,036	1,721	2,954	800	2,653
Seattle, WA	2,606	2,900	1,705	2,818	1,482	2,517
Washington, DC	1,098	229	1,170	140	2,278	241

Road Distances Between Selected U.S. Cities (in miles) (continued)

City	St. Louis	Salt Lake City	San Antonio	San Francisco	Seattle	Washington, DC
Atlanta, GA	553	1,976	1,022	2,563	2,843	640
Birmingham, AL	508	1,805	860	2,385	2,612	751
Boston, MA	1,179	2,425	2,108	3,179	3,043	440
Buffalo, NY	749	1,978	1,630	2,732	2,596	386
Chicago, IL	293	1,458	1,245	2,212	2,052	695
Cleveland, OH	567	1,786	1,488	2,540	2,404	369
Dallas, TX	638	1,239	271	1,765	2,122	1,372
Denver, CO	871	512	975	1,266	1,373	1,635
Detroit, MI	529	1,721	1,458	2,475	2,339	526
El Paso, TX	1,179	877	566	1,202	1,760	1,997
Houston, TX	799	1,465	193	1,958	2,348	1,443
Indianapolis, IN	239	1,545	1,192	2,299	2,241	565
Kansas City, MO	255	1,128	788	1,882	1,909	1,071
Los Angeles, CA	1,864	728	1,388	403	1,150	2,680
Louisville, KY	264	1,647	1,059	2,401	2,355	601
Memphis, TN	295	1,556	692	2,151	2,363	902
Miami, FL	1,241	2,571	1,435	3,097	3,389	1,101
Minneapolis, MN	559	1,243	1,188	1,997	1,641	1,114
New Orleans, LA	696	1,743	574	2,269	2,606	1,098
New York, NY	968	2,282	1,889	3,036	2,900	229
Omaha, NE	459	967	928	1,721	1,705	1,170
Philadelphia, PA	886	2,200	1,795	2,954	2,818	140
Phoenix, AZ	1,485	651	1,009	800	1,482	2,278
Pittsburgh, PA	599	1,899	1,528	2,653	2,517	241
St. Louis, MO	—	1,383	949	2,137	2,164	836
Salt Lake City, UT	1,383	—	1,364	754	883	2,110
San Antonio, TX	949	1,364	—	1,793	2,301	1,648
San Francisco, CA	2,137	754	1,793	—	817	2,864
Seattle, WA	2,164	883	2,301	817	—	2,755
Washington, DC	836	2,110	1,648	2,864	2,755	—

Traveling withYour Pet *The Pets Allowed Directory* lists hotels and motels that welcome small pets. Call (800) 569-7326 for ordering information. A newsletter, *Dog Gone*, describes pet-friendly destinations; call (407) 569-8434 for subscription information.

Air Distances Between Selected U.S. Cities (in miles)

City	Atlanta	Boston	Chicago	Cleveland	Dallas	Denver
Atlanta, GA	—	946	606	554	721	1,208
Boston, MA	946	—	851	551	1,551	1,769
Chicago, IL	606	851	—	308	803	920
Cleveland, OH	554	551	308	—	1,025	1,227
Dallas, TX	721	1,551	803	1,025	—	663
Denver, CO	1,208	1,769	920	1,227	663	—
Detroit, MI	595	613	238	90	999	1,156
El Paso, TX	1,291	2,072	1,252	1,525	572	557
Houston, TX	689	1,605	940	1,114	225	879
Indianapolis, IN	427	807	165	263	763	1,000
Kansas City, MO	681	1,251	414	700	451	558
Los Angeles, CA	1,946	2,596	1,745	2,049	1,240	831
Louisville, KY	319	826	269	311	726	1,038
Memphis, TN	337	1,137	482	630	420	879
Miami, FL	604	1,255	1,188	1,087	1,111	1,726
Minneapolis, MN	906	1,123	355	630	862	700
New Orleans, LA	425	1,359	833	924	443	1,082
New York, NY	760	188	713	405	1,374	1,631
Omaha, NE	821	1,282	432	739	586	488
Philadelphia, PA	665	271	666	360	1,299	1,579
Phoenix, AZ	1,587	2,300	1,453	1,749	887	586
Pittsburgh, PA	526	483	410	115	1,070	1,320
St. Louis, MO	484	1,038	262	492	547	796
Salt Lake City, UT	1,589	2,099	1,260	1,568	999	371
San Francisco, CA	2,139	2,699	1,858	2,166	1,483	949
Seattle, WA	2,182	2,493	1,737	2,026	1,681	1,021
Washington, DC	532	393	597	306	1,185	1,494

Air Distances Between Selected U.S. Cities (in miles) (continued)

City	Detroit	El Paso	Houston	Indianapolis	Kansas City	Los Angeles
Atlanta, GA	595	1,291	689	427	681	1,946
Boston, MA	613	2,072	1,605	807	1,251	2,596
Chicago, IL	238	1,252	940	165	414	1,745
Cleveland, OH	90	1,525	1,114	263	700	2,049
Dallas, TX	999	572	225	763	451	1,240
Denver, CO	1,156	557	879	1,000	558	831
Detroit, MI	—	1,479	1,105	240	645	1,983
El Paso, TX	1,479	—	676	1,264	839	701
Houston, TX	1,105	676	—	865	644	1,374
Indianapolis, IN	240	1,264	865	—	453	1,809
Kansas City, MO	645	839	644	453	—	1,356
Los Angeles, CA	1,983	701	1,374	1,809	1,356	—
Louisville, KY	316	1,254	803	107	480	1,829
Memphis, TN	623	976	484	384	369	1,603
Miami, FL	1,152	1,643	968	1,024	1,241	2,339
Minneapolis, MN	543	1,157	1,056	511	413	1,524
New Orleans, LA	939	983	318	712	680	1,673
New York, NY	482	1,905	1,420	646	1,097	2,451
Omaha, NE	669	878	794	525	166	1,315
Philadelphia, PA	443	1,836	1,341	585	1,038	2,394
Phoenix, AZ	1,690	346	1,017	1,499	1,049	357
Pittsburgh, PA	205	1,590	1,137	330	781	2,136
St. Louis, MO	455	1,034	679	231	238	1,589
Salt Lake City, UT	1,492	689	1,200	1,356	925	579
San Francisco, CA	2,091	995	1,645	1,949	1,506	347
Seattle, WA	1,938	1,376	1,891	1,872	1,506	959
Washington, DC	396	1,728	1,220	494	945	2,300

Air Distances Between Selected U.S. Cities (in miles) (continued)

City	Louisville	Memphis	Miami	Minneapolis	New Orleans	New York
Atlanta, GA	319	337	604	906	425	760
Boston, MA	826	1,137	1,255	1,123	1,359	188
Chicago, IL	269	482	1,188	355	833	713
Cleveland, OH	311	630	1,087	630	924	405
Dallas, TX	726	420	1,111	862	443	1,374
Denver, CO	1,038	879	1,726	700	1,082	1,631
Detroit, MI	316	623	1,152	543	939	482
El Paso, TX	1,254	976	1,643	1,157	983	1,905
Houston, TX	803	484	968	1,056	318	1,420
Indianapolis, IN	107	384	1,024	511	712	646
Kansas City, MO	480	369	1,241	413	680	1,097
Los Angeles, CA	1,829	1,603	2,339	1,524	1,673	2,451
Louisville, KY	—	320	919	605	623	652
Memphis, TN	320	—	872	699	358	957
Miami, FL	919	872	—	1,511	669	1,092
Minneapolis, MN	605	699	1,511	—	1,051	1,018
New Orleans, LA	623	358	669	1,051	—	1,171
New York, NY	652	957	1,092	1,018	1,171	—
Omaha, NE	580	529	1,397	290	847	1,144
Philadelphia, PA	582	881	1,019	985	1,089	83
Phoenix, AZ	1,508	1,263	1,982	1,260	1,316	2,145
Pittsburgh, PA	344	660	1,010	743	919	317
St. Louis, MO	242	240	1,061	466	598	875
Salt Lake City, UT	1,402	1,250	2,089	987	1,434	1,972
San Francisco, CA	1,986	1,802	2,594	1,584	1,926	2,571
Seattle, WA	1,943	1,867	2,734	1,395	2,101	2,408
Washington, DC	476	765	923	934	966	205

Air Distances Between Selected U.S. Cities (in miles) (continued)

City	Omaha	Philadelphia	Phoenix	Pittsburgh	St. Louis	Salt Lake City
Atlanta, GA	821	665	1,587	526	484	1,589
Boston, MA	1,282	271	2,300	483	1,038	2,099
Chicago, IL	432	666	1,453	410	262	1,260
Cleveland, OH	739	360	1,749	115	492	1,568
Dallas, TX	586	1,299	887	1,070	547	999
Denver, CO	488	1,579	586	1,320	796	371
Detroit, MI	669	443	1,690	205	455	1,492
El Paso, TX	878	1,836	346	1,590	1,034	689
Houston, TX	794	1,341	1,017	1,137	679	1,200
Indianapolis, IN	525	585	1,499	330	231	1,356
Kansas City, MO	166	1,038	1,049	781	238	925
Los Angeles, CA	1,315	2,394	357	2,136	1,589	579
Louisville, KY	580	582	1,508	344	242	1,402
Memphis, TN	529	881	1,263	660	240	1,250
Miami, FL	1,397	1,019	1,982	1,010	1,061	2,089
Minneapolis, MN	290	985	1,280	743	466	987
New Orleans, LA	847	1,089	1,316	919	598	1,434
New York, NY	1,144	83	2,145	317	875	1,972
Omaha, NE	—	1,094	1,036	836	354	833
Philadelphia, PA	1,094	—	2,083	259	811	1,925
Phoenix, AZ	1,036	2,083	—	1,828	1,272	504
Pittsburgh, PA	836	259	1,828	—	559	1,668
St. Louis, MO	354	811	1,272	559	—	1,162
Salt Lake City, UT	833	1,925	504	1,668	1,162	—
San Francisco, CA	1,429	2,523	653	2,264	1,744	600
Seattle, WA	1,369	2,380	1,114	2,138	1,724	701
Washington, DC	1,014	123	1,983	192	712	1,848

Air Distances Between Selected U.S. Cities (in miles) (continued)

City	San Francisco	Seattle	Washington, DC
Atlanta, GA	2,139	2,182	532
Boston, MA	2,699	2,493	393
Chicago, IL	1,858	1,737	597
Cleveland, OH	2,166	2,026	306
Dallas, TX	1,483	1,681	1,185
Denver, CO	949	1,021	1,494
Detroit, MI	2,091	1,938	396
El Paso, TX	995	1,376	1,728
Houston, TX	1,645	1,891	1,220
Indianapolis, IN	1,949	1,872	494
Kansas City, MO	1,506	1,506	945
Los Angeles, CA	347	959	2,300
Louisville, KY	1,986	1,943	476
Memphis, TN	1,802	1,867	765
Miami, FL	2,594	2,734	923
Minneapolis, MN	1,584	1,395	934
New Orleans, LA	1,926	2,101	966
New York, NY	2,571	2,408	205
Omaha, NE	1,429	1,369	1,014
Philadelphia, PA	2,523	2,380	123
Phoenix, AZ	653	1,114	1,983
Pittsburgh, PA	2,264	2,138	192
St. Louis, MO	1,744	1,724	712
Salt Lake City, UT	600	701	1,848
San Francisco, CA	—	678	2,442
Seattle, WA	678	—	2,329
Washington, DC	2,442	2,329	—

State Tourist Office Phone Numbers

State	Phone	State	Phone
Alabama	(800) 252-2262	Montana	(800) 541-1447
Alaska	(907) 465-2010	Nebraska	(800) 228-4307
Arizona	(800) 842-8257	Nevada	(800) 638-2328
Arkansas	(800) 628-8725	New Hampshire	(603) 271-2343
California	(800) 862-2543	New Jersey	(800) 537-7397
Colorado	(800) 265-6723	New Mexico	(800) 545-2040
Connecticut	(800) 282-6863	New York	(800) 225-5697
Delaware	(800) 441-8846	North Carolina	(800) 847-4862
District of Columbia	(202) 789-7000	North Dakota	(800) 435-5663
Florida	(904) 487-1462	Ohio	(800) 282-5393
Georgia	(800) 847-4842	Oklahoma	(800) 652-6552
Hawaii	(808) 923-1811	Oregon	(800) 547-7842
Idaho	(800) 635-7820	Pennsylvania	(800) 847-4872
Illinois	(800) 223-0121	Rhode Island	(800) 556-2484
Indiana	(800) 289-6646	South Carolina	(800) 872-3505
Iowa	(800) 345-4692	South Dakota	(800) 732-5682
Kansas	(800) 252-6727	Tennessee	(800) 836-6200
Kentucky	(800) 225-8747	Texas	(800) 888-8839
Louisiana	(800) 633-6970	Utah	(800) 200-1160
Maine	(800) 533-9595	Vermont	(800) 837-6668
Maryland	(800) 543-1036	Virginia	(800) 248-4833
Massachusetts	(800) 447-6277	Washington	(800) 544-1800
Michigan	(800) 543-2937	West Virginia	(800) 225-5982
Minnesota	(800) 657-3700	Wisconsin	(800) 432-8747
Mississippi	(800) 927-6378	Wyoming	(800) 225-5996
Missouri	(800) 877-1234		

National Park Information

The federal government offers several National Park publications at a modest cost, available by writing to the Consumer Information Center, P.O. Box 100, Pueblo, CO 81002 (include name and number of the publication):

- Lesser-Known Areas of the National Park System (No. 137A, $1.50)
- A Guide to Your National Forests (No. 136A, $1.00)
- National Trails System Map and Guide (No. 139A, $1.25)

To make free reservations at selected National Park Campgrounds, call (800) 365-2267. Reservations may be made up to 8 weeks in advance at Acadia, Assateague Island, Cape Hatteras, Death Valley, Great Smoky Mountains, Joshua Tree, Rocky Mountain, Sequoia and Kings Canyon, Shenandoah, Whiskeytown, Yellowstone, and Yosemite.

Most Visited Sites in the National Park System

Site	Millions of Visitors (1994)
National Parks	
Great Smoky Mountains	8.6
Grand Canyon	4.4
Yosemite	4.0
Olympic	3.4
Yellowstone	3.0
Rocky Mountain	2.9
Acadia	2.7
Grand Teton	2.5
Zion	2.3
Glacier	2.2

Most Visited Sites in the National Park System (continued)

Site	Millions of Visitors (1994)
National Park System	
Blue Ridge Parkway	17.0
Golden Gate National Recreation Area	14.7
Lake Mead National Recreation Area	9.6
George Washington Memorial Parkway	5.6
National Capital Parks	5.4
Natchez Trace Parkway	5.3
Cape Cod National Seashore	5.2
Gulf Islands National Seashore	5.1
Delaware Water Gap National Recreation Area	4.8

Rule of Thumb: U.S. Interstate Highway Numbers

North–south interstate highways have 1- or 2-digit *odd* numbers, beginning with Interstate 5 on the West Coast and ending with Interstate 95 on the East Coast; many of these highways end with the numeral 5 as you cross the country.

East–west interstate highways have 1- or 2-digit *even* numbers, beginning with Interstate 10 (Florida to Los Angeles) and increasing as you go north (I–20, I–40, I–70, etc.), ending with I–90 (Boston to Seattle).

An interstate highway number with 3 digits connects with a main interstate route; for example, the Washington, DC, Beltway is I–495, which connects with I–95.

U.S. routes are numbered in a similar way except that they have 1 to 3 digits. Route 1 runs north–south along the East Coast, and Route 101 runs north–south along the West Coast. Route 2 runs east–west along the Canadian border, and Route 90 runs east–west near the Gulf of Mexico and the Mexican border.

Major North–South U.S. Interstate Highways

Highway No.	From	To	Connects	Distance (miles)
I–5	Blaine, WA	San Diego, CA	Seattle, Portland, Eugene, Sacramento, Bakersfield, Los Angeles	1,382
I–15	Sweetgrass, MT	San Diego, CA	Great Falls, Butte, Idaho Falls, Salt Lake City, Las Vegas, Barstow, Anaheim	1,437
I–25	Buffalo, WY	Las Cruces, NM	Casper, Cheyenne, Boulder, Denver, Colorado Springs, Santa Fe, Albuquerque	1,062
I–35	Duluth, MN	Laredo, TX	Minneapolis/St. Paul, Des Moines, Kansas City, Wichita, Oklahoma City, Dallas/Ft. Worth, Austin, San Antonio	1,568
I–55	Chicago, IL	La Place, LA	St. Louis, Memphis, Jackson	944
I–65	Gary, IN	Mobile, AL	Indianapolis, Louisville, Nashville, Birmingham, Montgomery	888
I–75	Sault Ste. Marie, MI	Miami, FL	Detroit, Toledo, Cincinnati, Lexington, Knoxville, Chattanooga, Montgomery, Atlanta, Tampa, Ft. Meyers	1,787
I–95	Houlton, ME	Miami, FL	Bangor, Portland, Boston, Providence, New York, Philadelphia, Baltimore, Washington, Richmond, Fayetteville, Savannah, Jacksonville, Ft. Lauderdale	1,894

Major East–West U.S. Interstate Highways

Highway No.	From	To	Connects	Distances (miles)
I–10	Jacksonville, FL	Los Angeles, CA	Pensacola, Mobile, New Orleans, Baton Rouge, Houston, San Antonio, El Paso, Tucson, Phoenix, Palm Springs	2,460
I–20	Florence, SC	Kent, TX	Columbia, Atlanta, Birmingham, Jackson, Shreveport, Dallas/Ft. Worth	1,536
I–40	Wilmington, NC	Barstow, CA	Greensboro, Knoxville, Nashville, Memphis, Little Rock, Oklahoma City, Amarillo, Albuquerque, Flagstaff	2,463
I–70	Baltimore, MD	Cove Fort, UT	Pittsburgh, Columbus, Indianapolis, St. Louis, Kansas City, Topeka, Denver, Grand Junction	2,175
I–80	Ridgefield Park, NJ	San Francisco, CA	Cleveland, Chicago, Des Moines, Omaha, Cheyenne, Salt Lake City, Reno	2,907
I–90	Boston, MA	Seattle, WA	Albany, Niagara Falls, Cleveland, Chicago, Madison, Rochester (MN), Sioux Falls, Billings, Butte, Missoula, Spokane	3,163

World Travel

U.S. Passport Information

To obtain a passport for the first time

You must apply in person.

Use Form DSP-11 (from post office, courthouse, or passport agency); include evidence of citizenship, valid ID (such as driver's license), and 2 photos (2" x 2") taken in last 6 months.

Fees: $55 + $10 execution fee, adults (age 18 and over); $30 + $10 for minors (under age 18).

Valid for 10 years from date of issue (5 years for minors).

Where to apply: clerk of any federal or state court of record; judge or clerk of any probate court that accepts applications; any post office authorized to accept applications (ask at your local post office); or passport agency offices in Boston, Chicago, Honolulu, Houston, Los Angeles, Miami, New Orleans, New York, Philadelphia, San Francisco, Seattle, Stamford, CT, and Washington, DC.

To renew a passport

Can be done by mail if your last passport was issued within 12 years of the date of your current application, and if your last passport was issued after your 18th birthday.

Use Form DSP-82 (from post office, courthouse, or passport agency); include your most recent U.S. passport, and 2 photos (2" x 2") taken in last 6 months. Mailing instructions are on form.

Fee: $55 (no execution fee for renewals)

To make a name change to your valid passport

Use Form DSP-19 (from post office, courthouse, or passport agency); include your current valid passport and an original or certified copy of your marriage certificate, divorce decree, or court-ordered name change and send to—

Passport Lock Box
P.O. Box 371971
Pittsburgh, PA 15250-7971

No fee is charged to amend your passport.

To report loss or theft of your passport

Report the loss or theft when you apply in person for a new passport. If you are overseas, notify the nearest U.S. embassy or consulate and the local police.

U.S. Passport Information (continued)

To inquire about the status of your passport application	Write to— National Passport Center 31 Rochester Avenue Portsmouth, NH 03801-2900 Phone: (603) 334-0500
For expedited renewal service by mail	With application, include $30 expedited service fee in addition to other fees, plus a photocopy of your airline ticket. Mark the envelope "Expedited Service."

U.S. Customs Limits (for most countries)

Personal exemption: $400
Liquor: 1 liter
Tobacco: 200 cigarettes, 100 cigars

Articles in excess of personal exemption, up to $1,000 in value, are subject to 10 percent duty rate.

You must file a report with U.S. Customs if you carry more than $10,000 into or out of the United States, in any form of currency, negotiable instruments in bearer form, or traveler's checks.

Sales Taxes in Europe: The VAT

In most European countries, the sales tax is included in the retail price of an item, not added on at the time of purchase as with U.S. sales taxes. This tax is called the VAT—value-added tax. These countries will issue full or partial refunds to tourists for purchases that they send or take home with them; procedures vary by country, and a minimum purchase is required per store, per day.

Country	VAT (%)	VAT Refund (%)
Austria	20	16.7
Belgium	20.5	17
Denmark	25	20
Finland	22	18
France	18.6	15.7
Germany	15	13
Great Britain	17.5	14.9
Greece	13–18	11.5–15.3

Sales Taxes in Europe: The VAT (continued)

Country	VAT (%)	VAT Refund (%)
Hungary	10–25	9.1–20
Ireland	24.4	17.4
Italy	13–19	11.5–16
Luxembourg	15	13
Netherlands	17.5	14.9
Norway	22	18.7
Portugal	16–30	14.5
Slovenia	10–32	9.1–24.2
Spain	15	13.8
Sweden	25	20
Switzerland	6.5	—

Important U.S. State Department Phone Numbers

Passport information: (202) 647-0518

Overseas Citizens Emergency Center: (202) 647-5225; the computer bulletin board number is (202) 647-9225. Ask for Consular Information Sheets, which cover political climate, security and crime, health risks, embassy locations, and entry requirements.

Health Information Phone Numbers

Centers for Disease Control and Prevention traveler's hotline: (404) 332-4559

International Association for Medical Assistance to Travelers (IAMAT): (716) 754-4883 (provides lists of approved English-speaking doctors abroad)

International SOS Assistance: (800) 523-8930 (provides worldwide medical assistance and information)

Information for Travelers with Disabilities

Mobility International: (503) 343-1234

Information Center for Travelers with Disabilities: (617) 727-5540

Travel Industry and Disabled Exchange: (818) 788-8747

Travelin' Talk: (615) 552-6670

Information for Older Travelers

American Association of Retired Persons: (202) 434-2277

National Council of Senior Citizens: (202) 347-8800

Elderhostel: (617) 426-7788

Travel Insurance (medical and trip cancellation policies)

Carefree Travel Insurance: (800) 323-3149

Travel Guard: (800) 782-5151

International Tourist Offices and Embassies in the U.S.

Most countries operate 1 or more tourist offices in the United States that are a good source of up-to-date, free literature about the country. The tourist office can give you information about visa requirements, but you will usually need to apply for a visa through the country's embassy in the United States. Phone numbers may change; for directory assistance for toll-free numbers, call (800) 555-1212; for directory assistance for other numbers, dial the area code and 555-1212. Always dial 1 before any long-distance number.

Country	Tourist Office	Embassy
Argentina	(212) 603-0443	(202) 939-6400
Australia	(212) 687-6300 (310) 552-1988	(202) 797-3222
Austria	(800) 474-9696 (212) 944-6880	(202) 895-6700
Bahamas	(800) 422-4262	(202) 319-2660
Baltic States		
Estonia	—	(202) 588-0101
Latvia	—	(202) 726-8213
Lithuania	—	(202) 234-5860
Belgium	(212) 758-8130	(202) 333-6900
Bermuda	(212) 818-9800	—
Bolivia	(202) 483-4410	(202) 483-4410
Brazil	(800) 544-5503	(202) 745-2700
Bulgaria	(212) 573-5530	(202) 387-7969
Canada	—	(202) 682-1740
Alberta	(800) 661-8888	—
British Columbia	(800) 663-6000	—

International Tourist Offices and Embassies
in the U.S. (continued)

Country	Tourist Office	Embassy
Canada (continued)		
Manitoba	(800) 665-0040	—
New Brunswick	(800) 561-0123	—
Newfoundland	(800) 563-6353	—
Northwest Territories	(800) 661-0788	—
Nova Scotia	(800) 341-6096	—
Ontario	(800) 668-2746	—
Prince Edward Island	(800) 565-0267	—
Quebec	(800) 363-7777	—
Saskatchewan	(800) 667-7191	—
Yukon	(403) 667-5340	—
Caribbean		
Anguilla	(800) 553-4939	—
Antigua	(212) 541-4117	(202) 362-5211
Aruba	(800) 862-7822	—
Barbados	(800) 221-9831	(202) 939-9200
Belize	(212) 563-6011	(202) 332-9636
Bonaire	(800) 826-6247	—
British Virgin Islands	(800) 835-8530	—
Cayman Islands	(212) 682-5582	
Curaçao	(212) 683-7660	—
Dominican Republic	(212) 768-2482	(202) 332-6280
French West Indies*	(212) 838-7800	—
Grenada	(800) 927-9554	—
Jamaica	(212) 856-9727	—
Montserrat	(809) 491-2230	—
Puerto Rico	(800) 223-6530	—
Saba	(800) 722-2394	—
St. Eustatius	(800) 692-4106	—
St. Kitts/Nevis	(800) 582-6208	—
St. Lucia	(800) 456-3984	—
St. Vincent/Grenadines	(800) 729-1726	—
Trinidad/Tobago	(201) 662-3403	—
Turks & Caicos	(800) 241-0824	—
U.S. Virgin Islands	(212) 332-2222	—
Chile	(202) 785-1746	(202) 785-1746
China	(212) 868-7752	(202) 328-2500

International Tourist Offices and Embassies in the U.S. (continued)

Country	Tourist Office	Embassy
Colombia	(202) 868-7752	(202) 387-8338
Costa Rica	(800) 327-7033	(202) 234-2945
Czech Republic	—	(202) 363-6315
Denmark	(212) 949-2333	(202) 234-4300
Ecuador	(305) 447-6300	(202) 234-7200
Egypt	(212) 332-2570 (213) 653-8815	(202) 234-3903
Finland	(212) 949-2333	(202) 298-5800
France	(212) 315-0888 (310) 271-2358	(202) 944-6000
Germany	(212) 661-7200 (310) 575-9799	(202) 298-4000
Great Britain	(800) 462-2748 (212) 986-2200	(202) 462-1340
Greece	(212) 421-5777 (213) 626-6696	(202) 939-5800
Guatemala	(800) 742-4529	(202) 745-4952
Hong Kong	—	(202) 462-1340
Hungary	(212) 355-0240	(202) 362-6730
Iceland	(212) 949-2333	(202) 265-6653
India	(212) 586-4901 (213) 380-8855	(202) 939-9839
Indonesia	(213) 387-8309	(202) 775-5200
Ireland	(800) 223-6470	(202) 462-3939
Israel	(800) 596-1199	(202) 364-5500
Italy	(212) 245-4822 (310) 820-0098	(202) 328-5500
Japan	(212) 757-5640 (312) 222-0874 (213) 623-1952	(202) 939-6700
Kenya	(212) 486-1300	(202) 387-6101 (212) 486-1300
Korea (South)	(201) 585-0909	—
Luxembourg	(212) 935-8888	(202) 265-4171
Malaysia	(213) 689-9702	(202) 328-2700
Mexico	(212) 755-7261	(202) 728-1600
Monaco	(212) 759-5227	—
Morocco	—	(202) 462-7979
Netherlands	(312) 819-0300	(202) 244-5300
New Zealand	(800) 388-5494	(202) 328-4800

International Tourist Offices and Embassies in the U.S. (continued)

Country	Tourist Office	Embassy
Norway	(212) 949-2333	(202) 333-6000
Paraguay	(202) 483-6960	(202) 483-6960
Peru	(202) 833-9860	(202) 833-9860
Poland	(212) 338-9412	(202) 234-3800
Portugal	(212) 354-4403	(202) 328-8612
Romania	(212) 697-6971	(202) 332-4846
Russia	(212) 758-1162	(202) 298-5700
Scotland	(800) 234-9123	(see Great Britain)
Singapore	—	(202) 537-3100
South Africa	(800) 822-5368	(202) 232-4400
Spain	(212) 759-8822	(202) 452-0100
Sri Lanka	—	(202) 483-4025
Sweden	(212) 949-2333	(202) 467-2600
Switzerland	(800) 467-9477 (212) 757-5944	(202) 745-7900
Tanzania	—	(202) 939-6125
Thailand	(213) 382-2353	(202) 944-3600
Turkey	(202) 429-9844	(202) 659-8200
Ukraine	—	(202) 333-0606
Uruguay	(305) 443-7431	(202) 331-1313
Venezuela	(202) 342-2214	(202) 342-2214
Yugoslavia	(212) 757-2801	(202) 462-6566

*Guadeloupe, Martinique, St. Bart's, St. Martin.

Weather: Best Times to Go

One of the most important factors in planning a vacation trip is knowing the best months to visit a country, weatherwise. Generally that means sunny days and relatively warm weather. See Chapter 8, Weather, for tables showing monthly average highs, lows, and precipitation levels for cities worldwide. The table below lists the best months to travel both in terms of weather and certain other factors, such as peak vacation times, when crowds may make travel less pleasant. Once you have decided on your destination, check a recent travel guidebook for the country or countries you plan to visit to learn particular dates of school vacations, national holidays, and local celebrations to fine-tune your travel plans.

Country	Best Time to Go
Argentina	June – August
Australia	May (northern regions), October – December, February – April
Austria	May – June, September – October
Bahamas	December – April
Baltic States	May – September
Estonia	May – September
Latvia	May – September
Lithuania	May – September
Belgium	May – September
Bermuda	Year-round except September
Bolivia	April – October
Brazil	April – October
Bulgaria	June – August
Canada	April – November
Alberta	August – September
British Columbia	June – September
Manitoba	July – September
New Brunswick	July – August
Newfoundland	July – August
Northwest Territories	July – September
Nova Scotia	July – September
Ontario	July – September
Prince Edward Island	July – August

Weather: Best Times to Go (continued)

Country	Best Time to Go
Canada (continued)	
Quebec	July – August
Saskatchewan	August – September
Yukon	July – September
Caribbean	December – April
Chile	Varies by region
China	April – June, October – November
Colombia	December – February, June – September
Costa Rica	October – November
Czech Republic	May – June, September – October
Denmark	May – June
Ecuador	May – October
Egypt	December – February, April – June, September – November
Finland	June – August
France	June – September
Germany	May – October
Great Britain	April – October
Greece	May, June, September
Guatemala	October – November
Hong Kong	October – November, March – April
Hungary	May, September
India	November – February
Indonesia	April – May, September – October
Ireland	April – September
Israel	December – March
Italy	April – October
Japan	March – June, September – November
Kenya	January – February, June – September
Korea (South)	October – March
Luxembourg	May – October

Weather: Best Times to Go (continued)

Country	Best Time to Go
Malaysia	October – February (east coast), January – August (west coast)
Mexico	October – May
Morocco	April – May, September – October
Myanmar (Burma)	November – February
Netherlands	April – October
New Zealand	October – November, February – April
Norway	May – September
Paraguay	May – August
Peru	April – November
Poland	May – September
Portugal	April – October
Romania	May – September
Russia	May – September
Scotland	April – October
Singapore	Year-round; February best
South Africa	January – March
Spain	May – October
Sri Lanka	September – January
Sweden	May – July
Switzerland	Year-round, except November, May
Tanzania	June – September
Thailand	November – February
Turkey	April – October
Uruguay	December – March
Venezuela	December – April

Electrical Needs Worldwide

If you plan to take a hair dryer or other small electrical appliance that runs on 110V (the U.S. standard), you will probably need an electrical converter and a set of adapter plugs (although hotels may have 110V circuits for electric shavers). Some U.S. hair dryers are switchable to 220V, but depending on the country you visit, you may still need a plug adapter. Also note that the frequency of the current, expressed in hertz (Hz), varies from country to country even if voltage is the same. This means that a clock or tape recorder designed to run on 60 Hz (the U.S. standard) will not work properly on a 50-Hz current.

Location	Current	Plug Type
Australia	230V	3 slanted flat blades
Bermuda	110V	same as U.S.
Canada	110V	same as U.S.
Caribbean, Central America	110V or 220V	varies
China	220V	3 slanted flat blades
Europe (includes eastern Europe)	Mostly 220V	2 or 3 round pins
Great Britain, Ireland	220V	3 flat pins
Hong Kong	200V	3 round pins
India	220V	2 or 3 round pins
Israel	230V	2 round pins or 3 slanted flat pins
Japan	100V	same as U.S.
Kenya	240V	3 round pins; 3 flat blades
Mexico	120V	same as U.S. (need adapter for polarized plugs)
New Zealand	230V	3 slanted flat blades
Russia	220V	2 round pins
South Africa	220–230V	2 round pins; 3 flat blades
South America	110V, 220V, 240V	varies
Southeast Asia	110V or 220V	varies
Turkey	220V	2 round pins

Numbers to Know: European Customary Practices

Dates: In the United States, we express dates numerically in this order: month/day/year. For example, June 1, 1998, is expressed 6/1/98. But in Europe and other parts of the world, the day comes first, so that June 1, 1998, is expressed as 1/6/98. To be on the safe side, spell out the name of the month when making reservations in European destinations.

Decimal points and commas: Europeans reverse the U.S. use of these marks. In numbers over 1,000, the comma is replaced by either a thin space or a decimal point. In numbers in which we use decimal points, Europeans use commas: 1.50 becomes 1,50.

Floor numbers in buildings: A second floor in the United States is considered the first floor in Europe; Europeans call the first floor the ground floor and don't give it a number.

Sevens and ones: To distinguish between the number 7 and the number 1, Europeans put a slash through the 7, like this: 7 Their numeral 1 often has an upswing: 1; this can look like a 7 to U.S. eyes.

Using your fingers to convey "how many": Always start with your thumb to mean "1" in Europe; if you hold up only your index finger when ordering a drink, you will probably receive 2 drinks, since the thumb—even if you do not hold it out—also counts as 1. Be careful about using the peace sign—the index and middle fingers held in a V shape—to indicate 2; in some countries this gesture is considered obscene.

Telling Time with a 24-Hour Clock

In Europe and much of the world, hours are expressed using a 24-hour clock rather than the 12-hour a.m. and p.m. scheme used in the United States. In this system, the a.m. hours remain the same, but 1:00 p.m. is 1300 hours, 2:00 p.m. is 1400, on up to 2400, for 12:00 midnight. Here's a handy drawing to aid in quick translation. See Chapter 7, Time, for a map of worldwide time zones.

Before You Go: Where to Look for Current Exchange Rates

Exchange rates can vary daily. The newspapers of most major cities, as well as the *Wall Street Journal* and other business dailies, list foreign currency exchange rates in their financial sections.

Here is a sampling of currency rates as of mid-1995, shown in currency units per U.S. dollar:

Canada (dollar)	1.3563
France (franc)	4.8565
Germany (mark)	1.4085
Great Britain (pound)	0.6241
Italy (lira)	1,582.0
Japan (yen)	91.35
Mexico (peso)	6.145

Large city banks and some American Automobile Association offices also list rates and may sell selected foreign currencies. Although you may get a better conversion rate in the country once you arrive, it may be a good idea to get $50 or $60 worth of the currency of each country you'll be visiting *before* you leave the United States, just in case you arrive in a country when banks and exchange offices are closed (hours of operation vary widely).

International Currencies

Country	Basic Unit	Subunit
Afghanistan	afghani	100 puls
Albania	lek	100 qindarka
Algeria	dinar	100 centimes
Andorra	peseta	100 centimos
Angola	kwanza	100 lwei
Argentina	peso	100 centavos
Armenia	dram	100 louma
Australia	dollar	100 cents
Austria	schilling	100 groschen
Azerbaijan	manat	100 gopik
Bahamas	dollar	100 cents

International Currencies (continued)

Country	Basic Unit	Subunit
Bahrain	dinar	1,000 fils
Bangladesh	taka	100 paisa (poisha)
Barbados	dollar	100 cents
Belarus	ruble	100 kopecks
Belgium	franc	100 centimes
Belize	dollar	100 cents
Benin	franc	100 centimes
Bhutan	ngultrum	100 chetrum
Bolivia	boliviano	100 centavos
Bosnia-Herzegovina	dinar	100 paras
Botswana	pula	100 thebes
Brazil	real	100 centavos
Brunei	dollar	100 cents
Bulgaria	lev	100 stotinki
Burkina Faso	franc	100 centimes
Burundi	franc	100 centimes
Cambodia	riel	100 sen
Cameroon	franc	100 centimes
Canada	dollar	100 cents
Cape Verde	escudo	100 centavos
Cayman Islands	dollar	100 cents
Central African Republic	franc	100 centimes
Chad	franc	100 centimes
Chile	peso	100 centavos
China	yuan	100 fen
Colombia	peso	100 centavos
Comoros	franc	100 centimes
Congo	franc	100 centimes
Costa Rica	colón	100 centimos
Croatia	dinar	100 paras
Cuba	peso	100 centavos
Cyprus	pound	100 cents
Czech Republic	koruna	100 haler
Denmark	krone	100 öre
Djibouti	franc	100 centimes
Dominica	dollar	100 cents
Dominican Republic	peso	100 centavos
Ecuador	sucre	100 centavos

International Currencies (continued)

Country	Basic Unit	Subunit
Egypt	pound	100 piastres
El Salvador	colón	100 centavos
Equatorial Guinea	franc	100 centimes
Eritrea	birr	100 cents
Estonia	kroon	100 sents
Ethiopia	birr	100 cents
Fiji	dollar	100 cents
Finland	markka	100 penni
France	franc	100 centimes
Gabon	franc	100 centimes
Gambia	dalasi	100 butut
Georgia	maneti	100 kopecks
Germany	deutsche mark	100 pfennig
Ghana	cedi	100 pesewa
Great Britain	pound	100 pence
Greece	drachma	100 lepta
Greenland	krone	100 öre
Grenada	dollar	100 cents
Guatemala	quetzal	100 centavos
Guinea	franc	100 centimes
Guinea-Bissau	peso	100 centavos
Guyana	dollar	100 cents
Haiti	gourde	100 centimes
Honduras	lempira	100 centavos
Hong Kong	dollar	100 cents
Hungary	forint	100 fillér
Iceland	króna	100 aurar
India	rupee	100 paise
Indonesia	rupiah	100 sen
Iran	rial	100 dinars
Iraq	dinar	1,000 fils
Ireland	pound	100 pence
Israel	shekel	100 agorot
Italy	lira	100 centesimi
Ivory Coast	franc	100 centimes
Jamaica	dollar	100 cents
Japan	yen	100 sen
Jordan	dinar	1,000 fils

International Currencies (continued)

Country	Basic Unit	Subunit
Kazakhstan	tenge	—
Kenya	shilling	100 cents
Kiribati	dollar	100 cents
Kuwait	dinar	1,000 fils
Kyrgyzstan	som	100 kopecks
Laos	kip	100 at
Latvia	lat	100 santimi
Lebanon	pound	100 piastres
Lesotho	loti	100 lisente
Liberia	dollar	100 cents
Libya	dinar	100 dirhams
Liechtenstein	franc	100 centimes
Lithuania	litas	100 centas
Luxembourg	franc	100 centimes
Macao	pataca	100 avos
Macedonia	dinar	100 paras
Madagascar	franc	100 centimes
Malawi	kwacha	100 tambala
Malaysia	ringgit	100 sen
Maldives	rifiyaa	100 laari
Mali	franc	100 centimes
Malta	lira	100 cents
Mauritania	ouguiya	5 khoums
Mauritius	rupee	100 cents
Mexico	peso	100 centavos
Monaco	franc	100 centimes
Mongolia	tugrik	100 mongo
Morocco	dirham	100 centimes
Mozambique	metical	100 centavos
Myanmar (Burma)	kyat	100 pyas
Namibia	dollar	100 cents
Nauru	dollar	100 cents
Nepal	rupee	100 paise
Netherlands	guilder	100 cents
Netherlands Antilles	guilder	100 cents
New Zealand	dollar	100 cents
Nicaragua	cordoba	100 centavos
Niger	franc	100 centimes

International Currencies (continued)

Country	Basic Unit	Subunit
Nigeria	naira	100 kobos
North Korea	won	100 chon
Norway	krone	100 öre
Oman	riyal-omani	1,000 baizas
Pakistan	rupee	100 paisa
Panama	balboa	100 centesimos
Papua New Guinea	kina	100 toea
Paraguay	guarani	100 centimos
Peru	nuevo sol	100 centavos
Philippines	peso	100 centavos
Poland	zloty	100 groszy
Portugal	escudo	100 centavos
Qatar	riyal	100 dirhams
Romania	leu	100 bani
Russia	ruble	100 kopecks
Rwanda	franc	10 centimes
St. Kitts and Nevis	dollar	100 cents
St. Lucia	dollar	100 cents
St. Vincent and the Grenadines	dollar	100 cents
San Marino	lira	100 centesimi
São Tomé and Príncipe	dobra	100 centimos
Saudi Arabia	riyal	100 halalah
Senegal	franc	100 centimes
Seychelles	rupee	100 cents
Sierra Leone	leone	100 cents
Singapore	dollar	100 cents
Slovakia	koruna	—
Slovenia	tolar	100 stotins
Solomon Islands	dollar	100 cents
Somalia	shilling	100 cents
South Africa	rand	100 cents
South Korea	won	100 chon
Spain	peseta	100 centimos
Sri Lanka	rupee	100 cents
Sudan	pound	100 piasters
Suriname	guilder	100 cents
Swaziland	lilangeni	100 cents

International Currencies (continued)

Country	Basic Unit	Subunit
Sweden	krona	100 öre
Switzerland	franc	100 centimes
Syria	pound	100 piasters
Taiwan	dollar	100 cents
Tajikistan	ruble	100 kopecks
Tanzania	shilling	100 cents
Thailand	baht	100 satang
Togo	franc	100 centimes
Tonga	pa'anga	100 seniti
Trinidad/Tobago	dollar	100 cents
Tunisia	dinar	1,000 millimes
Turkey	lira	100 kurus
Turkmenistan	manat	—
Tuvalu	dollar	100 cents
Uganda	shilling	100 cents
United Arab Emirates	dirham	1,000 fils
United States	dollar	100 cents
Uruguay	peso	100 centavos
Uzbekistan	som	—
Vanuatu	vatu	100 centimes
Vatican City	lira	100 centesimi
Venezuela	bolivar	100 centimos
Vietnam	dong	10 hao, 100 xu
Western Samoa	tala	100 sene
Yemen	rial	100 fils
Yugoslavia	dinar	100 para
Zaire	zaire	100 makuta
Zambia	kwacha	100 ngwee
Zimbabwe	dollar	100 cents

How to Translate Foreign Prices

It's worthwhile to take a few minutes after you enter a country to come up with a quick way to translate prices to U.S. dollar equivalents. Here's one way:

- Note what the major unit of currency is worth in U.S. dollars.

  Example: Assume the British pound (£) is worth US$1.53.

- Round off that number (in the example, round down to 1.5) and use it as a multiplier.

  Example: If a book costs £10, to get dollars multiply 10 by 1.5 to get the equivalent of $15. If a meal costs £5, multiply by 1.5 to get $7.50.

For currencies like the Italian lira, where strings of zeros can make comprehension difficult, start off by eliminating the zeros.

  Example: Assume that the exchange rate is 1,600 Italian lire to the U.S. dollar.

- Divide 16 (you've dropped the last two zeros) into $1.00 to get a multiplier: 0.0625, which for rough estimates you will round off to 6.

- Use that number, 6, to multiply by the lire amount.

  Example: Admission to a museum is 10,000 lire. Mentally drop the last 4 digits and multiply by 6, to get the rough equivalent of $6.

  Example: A leather handbag costs 225,000 lire. Mentally drop the last 4 digits and multiply by 6, to get the rough equivalent of $132.

International Dialing Codes and Time Differences (using U.S. Eastern Standard Time)

To dial an international number—

1. Dial the International Access Code: 011. Use this code for *all* countries except Canada and for the Caribbean islands that use the 809 area code (see note at end of table).

2. Dial the country code (for example, for Italy dial 39).

3. Dial the city code, if one is listed (for example, Rome is 6).

To sum up, if you were dialing someone in Rome, the complete number would be—

011 + 39 + 6 + the local phone number

Country/City	Country/City Code	Time Difference
Albania	355	+7
Tirana	42	+7
Algeria[a]	213	+6
American Samoa[a]	684	6
Andorra	33	+6
Angola	244	+6
Argentina	54	+2
Buenos Aires	1	+2
Armenia	374	8
Aruba	297	+1[a]
Australia	61	+15[b]
Canberra	62	+15[b]
Austria	43	+6
Vienna	1 or 222	+6
Bahamas	809	0
Bahrain[a]	973	+8
Bangladesh	880	+11
Dhaka	2	+11
Belarus	372	+8
Belgium	32	+6
Brussels	2	+6
Belize	501	−1
Belize City	2	−1
Belmopan	8	−1
Benin[a]	229	+6
Bermuda	441	+1

International Dialing Codes and Time Differences (using U.S. Eastern Standard Time) (continued)

Country/City	Country/City Code	Time Difference
Bhutan[a]	975	+10.5
Bolivia	591	+1
La Paz	2	+1
Bosnia-Herzegovina	387	+6
Botswana	267	+7
Gaborone	31	+7
Brazil	55	+2
Brasilia	61	+2
Brunei	673	+13
Bandar Seri Begawan	2	+13
Bulgaria	359	+7
Sofia	2	+7
Burkina Faso[a]	226	+5
Burundi	257	+7
Bujumbura	22	+7
Cameroon[a]	237	+6
Canada	(c)	varies
Cape Verde Islands[a]	238	+4
Caribbean islands	(d)	+1
Central African Republic	236	+6
Chad	235	+6
N'Djamena	51	+6
Chile	56	+1
Santiago	2	+1
China (except Taiwan)	86	+13
Beijing	1	+13
Colombia	57	0
Bogotá	1	0
Comoros	269	+9
Moroni	73	+9
Congo[a]	242	+6
Cook Islands[a]	682	−5
Costa Rica[a]	506	−1
Croatia	385	+6
Cyprus	357	+7
Czech Republic	42	+6
Prague	2	+6

International Dialing Codes and Time Differences (using U.S. Eastern Standard Time) (continued)

Country/City	Country/City Code	Time Difference
Denmark[a]	45	+6
Djibouti[a]	253	+8
Ecuador	593	0
Quito	2	0
Egypt	20	+7
Cairo	2	+7
El Salvador[a]	503	−1
Equatorial Guinea	240	+6
Malabo	9	+6
Estonia	372	+7
Ethiopia	251	+8
Addis Ababa	1	+8
Fiji[a]	679	+17
Finland	358	+7
Helsinki	0	+7
France	33	+6
Paris	1	+6
French Antilles[a]	596	+1
French Guiana[a]	594	+2
French Polynesia[a]	689	−5
Gabon[a]	241	+6
Gambia[a]	220	+5
Germany	49	+6
Ghana	233	+5
Accra	21	+5
Gibraltar[a]	350	+6
Great Britain	44	+5
Belfast	1232	+5
Cardiff	1222	+5
Edinburgh	131	+5
London	171/181	+5
Greece	30	+7
Athens	1	+7
Greenland	299	+2[b]
Godthaab	2	+2[b]
Guadeloupe[a]	590	+1
Guam[a]	671	+15

International Dialing Codes and Time Differences (using U.S. Eastern Standard Time) (continued)

Country/City	Country/City Code	Time Difference
Guatemala	502	−1
Guatemala City	2	−1
Guinea	224	+5
Conakry	4	+5
Guinea-Bissau[a]	245	+5
Guyana	592	+2
Georgetown	02	+2
Haiti	509	0
Port-au-Prince	1	0
Honduras[a]	504	−1
Hong Kong	852	+13
Hungary	36	+6
Budapest	1	+6
Iceland	354	+5
Reykjavik	1	+5
India	91	+10.5
New Delhi	11	+10.5
Indonesia	62	+12[b]
Jakarta	21	+12[b]
Iran	98	+8.5
Tehran	21	+8.5
Iraq	964	+8
Baghdad	1	+8
Ireland	353	+5
Dublin	1	+5
Israel	972	+7
Jerusalem	3	+7
Italy	39	+6
Florence	55	+6
Rome	6	+6
Venice	41	+6
Ivory Coast[a]	225	+5
Japan	81	+14
Tokyo	3	+14
Jordan	962	+7
Amman	6	+7
Kazakhstan	7	+11

International Dialing Codes and Time Differences (using U.S. Eastern Standard Time) (continued)

Country/City	Country/City Code	Time Difference
Kenya	254	+8
Nairobi	2	+8
Korea, Republic of	82	+14
Seoul	2	+14
Kuwait[a]	965	+8
Kyrgystan	7	+10
Latvia	371	+7
Lesotho[a]	266	+7
Liberia[a]	231	+5
Libya	218	+7
Tripoli	21	+7
Liechtenstein	41	+6
All points	75	+6
Lithuania	370	+7
Luxembourg[a]	352	+6
Macao[a]	853	+13
Madagascar	261	+8
Antananarivo	2	+8
Malawi	265	+7
Malaysia	60	+13
Kuala Lumpur	3	+13
Maldives[a]	960	+10
Mali[a]	223	+5
Malta[a]	356	+6
Mauritania[a]	222	+5
Mauritius[a]	230	+5
Mexico	52	+1[b]
Mexico City	5	+1[b]
Monaco	33	+6
All points	93	+6
Morocco	212	+5
Rabat	7	+5
Mozambique	258	+7
Maputo	1	+7
Namibia	264	+7
Windhoek	61	+7
Nauru[a]	674	+17

International Dialing Codes and Time Differences (using U.S. Eastern Standard Time) (continued)

Country/City	Country/City Code	Time Difference
Nepal[a]	977	+10.5
Netherlands	31	+6
Amsterdam	20	+6
The Hague	70	+6
Netherlands Antilles	599	+1
New Caledonia[a]	687	+16
New Zealand	64	+17
Auckland	9	+17
Wellington	4	+17
Nicaragua	505	−1
Managua	2	−1
Niger Republic[a]	227	+6
Nigeria	234	+6
Lagos	1	+6
Norway	47	+6
Oslo	2	+6
Oman[a]	968	+9
Pakistan	92	+10
Islamabad	51	+10
Palau[a]	680	+14
Panama[a]	507	0
Papua New Guinea	675	+15
Paraguay	595	+2
Asunción	21	+2
Peru	51	0
Lima	14	0
Philippines	63	+13
Manila	2	+13
Poland	48	+6
Warsaw	22	+6
Portugal	351	+5
Lisbon	1	+5
Qatar[a]	974	+8
Reunion Island	262	+9
Romania	40	+7
Bucharest	0	+7

International Dialing Codes and Time Differences (using U.S. Eastern Standard Time) (continued)

Country/City	Country/City Code	Time Difference
Russia	7	+8[b]
Moscow	095	+8[b]
Rwanda[a]	250	+7
St. Helena	290	+5
St. Pierre/Miquelon	508	+2
Saipan	670	+15
San Marino	378	+6
Saudi Arabia	966	+8
Jeddah	2	+8
Riyadh	1	+8
Senegal[a]	221	+5
Sierra Leone	232	+5
Freetown	22	+5
Singapore[a]	65	+13
Slovenia	386	+6
South Africa	27	+7
Johannesburg	11	+7
Pretoria	12	+7
Spain	34	+6
Madrid	1	+6
Sri Lanka	94	+10.5
Colombo	1	+10.5
Suriname[a]	597	+2
Swaziland[a]	268	+7
Sweden	46	+6
Stockholm	8	+6
Switzerland	41	+6
Berne	31	+6
Zurich	1	+6
Syria	963	+8
Damascus	11	+8
Taiwan	886	+13
Taipei	2	+13
Tanzania	255	+8
Dar es Salaam	51	+8
Thailand	66	+12
Bangkok	2	+12

International Dialing Codes and Time Differences (using U.S. Eastern Standard Time) (continued)

Country/City	Country/City Code	Time Difference
Togo[a]	228	+5
Tunisia	216	+6
Tunis	1	+6
Turkey	90	+7
Ankara	4	+7
Uganda	256	+8
Kampala	41	+8
Ukraine	7	+8
Kiev	044	+8
United Arab Emirates	971	+9
Abu Dhabi	2	+9
United States	1	0[b]
Uruguay	598	+2
Montevideo	2	+2
Vatican City	39	+6
All points	6	+6
Venezuela	58	+1
Caracas	2	+1
Yemen	967	+8
Sanaa	2	+8
Yugoslavia	381	+6
Zaire	243	+6[b]
Kinshasa	12	+6[b]
Zambia	260	+7
Lusaka	1	+7
Zimbabwe	263	+7
Harare	4	+7

a. City code not required.

b. More than one time zone.

c. No country code is required for Canada; dial 1 and the area code (see Chapter 5, Daily Life, for a list of U.S. and Canadian area codes).

d. The following countries and islands use area code 809: Anguilla, Antigua, Bahamas, Barbuda, Barbados, British Virgin Islands, Carriacou, Cayman Islands, Dominica, Dominican Republic, Grenada, Jamaica, Montserrat, Puerto Rico, St. Kitts and Nevis, St. Lucia, St. Vincent and the Grenadines, Trinidad and Tobago, Turks and Caicos, and the U.S. Virgin Islands.

Phoning to the United States from Other Countries

If you think you will be calling the United States while you are visiting abroad, check with your long-distance carrier before you go; there may be direct-dial numbers you can use to avoid having to deal with local operators and high long-distance charges.

- AT&T, USA Direct and World Connect: (800) 331-1140
- MCI WorldPhone: (800) 444-4444
- Sprint Express: (800) 793-1153

Longest Flight The longest commercial airline flight available is from New York City to Johannesburg, South Africa, covering 7,967 miles in 14 hours, 24 minutes, according to Runzheimer International, a consulting firm.

Air Distances Between Selected World Cities (in miles)

City	Berlin	Buenos Aires	Cairo	Calcutta	Cape Town	Caracas
Berlin	—	7,402	1,795	4,368	5,981	5,247
Buenos Aires	7,376	—	7,345	10,265	4,269	3,168
Cairo	1,795	7,345	—	3,539	4,500	6,338
Calcutta	4,368	10,265	3,539	—	6,024	9,605
Cape Town	5,981	4,269	4,500	6,024	—	6,365
Caracas	5,247	3,168	6,338	9,605	6,365	—
Chicago	4,405	5,598	6,129	7,980	8,494	2,501
Hong Kong	5,440	11,472	5,061	1,648	7,375	10,167
Honolulu	7,309	7,561	8,838	7,047	11,534	6,013
Istanbul	1,078	7,611	768	3,638	5,154	6,048
Lisbon	1,436	5,956	2,363	5,638	5,325	4,041
London	579	6,916	2,181	4,947	6,012	4,660
Los Angeles	5,724	6,170	7,520	8,090	9,992	3,632
Manila	6,132	11,051	5,704	2,203	7,486	10,620
Mexico City	6,047	4,592	7,688	9,492	8,517	2,232
Montreal	3,729	5,615	5,414	7,607	7,931	2,449
Moscow	1,004	8,376	1,803	3,321	6,300	6,173
New York	3,965	5,297	5,602	7,918	7,764	2,132
Paris	545	6,870	1,995	4,883	5,807	4,736
Rio de Janeiro	6,220	1,200	6,146	9,377	3,773	2,810
Rome	734	6,929	1,320	4,482	5,249	5,196
San Francisco	5,661	6,467	7,364	7,814	10,247	3,904
Shanghai	5,218	12,201	5,183	2,117	8,061	9,501
Stockholm	504	7,808	2,111	4,195	6,444	5,420
Sydney	10,006	7,330	8,952	5,685	6,843	9,513
Tokyo	5,540	11,408	5,935	3,194	9,156	8,799
Warsaw	320	7,662	1,630	4,048	5,958	5,517
Washington, DC	4,169	5,218	5,800	8,084	7,901	2,059
Wellington	11,265	6,260	9,915	7,042	7,019	7,880

Air Distances Between Selected World Cities (in miles) (continued)

City	Chicago	Hong Kong	Honolulu	Istanbul	Lisbon	London
Berlin	4,405	5,440	7,309	1,078	1,436	579
Buenos Aires	5,598	11,472	7,561	7,611	5,956	6,916
Cairo	6,129	5,061	8,838	768	2,363	2,181
Calcutta	7,980	1,648	7,047	3,638	5,638	4,947
Cape Town	8,494	7,375	11,534	5,154	5,325	6,012
Caracas	2,501	10,167	6,013	6,048	4,041	4,660
Chicago	—	7,793	4,250	5,477	3,990	3,950
Hong Kong	7,793	—	5,549	4,984	6,853	5,982
Honolulu	4,250	5,549	—	8,109	7,820	7,228
Istanbul	5,477	4,984	8,109	—	2,012	1,552
Lisbon	3,990	6,853	7,820	2,012	—	985
London	3,950	5,982	7,228	1,552	985	—
Los Angeles	1,745	7,195	2,574	6,783	5,621	5,382
Manila	8,143	693	5,299	5,664	7,546	6,672
Mexico City	1,691	8,782	3,779	7,110	5,390	5,550
Montreal	744	7,729	4,910	4,789	3,246	3,282
Moscow	4,974	4,439	7,037	1,091	2,427	1,555
New York	713	8,054	4,964	4,975	3,364	3,458
Paris	4,134	5,985	7,438	1,400	904	213
Rio de Janeiro	5,296	11,021	8,285	6,389	4,796	5,766
Rome	4,808	5,768	8,022	843	1,161	887
San Francisco	1,858	6,897	2,393	6,703	5,666	5,357
Shanghai	7,061	764	4,941	4,962	6,654	5,715
Stockholm	4,278	5,113	6,862	1,348	1,856	890
Sydney	9,272	4,584	4,943	9,294	11,302	10,564
Tokyo	6,299	1,794	3,853	5,560	6,915	5,940
Warsaw	4,667	5,144	7,355	863	1,715	899
Washington, DC	597	8,147	4,519	5,215	3,562	3,663
Wellington	8,349	5,853	4,708	10,663	11,745	11,682

Air Distances Between Selected World Cities (in miles) (continued)

City	Los Angeles	Manila	Mexico City	Montreal	Moscow	New York
Berlin	5,724	6,132	6,047	3,729	1,004	3,965
Buenos Aires	6,170	11,051	4,592	5,615	8,376	5,297
Cairo	7,520	5,704	7,688	5,414	1,803	5,602
Calcutta	8,090	2,203	9,492	7,607	3,321	7,918
Cape Town	9,992	7,486	8,517	7,931	6,300	7,764
Caracas	3,632	10,620	2,232	2,449	6,173	2,132
Chicago	1,745	8,143	1,691	744	4,974	713
Hong Kong	7,195	693	8,782	7,729	4,439	8,054
Honolulu	2,574	5,299	3,779	4,910	7,037	4,964
Istanbul	6,783	5,664	7,110	4,789	1,091	4,975
Lisbon	5,621	7,546	5,390	3,246	2,427	3,364
London	5,382	6,672	5,550	3,282	1,555	3,458
Los Angeles	—	7,261	1,589	2,427	6,003	2,451
Manila	7,261	—	8,835	8,186	5,131	8,498
Mexico City	1,589	8,835	—	2,318	6,663	2,094
Montreal	2,427	8,186	2,318	—	4,386	320
Moscow	6,003	5,131	6,663	4,386	—	4,665
New York	2,451	8,498	2,094	320	4,665	—
Paris	5,588	6,677	5,716	3,422	1,544	3,624
Rio de Janeiro	6,331	11,259	4,771	5,097	7,175	4,817
Rome	6,732	6,457	6,366	4,080	1,474	4,281
San Francisco	347	6,967	1,887	2,539	5,871	2,571
Shanghai	6,438	1,150	8,022	7,053	4,235	7,371
Stockholm	5,454	5,797	5,959	3,667	762	3,924
Sydney	7,530	3,944	8,052	9,954	9,012	9,933
Tokyo	5,433	1,866	7,021	6,383	4,647	6,740
Warsaw	5,922	5,837	6,365	4,009	715	4,344
Washington, DC	2,300	8,562	1,887	488	4,858	205
Wellington	6,714	5,162	6,899	8,778	10,279	8,946

Air Distances Between Selected World Cities (in miles) (continued)

City	Paris	Rio de Janeiro	Rome	San Francisco	Shanghai	Stockholm
Berlin	545	6,220	734	5,661	5,218	504
Buenos Aires	6,870	1,200	6,929	6,467	12,201	7,808
Cairo	1,995	6,146	1,320	7,364	5,183	2,111
Calcutta	4,883	9,377	4,482	7,814	2,117	4,195
Cape Town	5,807	3,773	5,249	10,247	8,061	6,444
Caracas	4,736	2,810	5,196	3,904	9,501	5,420
Chicago	4,134	5,296	4,808	1,858	7,061	4,278
Hong Kong	5,985	11,021	5,768	6,897	764	5,113
Honolulu	7,438	8,285	8,022	2,393	4,941	6,862
Istanbul	1,400	6,389	843	6,703	4,962	1,348
Lisbon	904	4,796	1,161	5,666	6,654	1,856
London	213	5,766	887	5,357	5,715	890
Los Angeles	5,588	6,331	6,732	347	6,438	5,454
Manila	6,677	11,259	6,457	6,967	1,150	5,797
Mexico City	5,716	4,771	6,366	1,887	8,022	5,959
Montreal	3,422	5,097	4,080	2,539	7,053	3,667
Moscow	1,544	7,175	1,474	5,871	4,235	762
New York	3,624	4,817	4,281	2,571	7,371	3,924
Paris	—	5,699	697	5,558	5,754	958
Rio de Janeiro	5,699	—	5,684	6,621	11,336	6,651
Rome	697	5,684	—	6,240	5,677	1,234
San Francisco	5,558	6,621	6,240	—	6,140	5,361
Shanghai	5,754	11,336	5,677	6,140	—	4,825
Stockholm	958	6,651	1,234	5,361	4,825	—
Sydney	10,544	8,306	10,136	7,416	4,899	9,696
Tokyo	6,034	11,533	6,135	5,135	1,097	5,051
Warsaw	849	6,467	817	5,841	4,951	501
Washington, DC	3,829	4,796	4,434	2,442	7,448	4,123
Wellington	11,791	7,349	11,524	6,739	6,054	10,461

Air Distances Between Selected World Cities (in miles) (continued)

City	Sydney	Tokyo	Warsaw	Washington, DC	Wellington
Berlin	10,006	5,540	320	4,169	11,265
Buenos Aires	7,330	11,408	7,662	5,218	6,260
Cairo	8,952	5,935	1,630	5,800	9,915
Calcutta	5,685	3,194	4,048	8,084	7,042
Cape Town	6,843	9,156	5,958	7,901	7,019
Caracas	9,513	8,799	5,517	2,059	7,880
Chicago	9,272	6,299	4,667	597	8,349
Hong Kong	4,584	1,794	5,144	8,147	5,853
Honolulu	4,943	3,853	7,355	4,519	4,708
Istanbul	9,294	5,560	863	5,215	10,663
Lisbon	11,302	6,915	1,715	3,562	11,745
London	10,564	5,940	899	3,663	11,682
Los Angeles	7,530	5,433	5,922	2,300	6,714
Manila	3,944	1,866	5,837	8,562	5,162
Mexico City	8,052	7,021	6,365	1,887	6,899
Montreal	9,954	6,383	4,009	488	8,778
Moscow	9,012	4,647	715	4,858	10,279
New York	9,933	6,740	4,344	205	8,946
Paris	10,544	6,034	849	3,829	11,791
Rio de Janeiro	8,306	11,533	6,467	4,796	7,349
Rome	10,136	6,135	817	4,434	11,524
San Francisco	7,416	5,135	5,841	2,442	6,739
Shanghai	4,899	1,097	4,951	7,448	6,054
Stockholm	9,696	5,051	501	4,123	10,461
Sydney	—	4,866	9,696	9,758	1,338
Tokyo	4,866	—	5,249	6,772	5,760
Warsaw	9,696	5,249	—	4,457	10,615
Washington, DC	9,758	6,772	4,457	—	8,745
Wellington	1,338	5,760	10,615	8,745	—

Road Distances Between Selected Canadian Cities (in kilometers)

City	Calgary	Charlottetown	Edmonton	Fredericton	Halifax	Montreal
Calgary	—	4,917	299	4,558	5,042	3,743
Charlottetown	4,917	—	4,949	359	232	1,184
Edmonton	299	4,949	—	4,598	5,082	3,764
Fredericton	4,558	359	4,598	—	346	834
Halifax	5,042	232	5,082	346	—	1,318
Montreal	3,743	1,184	3,764	834	1,318	—
Ottawa	3,553	1,374	3,574	1,024	1,508	190
Quebec	4,014	945	4,035	586	912	270
Regina	764	4,163	785	3,813	4,297	2,979
St. John's	6,183	1,294	6,212	1,622	1,349	2,448
Saskatoon	620	4,421	528	4,070	4,554	3,236
Thunder Bay	2,050	2,878	2,071	2,527	3,011	1,693
Toronto	3,434	1,724	3,455	1,373	1,857	539
Vancouver	1,057	5,985	1,244	5,634	6,119	4,801
Victoria	1,123	6,051	1,310	5,700	6,185	4,867
Whitehorse	2,385	7,034	2,086	6,684	7,168	5,850
Winnipeg	1,336	3,592	1,357	3,241	3,726	2,408
Yellowknife	1,811	6,460	1,511	6,109	6,593	5,275

City	Ottawa	Quebec	Regina	St. John's	Saskatoon	Thunder Bay
Calgary	3,553	4,014	764	6,183	620	2,050
Charlottetown	1,374	945	4,163	1,294	4,421	2,878
Edmonton	3,574	4,035	785	6,212	528	2,071
Fredericton	1,024	586	3,813	1,622	4,070	2,527
Halifax	1,508	912	4,297	1,349	4,554	3,011
Montreal	190	270	2,979	2,448	3,236	1,693
Ottawa	—	460	2,789	2,638	3,046	1,503
Quebec	460	—	3,249	2,208	3,507	1,963
Regina	2,789	3,249	—	5,427	257	1,286
St. John's	2,638	2,208	5,427	—	5,684	4,141
Saskatoon	3,046	3,507	257	5,684	—	1,543
Thunder Bay	1,503	1,963	1,286	4,141	1,543	—
Toronto	399	810	2,670	2,987	2,927	1,384
Vancouver	4,611	5,071	1,822	7,248	1,677	3,108
Victoria	4,677	5,137	1,888	7,314	1,743	3,174
Whitehorse	5,660	6,120	2,871	8,298	2,614	4,157
Winnipeg	2,218	2,678	571	4,855	829	715
Yellowknife	5,086	5,546	2,297	7,723	2,039	3,582

Road Distances Between Selected Canadian Cities (in kilometers) (continued)

City	Toronto	Vancouver	Victoria	Whitehorse	Winnipeg	Yellowknife
Calgary	3,434	1,057	1,123	2,385	1,336	1,811
Charlottetown	1,724	5,985	6,051	7,034	3,592	6,460
Edmonton	3,455	1,244	1,310	2,086	1,357	1,511
Fredericton	1,373	5,634	5,700	6,684	3,241	6,109
Halifax	1,857	6,119	6,185	7,168	3,726	6,593
Montreal	539	4,801	4,867	5,850	2,408	5,275
Ottawa	399	4,611	4,677	5,660	2,218	5,086
Quebec	810	5,071	5,137	6,120	2,678	5,546
Regina	2,670	1,822	1,888	2,871	571	2,297
St. John's	2,987	7,248	7,314	8,298	4,855	7,723
Saskatoon	2,927	1,677	1,743	2,614	829	2,039
Thunder Bay	1,384	3,108	3,174	4,157	715	3,582
Toronto	—	4,492	4,558	5,528	2,099	4,966
Vancouver	4,492	—	66	2,697	2,232	2,411
Victoria	4,558	66	—	2,763	2,298	2,477
Whitehorse	5,528	2,697	2,763	—	3,524	2,704
Winnipeg	2,099	2,232	2,298	3,524	—	2,868
Yellowknife	4,966	2,411	2,477	2,704	2,868	—

Numbers to Know: Miles and Kilometers

A kilometer is about 0.6 mile. A quick way to convert kilometers to miles is to divide the distance in kilometers by 2, then add 10 percent of the original kilometer figure.

Example: Paris is 640 km away. To get miles, divide 640 by 2 to get 320; then take 10 percent of 640 (64) and add it to 320, to get 384 miles.

To convert miles to kilometers, multiply miles by 1.6.

To convert kilometers to miles, multiply kilometers by 0.6.

Here are some speed limits to remember:

30 mph = 48 km/h
55 mph = 88 km/h
65 mph = 104 km/h

Road Distances Between Selected British Cities (in miles)

City	Aberdeen	Birmingham	Cambridge	Cardiff	Edinburgh
Aberdeen	—	430	468	532	130
Birmingham	430	—	101	107	293
Cambridge	468	101	—	191	337
Cardiff	532	107	191	—	395
Edinburgh	130	293	337	395	—
Glasgow	149	291	349	393	45
Liverpool	361	98	195	200	222
London	543	118	60	155	405
Manchester	354	88	153	188	218
Newcastle	239	198	224	311	107
Oxford	497	63	82	109	361
Plymouth	624	199	275	164	488
Shrewsbury	412	48	142	110	276
Southampton	571	128	133	123	437
York	325	128	153	241	191

City	Glasgow	Liverpool	London	Manchester	Newcastle
Aberdeen	149	361	543	354	239
Birmingham	291	98	118	88	198
Cambridge	349	195	60	153	224
Cardiff	393	200	155	188	311
Edinburgh	45	222	405	218	107
Glasgow	—	220	402	214	150
Liverpool	220	—	210	34	170
London	402	210	—	199	280
Manchester	214	34	199	—	141
Newcastle	150	170	280	141	—
Oxford	354	164	56	153	253
Plymouth	486	294	215	281	410
Shrewsbury	272	64	162	69	216
Southampton	436	241	76	227	319
York	208	100	209	71	83

Road Distances Between Selected British Cities (in miles) (continued)

City	Oxford	Plymouth	Shrewsbury	Southampton	York
Aberdeen	497	624	412	571	325
Birmingham	63	199	48	128	128
Cambridge	82	275	142	133	153
Cardiff	109	164	110	123	241
Edinburgh	361	488	276	437	191
Glasgow	354	486	272	436	208
Liverpool	164	294	64	241	100
London	56	215	162	76	209
Manchester	153	281	69	227	71
Newcastle	253	410	216	319	83
Oxford	—	193	113	67	185
Plymouth	193	—	242	155	340
Shrewsbury	113	242	—	190	144
Southampton	67	155	190	—	252
York	185	340	144	252	—

Road Distances Between Selected European Cities (in kilometers)

City	Athens	Barcelona	Brussels	Calais	Cherbourg
Athens	—	3,313	2,963	3,175	3,339
Barcelona	3,313		1,318	1,326	1,294
Brussels	2,963	1,318	—	204	583
Calais	3,175	1,326	204	—	460
Cherbourg	3,339	1,294	583	460	—
Cologne	2,762	1,498	206	409	785
Copenhagen	3,276	2,218	966	1,136	1,545
Geneva	2,610	803	677	747	853
Gibraltar	4,485	1,172	2,256	2,224	2,047
Hamburg	2,977	2,018	597	714	1,115
Lisbon	4,532	1,304	2,084	2,052	1,827
Lyons	2,753	645	690	739	789
Madrid	3,949	636	1,558	1,550	1,347
Marseilles	2,865	521	1,011	1,059	1,101
Milan	2,282	1,014	925	1,077	1,209
Munich	2,179	1,365	747	977	1,160
Paris	3,000	1,033	285	280	340
Rome	817	1,460	1,511	1,662	1,794
Stockholm	3,927	2,868	1,616	1,786	2,196
Vienna	1,991	1,802	1,175	1,381	1,588

City	Cologne	Copenhagen	Geneva	Gibraltar	Hamburg
Athens	2,762	3,276	2,610	4,485	2,977
Barcelona	1,498	2,218	803	1,172	2,018
Brussels	206	966	677	2,256	597
Calais	409	1,136	747	2,224	714
Cherbourg	785	1,545	853	2,047	1,115
Cologne	—	760	1,662	2,436	460
Copenhagen	760	—	1,418	3,196	460
Geneva	1,662	1,418	—	1,975	1,118
Gibraltar	2,436	3,196	1,975	—	2,897
Hamburg	460	460	1,118	2,897	—
Lisbon	2,290	2,971	1,936	676	2,671
Lyons	714	1,458	158	1,817	1,159
Madrid	1,764	2,498	1,439	698	2,198
Marseilles	1,035	1,778	425	1,693	1,479
Milan	911	1,537	328	2,185	1,238
Munich	583	1,104	591	2,565	805
Paris	465	1,176	513	1,971	877
Rome	1,497	2,050	995	2,631	1,751
Stockholm	1,403	650	2,000	3,880	949
Vienna	937	1,455	1,019	2,974	1,155

Road Distances Between Selected European Cities (in kilometers) (continued)

City	Lisbon	Lyons	Madrid	Marseilles	Milan
Athens	4,532	2,753	3,949	2,865	2,282
Barcelona	1,304	645	636	521	1,014
Brussels	2,084	690	1,558	1,011	925
Calais	2,052	739	1,550	1,059	1,077
Cherbourg	1,827	789	1,347	1,101	1,209
Cologne	2,290	714	1,764	1,035	911
Copenhagen	2,971	1,458	2,498	1,778	1,537
Geneva	1,936	158	1,439	425	328
Gibraltar	676	1,817	698	1,693	2,185
Hamburg	2,671	1,159	2,198	1,479	1,238
Lisbon	—	1,778	668	1,762	2,250
Lyons	1,778	—	1,281	320	328
Madrid	668	1,281	—	1,157	1,724
Marseilles	1,762	320	1,157	—	618
Milan	2,250	328	1,724	618	—
Munich	2,507	724	2,010	1,109	331
Paris	1,799	471	1,273	792	856
Rome	2,700	1,048	2,097	1,011	586
Stockholm	3,231	2,108	3,188	2,428	2,187
Vienna	2,935	1,157	2,409	1,363	898

City	Munich	Paris	Rome	Stockholm	Vienna
Athens	2,179	3,000	817	3,927	1,991
Barcelona	1,365	1,033	1,460	2,868	1,802
Brussels	747	285	1,511	1,616	1,175
Calais	977	280	1,662	1,786	1,381
Cherbourg	1,160	340	1,794	2,196	1,588
Cologne	583	465	1,497	1,403	937
Copenhagen	1,104	1,176	2,050	650	1,455
Geneva	591	513	995	2,068	1,019
Gibraltar	2,565	1,971	2,631	3,886	2,974
Hamburg	805	877	1,751	949	1,155
Lisbon	2,507	1,799	2,700	3,231	2,935
Lyons	724	471	1,048	2,108	1,157
Madrid	2,010	1,273	2,097	3,188	2,409
Marseilles	1,109	792	1,011	2,428	1,363
Milan	331	856	586	2,187	898
Munich	—	821	946	1,754	428
Paris	821	—	1,476	1,827	1,249
Rome	946	1,476	—	2,707	1,209
Stockholm	1,754	1,827	2,707	—	2,105
Vienna	428	1,249	1,209	2,105	—

Using ATMs Abroad

If your ATM card works with the MasterCard Cirrus or the Visa Plus networks, you may be able to get cash from ATM machines abroad at near-wholesale exchange rates. Check with your bank before leaving—you may need to obtain a different PIN (personal identification number).

Tipping Practices in Restaurants, by Country

Country	Amount of Tip
Argentina	Leave 10%
Australia	Tipping not customary but 10% to 15% becoming standard
Austria	Service included; add 5%
Belgium	Service included; add 5%
Bolivia	Service included
Bulgaria	Service included; leave 2 to 5 leva
Caribbean	If service included, leave 5%; if not, leave 10% to 15%
Colombia	If service included, leave 5%; if not, leave 10%
Czech Republic	Leave 10% to 15%
Denmark	Service included; leave 10% to 15% for good service
Ecuador	If service included, leave 5%; if not, leave 10%
Egypt	Service included; leave extra 5% to 10%
Finland	Service included; round up to nearest 5 or 10 FIM
France	Service included; leave small change
Germany	Service included; round up to nearest DM
Great Britain	Service may be included; if not, leave 10% to 15%
Greece	Service may be included; if not, leave 10%
Hong Kong	Service included; leave extra 10%
Hungary	Leave 10% to 15%
Indonesia	Tipping not customary; discouraged
Ireland	Service included; tipping not expected
Italy	Service included; leave extra 5%
Japan	Service included in fine restaurants; tipping not customary
Luxembourg	Service included; round up to nearest franc
Malaysia	Service included; add 10%
Netherlands	Service included
New Zealand	Tipping not customary; service included in some fine restaurants
Norway	Service included; leave extra 5%

Tipping Practices in Restaurants, by Country (continued)

Country	Amount of Tip
Peru	Service included; leave extra 5%
Poland	Leave 10%
Portugal	Leave 5% to 10%
Romania	Service included
Spain	Leave 10%; service charge prohibited by law
Sweden	Tips not expected
Switzerland	Service included; leave small extra tip
Thailand	Service may be included; if not, leave 10%
Turkey	Service usually included; leave extra 10%
Uruguay	Leave 10%
Venezuela	Service included; leave extra 5%

Wiring Money Abroad

If you need to wire money overseas, try contacting these services:

American Express MoneyGram: (800) 926-9400 or call collect from abroad: (303) 980-3340

Western Union Financial Services: (800) 325-6000 or (800) 325-4176

Typical Flying Times* Between Selected World Cities

From	To	Flight Time (hours)
Auckland	Chicago	14.4
Berlin	Peking	7.9
Boston	Frankfurt	6.3
Buenos Aires	Singapore	17.0
Cairo	Singapore	8.8
Chicago	Calcutta	13.8
Copenhagen	Los Angeles	9.7
Copenhagen	Montreal	8.6
Denver	Buenos Aires	10.3
Denver	Moscow	9.5
Edmonton	Amsterdam	7.5
Hong Kong	Chicago	13.5
Honolulu	London	12.5
Honolulu	Mexico City	6.6
London	Wellington	20.2
Los Angeles	London	10.0
Madrid	Miami	7.6
Manila	Sydney	6.7
Melbourne	San Francisco	13.6
Melbourne	Wellington	4.5
Moscow	Washington, DC	10.9
New York	Honolulu	8.6
New York	London	6.5
Rio de Janeiro	Bombay	14.3
Rio de Janeiro	London	9.9
Rome	Delhi	6.4
Rome	Nairobi	5.8
Seattle	Istanbul	10.5
Tokyo	Buenos Aires	19.7
Tokyo	Seattle	8.3
Vancouver	Mexico City	4.3
Wellington	New Orleans	13.5
Winnipeg	Glasgow	6.2

*Does not include time on ground.

Useful General Information

U.S. Airline Phone Numbers

American	(800) 433-7300
America West	(800) 235-9292
Continental	(800) 525-0280
Delta	(800) 221-1212
Midwest Express	(800) 452-2022
Northwest	(800) 225-2525
Southwest Airlines	(800) 435-9792
Trans World Airlines (TWA)	(800) 221-2000
United Airlines	(800) 241-6522
USAir, Inc.	(800) 428-4322
Valujet Airlines	(800) 825-8538

Cheap Airline Tickets

Known in the travel business as "bucket shops," consolidators sell blocks of unfilled seats on airline flights. There is usually no advance-purchase restriction, and prices are usually less than the cheapest fare you'd be able to get from the airline directly. Here are three of the largest consolidators:

Council Charter: (800) 800-8222 or (212) 661-0311
Travac: (800) 872-8800 or (212) 563-3303
UniTravel: (800) 325-2222 or (314) 569-0900

International Airline Phone Numbers

Aer Lingus (Ireland)	(800) 223-6537
Air Canada	(800) 776-3000
Air France	(800) 237-2747
Air Jamaica	(800) 523-5585
Air Malta	(415) 362-2929
Air New Zealand	(800) 262-1234
Alitalia (Italy)	(800) 223-5730
All Nippon Airways	(800) 235-9262
Austrian Airlines	(800) 843-0002
Avianca (Colombia)	(800) 284-2622
Balkan Bulgarian Airlines	(800) 776-5706
British Airways	(800) 247-9297
Canadian Air International	(800) 776-3000
Cathay Pacific Airways	(800) 233-2742
China Airlines	(800) 227-5118
CSA/Czechoslovak Airlines	(800) 223-2365
Cyprus Airways	(800) 333-2977
El Al Israel Airlines	(800) 223-6700
Finnair	(800) 950-5000
Gulf Air	(800) 553-2824
Iberia Airlines (Spain)	(800) 772-4642
Icelandair	(800) 223-5500
Japan Air Lines	(800) 525-3663
KLM Royal Dutch Airlines	(800) 777-5553
Korean Air	(800) 438-5000
Ladeco Chilean Airlines	(800) 825-2332
LOT Polish Airlines	(800) 223-0593
LTU International Airways	(800) 888-0200
Lufthansa German Airlines	(800) 645-3880
Malév Hungarian Airlines	(800) 223-6884
Martinair Holland	(800) 627-8462
Olympic Airways (Greece)	(800) 223-1226
Qantas Airways (Australia)	(800) 227-4500
Sabena Belgian World Airways	(800) 955-2000
Saudi Arabian Airlines	(800) 472-8342
Scandinavian Airlines (SAS)	(800) 221-2350
Singapore Airlines	(800) 742-3333
Swissair	(800) 221-4750
TAP Air Portugal	(800) 221-7370

International Airline Phone Numbers (continued)

Tarom Romanian Airlines	(212) 687-6013
Thai Airways International	(800) 426-5204
THY Turkish Airlines	(800) 874-8875
Tower Air	(800) 348-6937
Virgin Atlantic Airways	(800) 862-8621

Hotel and Motel Chain Phone Numbers

Best Western	(800) 528-1234
Club Med	(800) 453-7447
Comfort Inns	(800) 424-6423
Days Inn	(800) 329-7466
Doubletree Hotels	(800) 222-8733
Econo Lodges	(800) 424-6423
Embassy Suites	(800) 362-2779
Forte Hotels	(800) 225-5843
Friendship Inns International	(800) 424-6423
Hampton Inn Hotels	(800) 426-7866
Harley Hotels	(800) 321-2323
Hilton Hotels	(800) 445-8667
Holiday Inns	(800) 465-4329
Howard Johnson	(800) 654-2000
Hyatt Hotels	(800) 233-1234
Marriott Courtyard	(800) 321-2211
Marriott Hotels	(800) 228-9290
Marriott Residence Inn	(800) 331-3131
Omni Hotels	(800) 843-6664
Radisson Hotels	(800) 333-3333
Ramada	(800) 228-2828
Red Carpet Inns	(800) 251-1962
Red Lion Hotels & Inns	(800) 547-8010
Red Roof Inns	(800) 843-7663
Resorts International	(800) 321-3000
Ritz-Carlton	(800) 241-3333
Sheraton Hotels & Inns	(800) 325-3535
Sonesta International Hotels	(800) 766-3782
Stouffer Hotels	(800) 468-3571
Westin Hotels	(800) 228-3000

Car Rental Company Phone Numbers

Alamo Rent-A-Car	(800) 327-9633
All England/All Ireland/ All Germany Car Rentals	(800) 241-3228
Auto Europe	(800) 223-5555 (800) 268-8810 (in Canada)
Avis	(800) 831-2847
Budget Rent-A-Car	(800) 527-0700
Cortell International	(800) 228-2535
Dollar Rent-A-Car	(800) 800-4000
Enterprise Rent-A-Car	(800) 325-8007
Eurodollar Rent-A-Car, Ltd.	(800) 800-6000
Europcar	(800) 227-3876
Europe By Car	(800) 223-1516
European Car Reservation	(800) 535-3303
Foremost	(800) 272-3299
Hertz	(800) 654-3001 (800) 263-0600 (in Canada)
Kemwel	(800) 678-0678
Meier's World Travel	(800) 937-0700
National	(800) 227-7368
Thrifty	(800) 367-2277
Tilden	(800) 227-7368

Cruise Line Phone Numbers

American Canadian Caribbean Line	(800) 556-7450
American Hawaii Cruises	(800) 765-7000
Carnival Cruise Lines	(800) 327-9501
Celebrity Cruises	(800) 235-3274
Classical Cruises	(800) 252-7745
Clipper Cruise Line	(800) 325-0010
Commodore Cruise Line	(800) 327-5617
Costa Cruises	(800) 462-6782
Crown Cruise Line	(800) 841-7447
Crystal Cruises	(213) 340-4121
Cunard Line Ltd.	(800) 221-8200
Dolphin/Majesty Cruise Lines	(800) 222-1003
Fantasy Cruises	(800) 437-3111
Holland America Line	(800) 426-0327
Norwegian Cruise Line	(800) 262-4625
Premier Cruise Lines	(800) 327-7113
Princess Cruises	(704) 274-2555
Renaissance Cruises	(800) 525-2450
Royal Caribbean Cruise Line	(800) 659-7225
Royal Cruise Line	(415) 956-7200
Royal Olympic Cruises	(800) 872-6400
Royal Viking Line	(800) 634-8000
Seabourn Cruise Line	(415) 391-7444
Windjammer Barefoot Cruises	(800) 327-2601
Windstar Cruises	(800) 258-7245
World Explorer Cruises	(800) 854-3835

Rule of Thumb: Tips on Ships

Plan on allowing $7 to $11 per passenger, per day, for all tipping while on board (including cabin stewards, waiters, busboys, wine stewards, and other service staff). Another way to calculate this amount is to figure on 10 percent of the total cruise cost. Practices vary; some cruise lines include a 15 percent service charge, and a few lines have banned tipping altogether.

Cruises: Best Times to Go

Destination	Best Times
Alaska	Mid-May — mid-September
Antarctica	December — February
Bermuda	Late April — September
Canada	June — October
Caribbean	November — April
Hawaii	June — August
Mexican Riviera	October — mid-May
Panama Canal	September — October, December — March, May
South America	December — March

Numbers to Know: Passenger/ Crew Ratios

One way to assess the level of service you are likely to receive on a cruise ship is to look at the ratio of crew to passengers. Guidebooks on cruise lines may list this ratio; if they don't, you can calculate it by dividing the passenger capacity (say, 100) by the number of crew (say, 20) to get the passenger/crew ratio of 5 to 1. When comparing cruise lines, remember that a lower ratio will probably mean better service.

European Railway Passes

Railway passes can be purchased through travel agents or through the services listed in the table that follows. Discounts are often available for youths, seniors, couples, groups, or travel that combines train and rental car use.

European Railway Pass Facts

Pass	Countries	Travel Days Allowed	Period Passes Available for—
Benelux	Belgium, Luxembourg, Netherlands	5	1 month
Britfrance	Britain and France	5	1 month
Britgerman	Britain and Germany	5 10	1 month 1 month
Central Europe Pass	Czech Republic, Germany, Poland, Slovakia	5	1 month
Eastern European	Austria, Czech Republic, Hungary, Poland, Slovakia	5 10	15 days 1 month
Eurailpass	Austria, Belgium, Denmark, Finland, France, Germany, Greece, Hungary, Ireland, Italy, Luxembourg, Netherlands, Norway, Portugal, Spain, Sweden, Switzerland	Unlimited	15 days 21 days 1 month 2 months 3 months
Eurail Flexipass	Same as Eurailpass	5 10 15	2 months 2 months 2 months
Europass	France, Germany, Italy, Spain, Switzerland; for additional fee, Austria, Belgium, Greece, Luxembourg, Portugal	5 8 11	3 countries, 2 months 4 countries, 2 months 5 countries, 2 months
Scanrail	Denmark, Finland, Norway, Sweden	5 10 Unlimited	15 days 1 month 1 month

World's Fastest Train The French TGV train *Atlantique* holds the world speed record for trains: 317 mph; its normal traveling speed is 186 mph. Each train set holds 485 passengers. The line runs to Switzerland, Belgium, the Netherlands, and England (through the Channel Tunnel).

Worldwide Railway Pass Phone Numbers

Australia	(800) 423-2880
Austrailpass	
Kangaroo Road 'n Rail Pass	
Austria	
Österreich Puzzle	(212) 944-6880
Bundesnetzkarte	(212) 944-6880
Belgium (and Luxembourg, the Netherlands)	
Belgian National Railways	(212) 758-8130
Benelux Tourrail Ticket (available at train stations)	
Canada	
Canrailpass	(800) 561-3949
Great Canadian Railtour Co.'s Rocky Mountaineer	(800) 665-7245
Europe	
Eurailpass, Europass	(800) 848-7245 (RailEurope)
	(800) 782-2424 (DER Tours)
Great Britain	
BritRail Pass	(212) 986-2200
Ireland	
Rambler ticket	(800) 243-7687
Japan	
Japan Railway Pass	(212) 698-4919
	(212) 944-8660
Spain	
RENFE Tourist Card	No U.S. representative
Switzerland	
Swiss Pass	(800) 438-7245
Swiss Card	(800) 438-7245
Scandinavia (Denmark, Norway, Sweden, Finland)	
Scanrail Pass	(800) 848-7245 (RailEurope)
	(800) 782-2424 (DER Tours)
United States	
Amtrak	(800) 872-7245

London to Paris, Through the Channel Tunnel

To make reservations on the Eurostar trains that make the trip between central London and central Paris in 3 hours, call RailEurope at (800) 848-7245. The "Chunnel" is 31 miles long. Trains also run from London to Brussels. Cars can be taken via car carrier from the Folkestone station, south of London, to Calais, France.

Reporting Lost or Stolen Credit Cards and Traveler's Checks

Type	In U.S./Caribbean	Outside U.S. (call collect)
Credit Card		
American Express		
Personal Card (Green)	(800) 528-4800	(919) 668-6668
Corporate Card	(800) 528-2122	(602) 492-5450
Gold Card	(800) 528-2121	(305) 476-2166
Platinum Card	(800) 525-3355	(602) 492-5450
Optima Card	(800) 635-5955	(602) 492-5450
Diners Club	(800) 234-6377	(303) 790-2433
MasterCard	(800) 826-2181	(314) 275-6690
Visa	(800) 336-8472	(314) 275-6690
Visa Gold	(800) 847-2911	(314) 275-6690
Traveler's Checks		
American Express	(800) 221-7282	(801) 964-6665
Mastercard	(800) 223-7373	(609) 987-7300
Thomas Cook	(800) 223-7373	(609) 987-7300
Visa	(800) 227-6811	(415) 574-7111

Vacation Photos: Choosing the Right Film Speed

The film speed—shown as the ISO number (formerly ASA) on the roll of film—indicates how the film responds to light. Here are suggested applications:

100	=	sunny days or with flash
200	=	all-purpose and with flash
400	=	low-light conditions or with telephoto lens
1,600	=	very dim light (e.g., in museums or buildings where no flash is allowed)

5

Daily Life

Home Decorating

Estimating Rolls of Wallpaper Needed

Wallpaper usually comes packaged in 2-roll bolts but is priced by the single roll. The average roll contains about 30 square feet; the width of a roll can vary.

Size of Room (ft.)	Ceiling Height (ft.)					Yards of Border
	8	9	10	11	12	
	Single Rolls Needed					
5 x 6	8	8	10	11	12	11
8 x 10	9	10	11	12	13	13
10 x 10	10	11	13	14	15	15
10 x 12	11	12	14	15	16	16
10 x 14	12	14	15	16	18	17
12 x 12	12	14	15	16	18	17
12 x 14	13	15	16	18	19	18
12 x 16	14	16	17	19	21	20
12 x 18	15	17	19	20	22	21
12 x 20	16	18	20	22	24	23
14 x 14	14	16	17	19	21	20
14 x 16	15	17	19	20	22	21
14 x 18	16	18	20	22	24	23
14 x 20	17	19	21	23	25	24
14 x 22	18	20	22	24	27	25
16 x 16	16	18	20	22	24	23
16 x 18	17	19	21	23	25	24
16 x 20	18	20	22	24	27	25
16 x 22	16	21	23	26	28	27
16 x 24	20	22	25	27	30	28
18 x 18	18	20	22	24	27	25
18 x 20	19	21	23	25	28	27
18 x 22	20	22	25	27	30	28
18 x 24	21	23	26	28	31	29

Rules of Thumb: Buying Wallpaper

After you've used the above table to get a general estimate of the number of rolls of wallpaper needed, you can refine your estimate by doing the following:

- Subtract 1 single roll for every 2 windows or 1 door in the room.
- Subtract 1 or 2 single rolls for a kitchen with many cabinets.
- Add 1 or 2 single rolls for patterns that have wide matches (called "pattern repeat" and given in inches on the label or in the wallpaper book).

One other number is important when buying wallpaper: the "run" number. Be sure before you start that all rolls have come from the same run, as marked on each roll's label. Rolls of different runs (meaning rolls that were printed at different times) may have slight color variations.

Estimating Gallons of Paint Needed—Interior

Walls

One gallon of paint can cover anywhere from 350 to 500 square feet, depending on the brand of paint and the condition of the surface. We use 350 in the calculations below to provide a conservative estimate.

1. To calculate the number of square feet, add the lengths of each wall and multiply your total by the ceiling height.

   Example: A 12-by-20-foot room that has 8-foot ceilings would amount to $12 + 12 + 20 + 20 = 64 \times 8 = 512$ square feet.

2. From that figure, subtract 20 square feet for each door or fireplace and 15 square feet for each window.

   Example: 1 door = 20 square feet, 3 windows = 45 square feet, for a total of 65 square feet

   $512 - 65 = 447$ square feet

3. Divide your result by 350.

   Example: $447 \div 350 = 1.277$ gallons of paint

There are 4 quarts in a gallon, so you could buy 1 gallon and 2 quarts, although 2 quarts may cost almost as much as a gallon; check prices before you decide.

4.   If you need to apply 2 coats, double the number of quarts or gallons.

Ceiling

Using wall or ceiling paint, which covers about 350 square feet per gallon, multiply the length by the width of the room to get the total number of square feet.

Example: For a 12-by-20-foot room, 12 × 20 = 240 square feet.

You would need a gallon of paint for 1 coat.

Trim

Trim paints cover between 350 and 400 square feet per gallon, or between 80 and 100 square feet per quart.

1.   Allow 8 square feet for each window and 25 square feet for each door.

Example: 1 door = 25 square feet, 3 windows = 24 square feet, for a total of 49 square feet

For the window trim and the door, you would need 1 quart of trim paint.

2.   To calculate base, chair rail, and ceiling molding, measure their lengths in inches and multiply by 6. Then divide that figure by 144 to get the number of square feet.

Example: Your 12-by-20-foot room has base and ceiling molding.

12 + 12 + 20 + 20 = 64 feet × 12 = 768 inches × 6 inches = 4,608 square inches

4,608 × 2 (base and ceiling molding) = 9,216 square inches

9,216 ÷ 144 = 64 square feet

For the moldings, you would need 1 quart of trim paint.

Estimating Gallons of Paint Needed—Exterior

Outside Walls

To calculate how much paint you will need for the exterior of your house, do the following:

1. Measure the length of each outside wall and add the lengths together. Then multiply by the height to the gables.

    Example: Your 2-story house is a rectangle 20 feet deep and 50 feet wide.

    20 + 20 + 50 + 50 = 140 × 18 feet high = 2,520 square feet

2. To calculate the area of the gables, multiply one-half their width by their height and multiply this figure by the number of gables.

    25 × 10 × 2 = 500 square feet

3. Add the results of steps 1 and 2.

    2,520 + 500 = 3,020

4. Divide that number by 400 (or the amount of coverage shown on the can label) to determine the number of gallons needed per coat.

    3,020 ÷ 400 = 7.55 gallons

Trim

To figure the amount of trim paint needed, count all the windows and multiply this number by 15 square feet, and count all the doors and multiply this number by 25 square feet.

Example: 10 windows × 15 = 150 square feet, 2 doors × 25 = 50 square feet, for a total of 200 square feet.

You would need about ½ gallon (2 quarts) of trim paint per coat.

Coverage Capacity of Various Finishes

Product	Square Feet per Gallon
Lacquer	200–300
Lacquer sealer	250–300
Stain (water-based)	350–400
Stain (oil-based)	300–350
Shellac	300–350
Varnish	300–350
Paste wax	125–175
Liquid wax	600–700
Paint	350–650

Light-Reflecting Power of Paint Colors

The color of your walls can affect the amount of reflective light available in a room; the greater the percentage of light-reflecting power, the better the visibility for close work such as reading and crafts.

Color	Percent of Light Reflected
White	70–90
Cream, ivory	55–90
Light yellow	65–70
Light green	40–50
Medium green	15–30
Medium gray	15–30
Orange	15–30
Medium blue	15–20
Dark blue	5–10
Red, maroon	3–18
Medium brown, dark brown	3–18

How Long Will It Last?

The estimates given assume normal wear, good initial product quality, careful upkeep, and careful surface preparation.

- Wallpaper: 10 years
- Interior Paint: 5 years
- Exterior Paint: 8 years

An unopened can of oil-based paint can last up to 50 years, but an unopened can of water-based (latex) paint may be good for only 7 or 8 years.

Furniture Spacing Guidelines

Location	Minimum Amount of Space to Allow (ft.)
Between most pieces of furniture	3
Between a chair or sofa and a coffee table	1$\frac{1}{2}$
Around a bed	2
In front of a chest of drawers	3
Behind a desk or dining chairs	3
Between a kitchen island and kitchen counters	3

Rule of Thumb: Typical Furniture Dimensions

Most desks measure 24 by 48 inches.

Most chairs and sofas are about 30 inches deep and have backs that are 33 to 36 inches high.

Most desks and dining tables are 29 to 30 inches high.

Mattress Sizes

Type	Width (in.)	Length (in.)
Standard		
Twin	39	75
Full	54	75
Queen	60	80
King	76	80
California King	72	84
Waterbed		
Single	48	84
Queen	60	84
King	72	84
Day Bed	39	75
Crib	27	52

Sizes of Sheets and Their Yardage Equivalents

| Size | Dimensions (in.) | Yardage Equivalent | | |
		36" Width	45" Width	54" Width
Twin	70 x 96	5	4	$3^1/_8$
Full	72 x 108	6	$4^1/_4$	$3^7/_8$
Queen	90 x 110	7	$5^7/_8$	$4^1/_2$
King	108 x 120	$8^3/_4$	7	$5^1/_8$

Pillow and Pillowcase Sizes

Type	Pillow (in.)	Pillowcase (in.)
Standard	20 x 26	20 x 30
Queen	20 x 30	20 x 34
King	20 x 36	20 x 40

Comforter/Duvet Sizes

Type	Size (in.)
Twin	60 x 86
Full/queen	86 x 86
King	102 x 86

Rule of Thumb: Hanging Pictures

Although the general rule is to hang pictures at eye level, here's a decorator's guideline that is a little more specific: The bottom of the picture frame should be no less than 5 inches and no more than 10 inches above a piece of furniture.

Estimating Slipcover Yardage

The estimates in this table are for 54-inch-wide fabric and do not allow for pattern matching. Add 2 to 3 yards to the estimates shown if you plan to add a tailored skirt.

Slipcover or upholstery fabric may be "railroaded," or run horizontally across the piece of furniture, if the fabric pattern is such that the direction in which it runs does not matter. Fabrics with prints that run in only one direction must be run vertically (from the top to the bottom of the piece of furniture).

	Yards Needed	
Piece	**Railroaded**	**Vertically Run**
Ottoman	$1^1/_2$	$1^1/_2$
Dining room chair	2	2
Armless boudoir chair	2	2
Armless chair	4	4
Barrelback chair	5	$6^1/_2$
Armchair	$5^1/_4$	$5^3/_4$
Armchair with back cushion	$6^3/_4$	$8^1/_4$
High-back wing chair	7	9
Love seat (to 60")	$9^3/_4$	12
Sofa (2-cushion, to 84")	$12^3/_4$	$16^1/_4$
with removable cushions	$15^3/_4$	19
Sofa (3-cushion, to 84")	$14^1/_4$	17
with removable cushions	$18^3/_4$	22
Wing sofa (3-cushion)	15	$17^3/_4$

Estimating Upholstery Yardage

Piece	Yards Needed	
	Railroaded	Vertically Run
Ottoman	$1^3/_4$	2
Dining chair seat pad	$2^1/_4$	$2^1/_2$
Armless boudoir chair	$2^1/_4$	3
Armless chair	4	4
Barrelback chair	$5^1/_4$	$5^1/_2$
Armchair	$5^1/_4$–7	6–7
Armchair with back cushion	$6^3/_4$	$7^1/_2$
High-back wing chair	6	$6^1/_2$
Love seat (to 60")	10	12
Sofa (2-cushion, to 84")	13	15
with removable cushions	$15^3/_4$	$17^3/_4$
Sofa (3-cushion, to 84")	$14^1/_4$	$16^1/_2$
with removable cushions	16	18
Wing sofa (3-cushion)	$15^1/_2$	17

Fabric Yardage Equivalents

If your sewing project calls for a fabric of a particular width (say, 45 inches) but your fabric is of a different width (say, 60 inches), use this table to determine the required yardage.

Fabric Width (in.)			
36	44–45	52–54	58–60
Yards			
$1^3/_4$	$1^3/_8$	$1^1/_8$	1
2	$1^5/_8$	$1^3/_8$	$1^1/_4$
$2^1/_4$	$1^3/_4$	$1^5/_8$	$1^3/_8$
$2^1/_2$	$2^1/_8$	$1^3/_4$	$1^5/_8$
$2^7/_8$	$2^1/_4$	$2^1/_4$	$1^3/_4$
$3^1/_8$	$2^1/_2$	$2^1/_4$	$1^7/_8$
$3^3/_8$	$2^3/_4$	$2^1/_4$	2
$3^3/_4$	$2^7/_8$	$2^3/_8$	$2^1/_4$
$4^1/_4$	$3^1/_8$	$2^5/_8$	$2^3/_8$
$4^1/_2$	$3^3/_8$	$2^3/_4$	$2^5/_8$
$4^3/_4$	$3^5/_8$	$2^7/_8$	$2^3/_4$
5	$3^7/_8$	$3^1/_0$	$2^7/_8$

Home Improvement and Remodeling

Which Home Improvements Pay Off?

Improvement	Percent of Return on Investment When House Is Sold
Kitchen remodeling, major	90–100
Kitchen remodeling, minor	85
New second full bath	85
New master suite	80
New family room with fireplace	80
Bath remodeling	75
New exterior siding	75
New sunroom	60
New windows	55–60
New deck or patio	50–70
New doors	40
New swimming pool	30

Rule of Thumb: Remodeling Costs Don't remodel to the point where you increase the value of your property to more than 20 percent of that of similar homes in your neighborhood. Also, keep your total remodeling expenses to no more than 10 percent of the value of your home.

Typical Home Improvements: What Will It Cost?

All prices listed are approximate and include labor or installation; they can vary greatly depending on the quality of products used and the local labor costs. Do-it-yourself costs will be about 20 to 30 percent less than the figures shown.

Item	Approximate Cost ($)
Air conditioning	
Complete central	2,000–3,000
Compressor	800–1,000
Attic conversion	18,000
Basement conversion	7,000
Bathroom remodeling	
Master bathroom	4,000–14,000
Powder room	3,000
Crown molding, 3-part (12' x 15' room)	600
Deck (10' x 20')	
Pressure-treated wood	3,000
Redwood	4,500
Electrical service upgrade	600–1,200
Family room addition	35,000
Fireplace	3,000–4,000
Flooring (12' x 15' room)	
Ceramic tile, unglazed	1,900
Oak strip	1,600
Vinyl	1,000
French doors	3,300
Front door, metal, with sidelights	1,400
Furnace, warm air	1,800–3,000
Garage	10,000–20,000
Kitchen island	1,000
Kitchen remodeling	7,000–20,000
Roof	
Asphalt shingle	1,500–2,200
Wood shingle	3,000–4,000
Septic tank	3,000–6,000
Water heater	350–750
Window	
New (3' x 5')	1,000
Replacement (2$\frac{1}{2}$' x 5')	500
Bay	3,000

**Rule of
Thumb:
Working with
a Contractor**

In your written contract with a construction contractor,
you should include these numbers:

- 10 percent: the limit on cost overruns
- 30 percent: the maximum amount of your down payment
- 10 percent: the difference between the amount of work paid
  for and the amount of work completed at any given time
- 15 percent: the amount to hold back until the job is done

Also ask for your contractor's license number; make sure he
or she is licensed to do work in the jurisdiction where your
property is located.

To purchase a standard form contract, call the American Institute
of Architects at (800) 365-2724; ask for Document A-107.

Estimating Square Footage in a House or a Room

Practice varies as to what constitutes the square footage in a house,
but generally the number is meant to indicate livable space—
not attics, unfinished basements, garages, or closets. The most
conservative system is to determine the square footage in each
room and hallway of the house and add the figures together.

To determine a room's square footage, multiply the length by the
width.

12 feet × 14 feet = 168 square feet

For odd-shaped rooms—say, a living room/dining room combination—
divide the area into two or more rectangles and calculate the square
footage of each.

Estimating Square Yardage Needed for Floor Coverings

First determine the room's square footage, as shown in the paragraphs above. Carpeting and vinyl floor covering are generally sold in square yards, so you must divide the square footage by 9, the number of square feet in a square yard.

168 square feet ÷ 9 = 18.67 square yards

Most carpeting comes in rolls that are 12 feet wide; if your room is wider than 12 feet, the carpet will have to be seamed at some point, and the calculations become a bit more complex. It's often best to let the carpet store determine the proper amount; there will probably be some waste that you will end up paying for.

Estimating Wood Flooring Needed

Unfinished flooring is usually sold in 12-board bundles that are 4 boards deep and 3 boards wide; the lengths of the pieces may vary by 6 inches either over or under the nominal length—that is, the stated length of the bundle. After you've determined the square footage of the area to be covered (as explained on the previous page), you'll need to determine the square footage of a bundle. The quickest way to do this is to multiply its nominal length (in feet) by the width of the wood strip (in inches). For example, if the nominal length of the bundle is 6 feet and the strips or planks are $2^{1/4}$ inches wide, you would multiply those 2 numbers:

6 × 2.25 = 13.5 square feet per bundle

Assume that the room size is 11 × 14 feet, or 154 square feet. Divide the square footage you get by the number of square feet in a bundle:

154 ÷ 13.5 = 11.407 bundles

Then multiply that result by the nominal length of the bundles— 6 feet—to get the total bundle feet needed:

11.407 × 6 = 66.44 bundle feet

Finally, add at least 5 percent to that amount to allow for waste.

66.44 × 0.05 = 3.422

66.44 + 3.422 = 69.86 (round up to 70) bundle feet

Estimating Tile Flooring Needed

To allow for waste, multiply the number of tiles needed by 7 percent (0.07) to arrive at the number of extra tiles you should buy.

Square Feet of Floor Space	Number of Tiles Needed	
	9" x 9"	12" x 12"
10	18	10
20	36	20
30	54	30
40	72	40
50	89	50
60	107	60
70	125	70
80	143	80
90	160	90
100	178	100
200	356	200
300	534	300
400	712	400
500	890	500

The Rule of 2: Estimating the Time Needed to Complete Home Improvement Projects

1. Determine the time needed to shop for the necessary materials; then double that time to allow for at least 1 return trip to the building supplier or hardware store for a forgotten item or to get a different size or color.

2. Determine the time needed to complete each subtask of a project (surface preparation, sanding, etc.) and add this figure to the time in step 1 to get a total.

3. Multiply that total by 2 to get a realistic estimate of the time required.

Standard Versus Actual Lumber Sizes

Lumber is measured in terms of its rough size; when "surfaced," it loses about $1/4$ inch (for 1-inch boards) and $1/2$ inch (for 2-inch or 4-inch boards) in thickness, and it loses about $1/2$ inch to $3/4$ inch in width.

Standard Size (in.)	Actual Size (seasoned wood) (in.)
1 x 4	$3/4$ x $3^1/2$
1 x 6	$3/4$ x $5^1/2$
1 x 8	$3/4$ x $7^1/4$
1 x 10	$3/4$ x $9^1/4$
1 x 12	$3/4$ x $11^1/4$
2 x 4	$1^1/2$ x $3^1/2$
2 x 6	$1^1/2$ x $5^1/2$
2 x 8	$1^1/2$ x $7^1/4$
2 x 10	$1^1/2$ x $9^1/4$
2 x 12	$1^1/2$ x $11^1/4$
4 x 4	$3^1/2$ x $3^1/2$
4 x 6	$3^1/2$ x $5^1/2$
4 x 8	$3^1/2$ x $7^1/4$
4 x 10	$3^1/2$ x $9^1/4$
4 x 12	$3^1/2$ x $11^1/4$

Quick Calculations to Determine Board Feet

Lumber is sold by the board foot, which is calculated by multiplying the length by the width by the thickness; 1 board foot equals 144 cubic inches.

Lumber Size (in.)	Do This—
1 x 4	Divide length by 3
1 x 6	Divide length by 2
1 x 8	Multiply length by 0.66
1 x 12	Linear equals board feet
2 x 4	Multiply length by 0.66
2 x 6	Linear equals board feet
2 x 8	Multiply length by 1.33
2 x 12	Multiply length by 2

Rule of Thumb: Saw Blade Teeth

If you use a blade with too few teeth, rough cuts can result; if you use a blade with too many teeth, you might burn the wood. If you are using a 10-inch table saw, follow these recommendations (if you are using a smaller portable saw, use a blade with proportionally fewer teeth):

- For ripping, or cutting with the grain of the wood, a 40-tooth blade is best for 1-inch-thick lumber; for 2-inch-thick lumber, use a 24-tooth blade.

- For cutting across the grain of the wood, use an 80-tooth blade for 1-inch boards and a 60-tooth blade for 2-inch boards.

Standard Sizes of Interior Materials

Item	Standard Width	Standard Height/Length
Doors		
Bifold (2-door)	2', 2'8", 3'	6'8"
Double-entry	60", 64", 72"	6'8"
Sliding glass	60", 72"	80"
Front	36"	6'8"
Room	30"	6'8"
Bathroom-closet	24"–30"	6'8"
Louvered	1'3"–3'	6'8"
Garage, single	9'	$6^1/_2$', 7'
Garage, double	16'	$6^1/_2$', 7'
Windows		
Bay	6'–12'	$3^1/_2$'–7'
Bow	6'–15'	3'–7'
Sash	20"–40"	34"–64"
Sliding glass	36", 48"	24", 34", 48", 56"
Bathroom fixtures		
Bathtub	2', 2'6"–2'8"	4'6", 5', 5'6"
Shower stall	2'6"–3'6"	2'6"–3'6"
Sink	30"–40"	22"
Toilet	19"–20"	26" (depth)
Kitchen equipment		
Range/oven	30"–36"	$23^1/_2$"–25" (depth)
Dishwasher	24"	$24^1/_4$"–30" (depth)
Refrigerator	28"–35"	25"–30" (depth)
Sink, single	24", 30"	21" (depth)
Sink, double	32"	20" (depth)

Weights of Common Substances

Substance	Approximate Weight (lb./cu. ft.)
Books, hardback	30–40
Brick, common	120
Clothing, firmly packed	10–15
Concrete	145
Earth, moist and loose	70–80
Gasoline	45–50
Gold	1,204
Ice (crushed)	37
Ice (solid)	57
Iron, cast	450
Lead	708
Mud	119
Oils (vegetable, mineral)	55–60
Sand (dry)	100
Sand (wet)	120
Silver	653
Snow (compacted)	30
Snow (freshly fallen)	10
Steel	490
Stone (crushed)	100
Water, fresh (1 gallon = 8.336 lbs.)	62.4
Water, sea	64.08

Mortar Requirements

Here's the formula for making 1 cubic foot of general-use mortar: Mix together—

- 16 pounds Portland cement
- 8½ pounds hydrated lime
- 100 pounds dry sand
- 2 to 3 gallons of water

Mortar Required for Common Bricks
(8" x 3¾" x 2¼")

Joint Thickness (in.)	Cubic Feet of Mortar per 1,000 Bricks
¹/₄	9
³/₈	14
¹/₂	20

Estimating Number of Bricks Needed

Joint Thickness (in.)	Number of Bricks per Square Foot
¹/₄	7
³/₈	6.5
¹/₂	6.17
⁵/₈	5.8
³/₄	5.5

Typical Concrete Mixtures by Volume

Cement:Sand:Gravel Ratio	Applications
1:3:6	Normal static loads, no rebar (not exposed)
1:2.5:5	Normal foundations and walls (exposed)
1:2.5:4	Basement walls
1:2.5:3.5	Waterproof basement walls
1:2.5:3	Light-duty floors, driveways
1:2.25:3	Steps, driveways, sidewalks
1:2:4	Lintels
1:2:4	Reinforced roads, buildings, walls (exposed)
1:2:3.5	Retaining walls, driveways
1:2:3	Swimming pools, fence posts
1:1.75:4	Light-duty floors
1:1.5:3	Watertight, reinforced tanks and columns
1:1:2	High-strength columns, girders, floors
1:1:1.5	Fence posts

Recommended Thicknesses of Concrete Slabs

Thickness (in.)	Applications
4	Home basement floors, farm building floors
4–5	Home garage floors, porches
5–6	Sidewalks, barn and granary floors, small shed floors
6–8	Driveways

Rule of Thumb: Estimating Concrete Blocks Needed to Build a Wall

Multiply the wall's length by its height; then multiply your result by 1.2 to get the number of concrete blocks you'll need.

You'll also need about 6 cubic feet of mortar for each 100 concrete blocks.

Finishing Nail Sizes: Length and Number per Pound

Size	Length (In.)	Approximate Number per Pound
2d	1	1,351
3d	1¼	807
4d	1½	584
5d	1¾	500
6d	2	309
8d	2½	189
10d	3	121
16d	3½	90
20d	4	62

Rule of Thumb: Nail Sizes

The "d" in nail sizes means "penny" and is the abbreviation for the Latin *denarius*, an ancient Roman coin. Originally, "2d," "10d," etc. referred to the cost in pennies for 100 nails. Now it refers to a definite size.

To determine the length of nails up to 3 inches (10d), divide the penny size of the nail by 4 and add ½ inch to get its length.

If you want to know the penny size of the nail (up to 3 inches in length), subtract ½ inch from the length of the nail and multiply by 4.

Common Nail Sizes: Length and Number per Pound

Size	Length (in.)	Approximate Number per Pound
2d	1	830
3d	1¼	528
4d	1½	316
5d	1¾	271
6d	2	168
7d	2¼	150
8d	2½	106
9d	2¾	96
10d	3	69
12d	3¼	63
16d	3½	49
20d	4	31
30d	4½	24
40d	5	18
50d	5½	14
60d	6	11

Thicknesses Needed for Various Types of Insulation to Obtain R-Values

The R-value measures a material's ability to resist heat conduction. The higher the R-value, the better a material's insulating ability. It is defined as:

$$\frac{\text{temperature difference} \times \text{area} \times \text{time}}{\text{heat loss}}$$

| | | Inches of Thickness Needed | | | |
| | | Loose and Blown Fill | | | |
R-Value	Mineral Fiber Blanket or Batts	Fiberglass	Rock Wool	Cellulosic Fiber	Perlite or Vermiculite
R-11	$3^1/_4$–$3^3/_4$	4–$5^1/_4$	$3^1/_2$	$3^3/_4$	3–$4^1/_2$
R-19	$5^3/_4$–$6^1/_4$	7–$8^3/_4$	$6^1/_4$	$6^1/_2$	$5^1/_2$–$7^3/_4$
R-30	9–$9^1/_2$	11–14	$9^3/_4$	$10^1/_2$	$8^1/_2$–$12^1/_4$
R-38	$11^1/_2$–12	14–$17^3/_4$	$12^1/_4$	13	$10^1/_2$–$15^1/_2$
R-49	15–$15^1/_2$	18–23	16	17	$13^3/_4$–20

Source: U.S. Department of Energy.
Note: Rigid cellular insulating boards have R-values of 4 to 8, depending on type, for 1 inch of thickness.

Thicknesses Needed to Meet Recommended Standard of R-19 for Exterior Walls

Material	Thickness to Equal R-19 (in.)
Brick	96
Concrete blocks	72
Wallboard	24
1/2" plywood	18
Urethane foam	3

Recommended R-Values for Insulation in Existing Houses in 8 Insulation Zones

Insula-tion Zone	Ceilings Below Ventilated Attics		Floors over Unheated Crawlspaces, Basements		Exterior Walls[a] (wood frame)		Crawlspace Walls[b]	
	Oil, Gas, Heat Pump	Electric Resistance	Oil, Gas, Heat Pump	Electric Resistance	Oil, Gas, Heat Pump	Electric Resistance	Oil, Gas, Heat Pump	Electric Resistance
1	19	30	0	0	0	11	11	11
2	30	30	0	0	11	11	19	19
3	30	38	0	19	11	11	19	19
4	30	38	19	19	11	11	19	19
5	38	38	19	19	11	11	19	19
6	38	38	19	19	11	11	19	19
7	38	49	19	19	11	11	19	19
8	49	49	19	19	11	11	19	19

Source: U.S. Department of Energy.

Note: These recommendations are based on the assumption that no structural modifications are needed to accommodate the added insulation.

a. The R-value of full wall insulation, which is $3^1/_2$ inches thick, will depend on the material used. The range is R-11 to R-13. For new construction, R-19 is recommended for exterior walls. Jamming an R-19 batt in a $3^1/_2$-inch cavity will not yield R-19.

b. Insulate crawlspace walls only if the crawlspace is dry all year, the floor above is not insulated, and all ventilation to the crawlspace is blocked. A vapor barrier (e.g., 4-mil or 6-mil polyethylene film) should be installed on the ground to reduce moisture migration into the crawlspace.

U.S. Department of Energy Insulation Zones

Roof Color Reflectivity

Type of Roof	Percent of Solar Heat Absorbed
Asphalt shingle	
Dark	90
Medium	85
Light	75
Gravel	
Medium	70
Light	50
Tile, medium	70–80
Cedar shake	80

British Thermal Unit (Btu) Equivalents

A Btu is the quantity of heat required to raise the temperature of 1 pound of water from 60° to 61°F at a constant pressure of 1 atmosphere.

To convert other forms of energy measurement to Btu's, multiply—

Cubic feet of gas × 1,031

Kilowatt hours of electricity × 3,412

Gallons of oil × 139,000

One therm equals 100,000 Btu's; 252 small calories (the metric system measurement of energy), or about 4 big calories (kilocalories), equal 1 Btu.

Estimating Room Air Conditioner Cooling Requirements

Air conditioners are rated in terms of "tons" of cooling capacity, with 1 ton equaling 12,000 Btu's per hour. The higher the Btu rating, the more heat the air conditioner can remove from the room. For a central air system, your house will need 1 ton of air-conditioning capacity for every 400 square feet of floor space if it was built before 1975; you will need 1 ton for every 650 square feet if your house was built after that date. If your house is highly energy efficient, with ceiling fans, adequate window shading, and light-colored walls and roof, you may need only 1 ton for every 1,000 square feet.

To determine the proper size of a room air conditioner, you must consider several factors: the room size, the amount of heat loss through the outside walls and ceiling, the amount of glass in the room, and the amount of sun that hits the outside walls of the room. The table below shows the estimated Btu requirements for rooms (size is expressed in square feet; multiply room length by width) with an attic above. If your room has another occupied room above it, the Btu capacity requirement will be less than that shown.

| | Room Size (sq. ft.) | | |
Btu Capacity	Exposed Walls Facing North or East	Exposed Wall Facing South	Exposed Wall Facing West
6,000	200	100	64
7,000	235	130	97
8,000	270	160	130
10,000	340	270	200
12,000	410	320	225

Rule of Thumb: Sizing Room Air Conditioners

To get a rough estimate of the size of room air conditioner needed, multiply the room's square footage by 27 to get the estimated Btu's per hour. A room that is poorly insulated or gets a lot of direct sun will need a larger unit.

Sizing Ceiling Fans

Room Size (ft.)	Minimum Fan Diameter (in.)
10 x 10	36
10 x 15	42
12 x 14	42
12 x 18	48

Sizing Bathroom Fans

To select the right size of ventilating fan for a bathroom, multiply the width of the room by its length, in feet. Then multiply the total floor area by 1.1 to get the correct measurement in cubic feet per minute (cfm). The cfm rating should appear on the fan. Also look at the sone rating, which measures the noise level of the fan—the lower the sone rating, the quieter the fan.

Comparing Light Bulb Brightness

The brightness of a light bulb is measured in lumens (1 lumen is equal to 1 standard candle); energy consumed is measured in watts. Here are some general equivalents; brands may vary.

Watts	Lumens
15	125
25	215
40	480–510
60	880
75	855–1,210
100	1,600–1,750
150	2,790

Comparing Types of Light Bulbs

Bulb Type	Lumens per Watt	Life (hrs.)
Incandescent	14–18	750–1,000+
Fluorescent (tube)	Up to 105	6,000–20,000
Compact fluorescent	Up to 105	10,000
Halogen	15–22	2,500–3,500

Energy Facts
- Only 10 to 15 percent of your electric bill goes to pay for the cost of lighting.
- The United States consumes 24 percent of the world's energy.
- In 1993, the major sources of home heating fuel in the United States were gas, 51 percent; electricity, 26 percent; oil, 12 percent; wood, 5 percent.
- The U.S. annual per capita consumption of energy is 326 million Btu's.

Appliance Energy Consumption

To determine how much energy household appliances consume, divide 1,000 by the wattage of the appliance (check the product label for the wattage). The result gives you the number of hours you can run that appliance to equal 1 kilowatt. To get an idea of how much this costs you, check a recent electricity bill to see what your utility company charges per kilowatt hour.

Here are some general ranges of electricity consumption:

Appliance	Watts
Clothes dryer	4,000–8,700
Clothes washer	400–800
Dishwasher	600–1,300
Iron	1,200–1,650
Microwave oven	800–1,500
Radio	6–12
Range/oven	8,000–16,000
Refrigerator/freezer	150–300
Television	200–400
Toaster	550–1,200
Water heater	2,000–5,000

Appliance Life Expectancies

Appliance	Average Life Expectancy (yrs.)
Air conditioner, central	12–15
Air conditioner, room	12
Blender	10
Clothes dryer	14–18
Clothes washer	11–13
Coffee maker, drip	6–8
Dishwasher	11–13
Food processor	8–10
Freezer	15–20
Furnace (gas)	20–30
Microwave oven	10–12
Range/oven	16–20
Refrigerator/freezer	16–20
Television, color	8–12
Vacuum cleaner	15
Water heater	10–12

Sizing a Water Heater

To estimate the size of water heater your family needs, use the following worksheet. First determine the *single hour of the day* when the water heater will be in most use; then fill in the blanks. The total number is the size of water heater, in gallons, that you will need.

Activity	Number		Gallons		Total Gallons
Shower	_____	x	20	=	_____
Bath	_____	x	20	=	_____
Washing hair	_____	x	4	=	_____
Shaving	_____	x	2	=	_____
Washing face and hands	_____	x	4	=	_____
Washing dishes					
by hand	_____	x	4	=	_____
by dishwasher	_____	x	14	=	_____
Total					_____

Numbers to Know: Saving on Heating and Cooling Costs

Reducing your home thermostat by 5°F in the winter for 8 hours can reduce your heating costs by up to 10 percent.

Raising your home thermostat by 1°F in the summer can reduce the energy needed for air conditioning by 3 percent.

Appliance Repair Phone Assistance

The following appliance manufacturers offer toll-free hotlines that you can call to get help with minor appliance repairs. All are free except for Sears, which charges a small fee. Have your appliance model number and date of purchase ready when you call.

Manufacturer	Phone Number	Hours
Sears	(800) 473-7247	24 hours a day, 7 days a week
Frigidaire	(800) 777-8349	8 a.m. to 8 p.m. EST, Monday–Saturday
General Electric	(800) 626-2000	24 hours a day, 7 days a week
Maytag	(800) 688-9900	8 a.m. to 5 p.m. EST, Monday–Friday
Whirlpool	(800) 253-1301	7 a.m. to 11 p.m. EST, 7 days a week

Home Hazards

Microwave Ovens. The government has set safety standards that limit levels of radiation outside a microwave oven to no more than 5 milliwatts per square centimeter at a distance of 5 centimeters from the oven door. Low-cost testers are available at electronics and some hardware stores.

Radon. If the radon level in your house is 4 pCi/L (0.02 WL) or higher, you should take steps to lower the level; call your state Radon Office for details, or call the U.S. Environmental Protection Agency's Radon Hotline at (800) SOS-RADON or its Radon Information Line at (703) 356-5346. Home testers are available at hardware stores and are easy to use. Most homes can be easily fixed to limit radon to 2 pCi/L (0.01 WL) or less.

Radon Risk Evaluation Chart

pCi/L	WL	Estimated Lung Cancer Deaths Due to Radon Exposure (per 1,000 such deaths)	Comparable Exposure Levels
200	1	440–770	1,000 times average outdoor level
100	0.5	270–360	100 times average indoor level
40	0.2	120–380	
20	0.1	60–210	
10	0.05	30–120	10 times average indoor level
4	0.02	13–15	
2	0.01	7–30	
1	0.005	3–13	Average indoor level
0.2	0.001	1–3	Average outdoor level

Source: Environmental Protection Agency, Office of Air and Radiation Programs.

Note: Measurement results are reported in either pCi/L (picocuries per liter), which measure radon gas; or WL (working levels), which measure radon decay products.

Lawn and Garden

Correct Mowing Heights

The right mowing height can make all the difference in your lawn. If you cut your grass too short, more sun can get to the surface, encouraging weeds to thrive and leaving the grass itself with insufficient food-making capacity. Cutting your grass taller encourages deeper rooting and reduces surface water evaporation. Here's how tall various types of grass should be after mowing.

Variety	Mowing Height (in.)
Bahia	$1\frac{1}{2}$–3
Bermuda	
Common	$\frac{1}{2}$–$1\frac{1}{2}$
Improved	$\frac{1}{2}$–1
Fine fescue	1–2
Kentucky bluegrass	1–2
Merion Kentucky bluegrass	$\frac{1}{2}$–1
Perennial ryegrass	$1\frac{1}{2}$–$2\frac{1}{2}$
Red fescue	$1\frac{1}{2}$–2
St. Augustine	2–3
Tall fescue	2–3
Zoysia	1–2

Rule of Thumb: Grass Mowing

Mow each time your grass grows one-third again its normal cutting height. In the hottest part of the summer, increase that to one-half again its normal cutting height.

Experts say that even during periods of slow growth, you should cut your lawn at least once every 10 days.

Plant Zone Hardiness Map

This map shows average minimum temperature ranges; the plants you buy for your yard are usually rated using these zone numbers to show their ability to withstand cold temperatures.

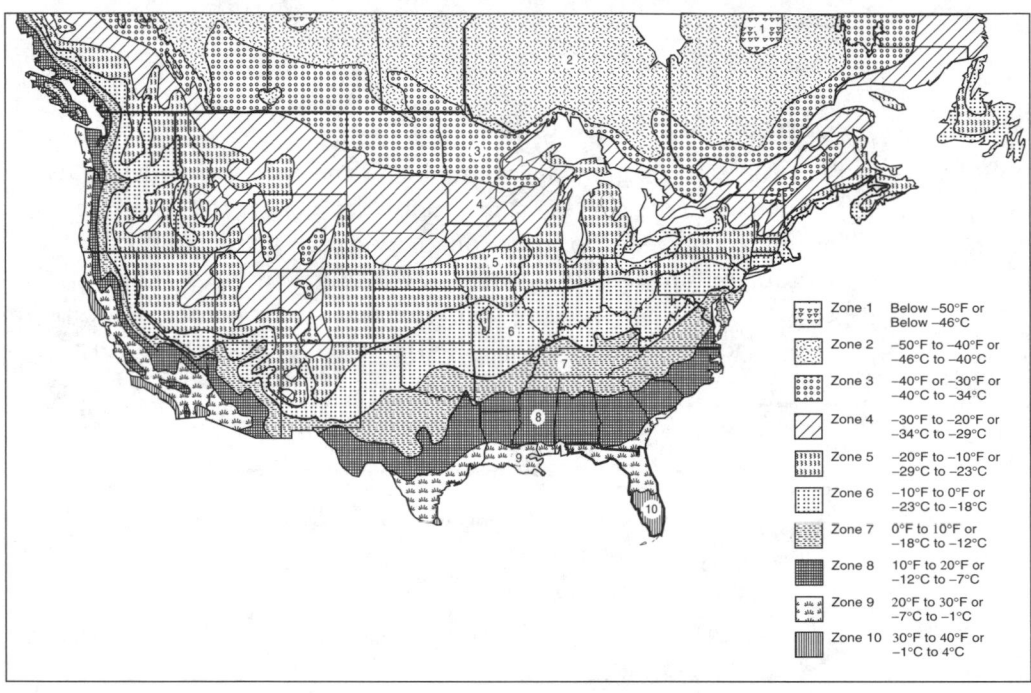

Zone 1 — Below −50°F or Below −46°C

Zone 2 — −50°F to −40°F or −46°C to −40°C

Zone 3 — −40°F or −30°F or −40°C to −34°C

Zone 4 — −30°F to −20°F or −34°C to −29°C

Zone 5 — −20°F to −10°F or −29°C to −23°C

Zone 6 — −10°F to 0°F or −23°C to −18°C

Zone 7 — 0°F to 10°F or −18°C to −12°C

Zone 8 — 10°F to 20°F or −12°C to −7°C

Zone 9 — 20°F to 30°F or −7°C to −1°C

Zone 10 — 30°F to 40°F or −1°C to 4°C

Buying by the "Yard": Mulch, Sand, and Topsoil

Many items bought for home and yard use are sold by the cubic yard. Here's how to calculate how much you'll need.

1. Find the volume of the area to be covered: Multiply length by width by depth. The result is the volume in cubic feet.

    Example: You want mulch to cover an area 10 feet long by 12 feet wide to a depth of 6 inches (or $1/2$ foot).

    $10 \times 12 \times 0.5 = 60$ cubic feet

2. Divide your result by the number of cubic feet in a cubic yard (27).

    $60 \div 27 = 2.22$ cubic yards

Thickness of Mulch Needed to Control Weeds

Material	Thickness (in.)
Compost	2–4
Grass clippings	2–3
Hay	6–8
Leaves, shredded	2–3
Peat moss	1–2
Pine bark chips	3–4
Pine needles	3–4
Newspapers, shredded	$1/2$–2
Straw	6–8

Fertilizer Numbers: What They Mean

The numbers on a bag of fertilizer represent the percentage, by weight, of the three chemical elements essential to plant growth:

First number = nitrogen (stem and leaf growth)

Second number = phosphorus (root growth)

Third number = potassium (flower, fruit, sturdiness)

For example, a 10-10-5 fertilizer contains 10 percent nitrogen, 10 percent phosphorus, and 5 percent potassium. The remaining contents are filler material.

The balance is therefore the important factor when buying fertilizer. If you want to encourage flowering, you should select a fertilizer with a high potassium percentage (the third number).

The difference between a 10-10-10 fertilizer and a 20-20-20 one is the total amount of nutrients in the bag, which is reflected in the number and frequency of applications needed and the price.

Estimating Bags of Fertilizer Needed

First determine the square footage of your lawn.

1. Start with your lot size.

   80 feet × 100 feet = 8,000 square feet

2. Determine the number of square feet that your house, other structures, driveway, and planting areas occupy.

   House = 30 × 50 feet = 1,500 square feet

   Driveway = 15 × 30 feet = 450 square feet

   Total = 1,950 square feet.

3. Subtract that result from the lot size in item 1 above.

   8,000 − 1,950 = 6,050 square feet of lawn

4. Most bags of fertilizer contain enough to cover about 5,000 square feet; check the specifications on the bag to be sure. You'll need more than one bag in this example. Calculations for other lawn products, such as mulch, lime, and weed killers, can be made in this way as well.

Soil pH

Acidity and alkalinity are measured in terms of the pH scale, which runs from 0 to 14, with 7 being neutral. Anything above 7 is alkaline; anything below 7 is acidic. Flower and vegetable gardens do well in a soil pH of 6.5; lawns and most other plants prefer a range of 6.0 to 7.0. Rhododendrons and azaleas like acid soil of 4.5 to 5.5 pH.

Tree Heights at Maturity

When selecting a tree, be sure to consider its height at maturity to prevent the problem of obstructed views or excessive shading in later years.

Tree	Mature Height (ft.)	Tree	Mature Height (ft.)
Alder	40–70	Japanese maple	15–20
American arborvitae	40–60	Lilac	20–30
American beech	40–60	Live oak	40–80
American elm	60–80	Lombardy poplar	50–80
Ash	80	Magnolia	20–25
Bailey's acacia	25–30	Mulberry	20–30
Bald cypress	50–80	Norway maple	70 90
Basswood	60–80	Norway spruce	40–80
Beech	50–100	Ohio buckeye	50–75
Black gum	40–60	Olive	25–30
California buckeye	40	Pear	20–60
Canadian hemlock	40–70	Pin oak	60–80
Catalpa	30–45	Pistachio	30–60
Cedar	40–60	Poplar	40–100
Chestnut	40–100	Red cedar	40–50
Chinese elm	40–50	Red maple	40–60
Colorado blue spruce	30–50	Red oak	60–80
Crab apple	15–25	Silver maple	60–100
Crape myrtle	15–30	Sugar maple	50–80
Dogwood	15–40	Sweet gum	25–80
Ginkgo	50–80	Sycamore	75–100
Hackberry	70–80	Tulip	100
Hawthorn	30	Walnut	30–70
Hemlock	30–70	Weeping willow	50
Hickory	60–80	White ash	60–100
Holly	10–50	White mulberry	45
Horse chestnut	20–50	White oak	60–100
Japanese cherry	25–35	Willow oak	60–80
Japanese crab	20–30	Yew	10–60

The Oldest Trees Of the 850 species of trees in the United States, the oldest is the bristlecone pine of Nevada and southern California; these trees can live for more than 4,600 years. Giant sequoias can reach an age of 2,500 years.

The oldest existing tree species in the world is the Chinese ginkgo (also known as the maidenhair), believed to have first appeared 160 million years ago.

Clothing Sizes

Babies and Children

Three kinds of sizing systems are used for babies' clothes: by age, by weight, and by general size (newborn to extra large). But since babies can vary so much in size for their age, buying by weight is the safest bet (and fit can vary by manufacturer). Toddlers' clothes come in sizes 1T to 4T, with the numbers referring generally to age.

Children's clothes come in sizes 2 to 6X; after that, boys' and girls' clothes are sized differently. Girls' clothes come in even numbers after 7/8 up to 16. Boys' sizes go from 6 to 24, in even numbers.

Women's and Men's Sizes: United States, United Kingdom, and Europe

Item	Women			Men		
	U.S.	U.K.	Europe	U.S.	U.K.	Europe
Dresses/Suits	6	8	36	38	38	51
	8	10	38	39	39	$52^1/_2$
	10	12	40	40	40	54
	12	14	42	41	41	$55^1/_2$
	14	16	44	42	42	57
Blouses/Shirts	6	30	36	15	15	38
	8	32	38	$15^1/_2$	$15^1/_2$	39
	10	34	40	16	16	41
	12	36	42	$16^1/_2$	$16^1/_2$	42
	14	38	44	17	17	43
	16	40	46	18	18	45
Shoes	5	$3^1/_2$	36	6	5	38
	$5^1/_2$	4	$36^1/_2$	7	6	$39^1/_2$
	6	$4^1/_2$	37	8	7	41
	$6^1/_2$	5	$37^1/_2$	9	8	42
	7	$5^1/_2$	38	10	9	43
	$7^1/_2$	6	$38^1/_2$	11	10	$44^1/_2$
	8	$6^1/_2$	39	12	11	46
	$8^1/_2$	7	$39^1/_2$	13	12	47
	9	$7^1/_2$	40	14	13	48

Women's Clothes: Five Kinds of Sizes

Category	Height	Figure Type	Size Range
Junior	5'2"–5'5"	Slender	3–15
Misses	5'5"–5'7"	Developed, well proportioned	4–18
Misses petite	4'8"–5'4"	Developed, well proportioned	2–16
Half	5'2"–5'4"	Fuller, rounder than misses	10½–26½
Women's	5'5"–5'8"	Fuller, rounder than misses	34–52

Shoe Size Logic

Children's shoe sizes were first based on the width of an adult hand (about 4 inches) to make the smallest size, 0; the length of the hand (a large one, obviously) of 9 inches gave the largest size, 13.

European shoe sizes are based on the metric system—each size unit equals two-thirds of a centimeter. Sizes for men and women are the same. In fact, most of the world uses this system.

Bar Codes

The Universal Product Code (UPC) is a 12-digit computer code, grouped into 4 parts. Here's what the parts mean:

The first digit identifies the product. All nationally branded products are coded 0, with the following exceptions: 2 = random weight items (e.g., cheese, meat); 3 = drugs and some health products; 4 = retailer sale items; and 5 = cents-off coupons.

The next 5 digits refer to the manufacturer, a code assigned by the Uniform Code Council in Dayton, OH.

The next 5 digits refer to the manufacturer's particular product and can include size, color, and other information.

The last digit is the check digit and is used by the computer to check the accuracy of the other digits.

Paper and Books

International Paper Sizes

Type	Inches	Millimeters
A series (writing paper, books, magazines)		
A0	33.11 x 46.81	841 x 1189
A1	23.39 x 33.1	594 x 841
A2	16.54 x 23.29	420 x 594
A3	11.69 x 16.54	297 x 420
A4	8.27 x 11.69	210 x 297
A5	5.83 x 8.27	148 x 210
A6	4.13 x 5.83	105 x 148
A7	2.91 x 4.13	74 x 105
A8	2.05 x 2.91	52 x 74
A9	1.46 x 2.05	37 x 52
A10	1.02 x 1.46	26 x 37
B series (posters)		
B0	39.37 x 55.67	1000 x 1414
B1	27.83 x 39.37	707 x 1000
B2	19.68 x 27.83	500 x 707
B3	13.90 x 19.68	353 x 500
B4	9.84 x 13.90	250 x 353
B5	6.93 x 9.84	176 x 250
B6	4.92 x 6.93	125 x 176
B7	3.46 x 4.92	88 x 125
B8	2.44 x 3.46	62 x 88
B9	1.73 x 2.44	44 x 62
B10	1.22 x 1.73	31 x 44

International Paper Sizes (continued)

Type	Inches	Millimeters
C series (envelopes)		
C0	36.00 x 51.20	917 x 1297
C1	25.60 x 36.00	648 x 917
C2	18.00 x 25.60	458 x 648
C3	12.80 x 18.00	324 x 458
C4	9.00 x 12.80	229 x 324
C5	6.40 x 9.00	162 x 229
C6	4.50 x 6.40	114 x 162
C7	3.20 x 4.50	81 x 114
DL	4.33 x 8.66	110 x 220
C7/6	3.19 x 6.38	81 x 162

U.S. Printing Paper Sizes and Weights

Grade and Basic Size (in.)	Standard Sizes (in.)	Weights (lbs.)	Thickness Range (in.)
Bond 17 x 22	8^1/$_2$ x 11, 8^1/$_2$ x 14, 11 x 17, 17 x 22, 17 x 28, 19 x 24, 19 x 28, 22 x 34, rolls	9, 12, 16, 20, 24, 28	.002–.006
Uncoated Book 25 x 38	17^1/$_2$ x 22^1/$_2$, 23 x 29, 23 x 35, 25 x 38, 35 x 45, 38 x 50, rolls	30, 32, 35, 40, 45, 50, 60, 70, 80	.003–.006
Text 25 x 38	17^1/$_2$ x 22^1/$_2$, 23 x 35, 25 x 38, 26 x 40	70, 75, 80, 100	.005–.008
Coated Book 25 x 38	19 x 25, 23 x 29, 23 x 35, 25 x 38, 35 x 45, 38 x 50, rolls	*Sheets:* 60, 70, 80, 100; *Rolls:* 40, 45, 50, 60, 70, 80, 100	.003–.007
Cover 20 x 26	20 x 26, 23 x 35, 25 x 38, 26 x 40	65, 80, 100; *Calipers:* .007, .008, .010, .012, .015	.006–.015

Estimating the Number of Words in a Manuscript

A standard $8^{1}/_{2}$-by-11-inch typewritten page, double-spaced, contains about 250 words; a single-spaced page contains close to 500. Most of today's word processing programs provide a word count. In WordPerfect, for example, the word count for a page or document is given at the end of the spell-checking program; in Microsoft Word, pressing the spelling button on the standard toolbar starts the program.

Readability Formulas

Readability formulas have been developed to measure how easily a reader can understand a document. Three types of formulas are used: the Flesch Reading Ease Formula, the Flesch-Kincaid Index, and the Gunning Fog Index. Each takes representative samples of at least 100 words from various parts of a document and measures the number of polysyllabic words and the average sentence length.

- The Flesch Reading Ease Formula: Reading ease = 206.835 − (1.015ASL + 0.846ASW), where ASL = average sentence length and ASW = average number of syllables per 100 words. A score of 90 to 100 is rated very easy (fourth-grade level), 50 to 60 is fairly difficult (some high school), and 0 to 30 is very difficult (college education).

- Flesch-Kincaid Index: GL (grade level) = 0.39NWS = 11.80NSW, where NWS = average number of words per sentence and NSW = average number of syllables per word. The result is the educational grade level required to understand the document.

- The Gunning Fog Index: Fog index = 0.4 (ASL + TSW), where ASL = average sentence length and TSW = number of trisyllabic words per 100 ("difficult" words, not proper nouns or combined words). The resulting score reflects the grade level of education that the reader would need to understand the document—that is, a score of 12 would mean that the reader would need a 12th-grade education.

The Dewey Decimal System

000	**Generalities**		**400**	**Language**
010	Bibliography		410	Linguistics
020	Library and information sciences		420	English and Old English
030	General encyclopedic works		430	Germanic languages; German
040	—		440	Romance languages; French
050	General serials and their indexes		450	Italian, Romanian, Rhaeto-Romanic
060	General organizations and museology		460	Spanish and Portuguese languages
070	News media, journalism, publishing		470	Italic languages; Latin
080	General collections		480	Hellenic languages; classical Greek
090	Manuscripts and rare books		490	Other languages
100	**Philosophy and psychology**		**500**	**Natural sciences and mathematics**
110	Metaphysics		510	Mathematics
120	Epistemology, causation, humankind		520	Astronomy and allied sciences
130	Paranormal phenomena		530	Physics
140	Specific philosophical schools		540	Chemistry and allied sciences
150	Psychology		550	Earth sciences
160	Logic		560	Paleontology; paleozoology
170	Ethics (moral philosophy)		570	Life sciences
180	Ancient, medieval, Oriental philosophy		580	Botanical sciences
190	Modern Western philosophy		590	Zoological sciences
200	**Religion**		**600**	**Technology (applied sciences)**
210	Natural theology		610	Medical sciences; medicine
220	Bible		620	Engineering and allied operations
230	Christian theology		630	Agriculture
240	Christian moral and devotional theology		640	Home economics and family living
250	Christian orders and local church		650	Management and auxiliary services
260	Christian social theology		660	Chemical engineering
270	Christian church history		670	Manufacturing
280	Christian denominations and sects		680	Manufacture for specific uses
290	Other and comparative religions		690	Buildings
300	**Social sciences**		**700**	**The arts**
310	General statistics		710	Civic and landscape art
320	Political science		720	Architecture
330	Economics		730	Plastic arts; sculpture
340	Law		740	Drawing and decorative arts
350	Public administration		750	Painting and paintings
360	Social services; association		760	Graphic arts; printmaking and prints
370	Education		770	Photography and photographs
380	Commerce, communications, transport		780	Music
390	Customs, etiquette, folklore		790	Recreational and performing arts

The Dewey Decimal System (continued)

800	Literature and rhetoric	900	Geography and history
810	American literature in English	910	Geography and travel
820	English and Old English literatures	920	Biography, genealogy, insignia
830	Literatures of Germanic languages	930	History of ancient world
840	Literatures of Romance languages	940	General history of Europe
850	Italian, Romanian, Rhaeto-Romanic literatures	950	General history of Asia; Far East
860	Spanish and Portuguese literatures	960	General history of Africa
870	Italic literatures; Latin	970	General history of North America
880	Hellenic literatures; classical Greek	980	General history of South America
890	Literatures of other languages	990	General history of other areas

The International Standard Book Number (ISBN)

The ISBN (International Standard Book Number) is used by booksellers and librarians to provide a standard code for ordering books. The number usually appears on the back cover and on the copyright page. Here is how to translate, for example, ISBN 0-06-270064-2:

Part of Code	Meaning
0	Area of origin, with 0 meaning that the book was published in an English-speaking country.
06	The publisher's assigned code. (Large publishers have small numbers, and small publishers have large numbers.)
270064	The book's title, assigned by the publisher.
2	The check number, which serves to guard against a computer's accepting an invalid number.

Music

Frequency Ranges of Instruments and Voices

Instrument	Frequency in Hz (cycles per second)		Instrument	Frequency in Hz (cycles per second)	
	Lower Limit	Upper Limit		Lower Limit	Upper Limit
Organ	32	4,186	English horn	164	934
Piano	27	4,186	Tenor trombone	82	466
Contrabassoon	29	311	Guitar	82	698
Harp	32	3,136	Clarinet	147	1,568
Double bass	41	247	Trumpet	165	932
Bass violin	41	262	Violin	196	2,093
Bass tuba	55	311	Oboe	233	1,397
Trombone	82	524	Flute	262	2,043
Bassoon	58	622	Piccolo	587	3,729
French horn	61	698			
Cello	65	659	Voice		
Bass clarinet	73	622	Bass	82	294
E-flat baritone saxophone	69	416	Baritone	110	392
B-flat tenor saxophone	104	622	Tenor	147	466
Viola	131	1,046	Alto	196	698
E-flat alto saxophone	138	831	Soprano	262	1,046

Frequencies of the Musical Scale

Note	Frequency (Hz)
C♭	261.63
C#	277.18
D	293.67
D#	311.13
E	329.63
F	349.23
F#	369.99
G	392.00
G#	415.31
A	440.00
A#	466.16
B	493.88

Musical Group Definitions

A *duet* is a composition for 2 performers who are of equal importance. A piano duet is a composition for 2 pianists, playing either on 1 piano or on 2 pianos. An instrumental duet is a composition for 2 instrumentalists, such as 2 violinists or a flutist and an oboist.

A *trio* is any composition in 3 parts. A piano trio is for 1 piano, 1 violin, and 1 cello.

A *quartet* is a composition for 4 instruments. A string quartet is for 2 violins, 1 viola, and 1 cello. A piano quartet is for a piano, a violin, a viola, and a cello. A woodwind quartet is usually for a flute, an oboe, a clarinet, and a bassoon. A brass quartet is for either 2 trumpets, a horn, and a trombone; 2 trumpets and 2 trombones; or some other combination of brass instruments. A wind quartet is for mixed brasses and woodwinds.

A *quintet* is a composition for 5 instruments. A string quintet is usually for 2 violins, 2 violas, and 1 cello. A piano quintet is for a piano combined with 2 violins, a viola, and a cello. A wind quintet is usually for a flute, an oboe, a clarinet, a horn, and a bassoon. A brass quintet is for 2 trumpets, a horn, a trombone, and a tuba.

The terms "duet," "trio," "quartet," and "quintet" may also refer to the groups who perform these compositions.

Number and Type of Instruments in a Symphony Orchestra

A symphony orchestra is made up of 98 to 112 performers, who play the following instruments:

Section/No. of Instruments	Composition
Woodwinds (16)	1 piccolo
	3 flutes
	3 oboes
	1 English horn
	3 clarinets
	1 bass clarinet
	3 bassoons
	1 contrabassoon
Brass (13 to 18)	6 to 8 French horns
	3 to 5 trumpets
	3 or 4 trombones
	1 tuba
Percussion (5 to 9)	2 to 4 timpani
	3 to 5 other percussion instruments (e.g., other drums, chimes, xylophone, cymbals)
Unbowed string instruments (2 or 3)	1 or 2 harps
	1 piano
Bowed strings (62 to 66)	16 to 18 first violins
	16 second violins
	12 violas
	10 cellos
	8 to 10 double basses

Letter and Package Size and Weight Requirements

For further information, contact the carriers at the phone numbers listed in the left-hand column.

Carrier	Size/Weight Requirements
U.S. Postal Service (800) 222-1811	Pieces must be at least 0.007" thick.
	Pieces less than $1/4$" thick must be rectangular in shape, at least $3^1/2$" high, and at least 5" long.
	Pieces greater than $1/4$" thick can be mailed even if they are less than $3^1/2$" by 5".
	Nonstandard mail is any piece that weighs 1 ounce or more and is longer than $11^1/2$", taller than $6^1/8$", and/or thicker than $1/4$", or any piece whose length divided by its height is less than 1.3 or more than 2.5. A 10-cent surcharge applies per piece, except for presort first-class and carrier route first-class pieces, which are assessed a 5-cent surcharge.
	Maximum weight for Express Mail service is 70 pounds, with maximum size of 108" in combined width and length.
United Parcel Service (800) 742-5877	Maximum size is 130", using this formula: [longest side] + [next longest side x 2] + [shortest side x 2]. Maximum weight is 70 pounds per package.
Federal Express (800) 238-5355	Maximum size is 160", using this formula: [longest side] + [next longest side x 2] + [shortest side x 2]. Maximum weight is 150 pounds per package.
Airborne Express (800) 426-2323	Maximum weight is 150 pounds per package for overnight service; there is no limit on regular service.
DHL (800) 225-5345	Maximum weight is 110 pounds per package for regular service, using DHL's formulas for actual or dimensional weight; call for information on shipping heavier packages.

Telephone Numbers and ZIP Codes

Useful Federal Government Phone Numbers

Here are some frequently needed federal office phone numbers; other chapters in this book give additional phone numbers for specific offices and agencies in particular subject areas. The General Services Administration offers a clearinghouse, the Federal Information Center, to help consumers locate the right federal agency. Call (301) 722-9000 or (800) 688-9889 for assistance.

Federal Office	Telephone
Consumer Product Safety Commission Washington, DC 20207	(800) 638-2772 (800) 638-8270 (TTY)
Environmental Protection Agency 401 M Street, SW Washington, DC 20460	(202) 260-2090
Equal Employment Opportunity Commission Office of Communications and Legislative Affairs 1801 L Street, NW Washington, DC 20507	(202) 663-4264 (800) 669-4000 (to contact local EEOC office)
Federal Communications Commission 1919 M Street, NW Washington, DC 20554	(202) 418-0200
Federal Deposit Insurance Corporation 550 17th Street, NW Washington, DC 20429	(202) 393-8400
Federal Energy Regulatory Commission 825 North Capitol Street, NE Washington, DC 20426	(202) 208-0200
Federal Reserve System 20th and C Streets, NW Washington, DC 20551	(202) 452-3000

Useful Federal Government Phone Numbers (continued)

Federal Office	Telephone
Federal Trade Commission 6th Street and Pennsylvania Avenue, NW Washington, DC 20580	(202) 326-2222
Food and Drug Administration 5600 Fishers Lane Rockville, MD 20857	(301) 443-3170
Government Printing Office Superintendent of Documents Washington, DC 20402	(202) 512-1800
Immigration and Naturalization Service U.S. Department of Justice 425 I Street, NW, Room 7116 Washington, DC 20536	(800) 755-0777
Internal Revenue Service 1111 Constitution Avenue, NW Washington, DC 20423	(800) 829-3676 (tax forms) (800) 829-1040 (information and assistance)
Interstate Commerce Commission 12th Street and Constitution Avenue, NW Washington, DC 20423	(202) 927-7119
National Labor Relations Board 1099 14th Street, NW Washington, DC 20570	(202) 273-1000
National Transportation Safety Board 490 L'Enfant Plaza, SW Washington, DC 20594	(202) 382-6600
Occupational Safety and Health Administration 200 Constitution Avenue, NW Washington, DC 20210	(202) 219-8148
Pension Benefit Guaranty Corporation Office of Coverage and Inquiries 1200 K Street, NW Washington, DC 20005	(202) 326-4000
Securities and Exchange Commission 450 5th Street, NW Washington, DC 20549	(202) 272-3100
Small Business Administration 409 3rd Street, SW Washington, DC 20416	(800) 827-5722
U.S. Postal Service 475 L'Enfant Plaza, SW Washington, DC 20260	(202) 635-5300 (current rates, fees, and services)

How to Call the President and Your U.S. Senators or Representative

The main switchboard at the White House is (202) 456-1414. The main information phone number at the U.S. Capitol for both the Senate and the House of Representatives is (202) 224-3121.

Area Codes of the United States, Canada, and Caribbean Islands by Number Code

In the early 1990s, phone companies in the United States began to face a shortage of available phone numbers, caused by the need to assign new phone numbers to computer modems, fax machines, cellular and car phones, and pagers. The solution to the problem has been to add more area codes. In the early days of long-distance calling, the middle number of every area code was either a 0 or a 1. To increase the number of new combinations, the system was modified so that area codes could have any number as a middle number; the requirement to dial "1" before any long distance number was introduced years ago to distinguish area codes from the first 3 digits of a local phone number.

The tables below include changes made in 1995; new area codes will continue to be added in the years ahead.

In the next table, if no city is listed, the area code given is for the entire state, territory, or province. To reach the information operator for a specific area code, dial 1 + the area code + 555-1212. To reach the information operator in your local area, dial 411. The table that begins on page 365 gives area codes for hundreds of towns and cities in the United States.

Area Codes of the United States, Canada, and Caribbean Islands, by Number Code

Area Code	Location	Area Code	Location
201	New Jersey (Jersey City, Newark)	316	Kansas (Emporia, Wichita)
202	District of Columbia	317	Indiana (Indianapolis, Muncie)
203	Connecticut (Fairfield and New Haven Counties)	318	Louisiana (Lafayette, Shreveport)
204	Manitoba, Canada (plus Creighton, Saskatchewan, Canada)	319	Iowa (Cedar Rapids, Dubuque)
		334	Alabama (southern)
205	Alabama (northern)	352	Florida (Gainesville)
206	Washington (Seattle)	360	Washington (western)
207	Maine	401	Rhode Island
208	Idaho	402	Nebraska (Lincoln, Omaha)
209	California (Fresno)	403	Alberta, British Columbia (Alaskan Highway), Northwest Territories, and Yukon, Canada (plus Aberfeldy, Greenstreet, and Tangleflag, Saskatchewan, Canada)
210	Texas (Laredo, San Antonio)		
212	New York City (Manhattan)		
213	California (Los Angeles)		
214	Texas (Dallas)		
215	Pennsylvania (Philadelphia)	404	Georgia (Atlanta)
216	Ohio (Akron, Cleveland)	405	Oklahoma (Lawton, Oklahoma City)
217	Illinois (Decatur, Springfield)	406	Montana
218	Minnesota (Duluth)	407	Florida (Melbourne, Orlando)
219	Indiana (Ft. Wayne, Gary)	408	California (San Jose)
281	Texas (Houston—new numbers)	409	Texas (Beaumont, Galveston)
301	Maryland (Bethesda, Frederick, Rockville)	410	Maryland (Annapolis, Baltimore)
		412	Pennsylvania (Pittsburgh)
302	Delaware	413	Massachusetts (Amherst, Springfield)
303	Colorado (Denver)	414	Wisconsin (Green Bay, Milwaukee)
304	West Virginia	415	California (San Francisco)
305	Florida (Miami)	416	Ontario, Canada (Toronto)
306	Saskatchewan, Canada (Regina, Saskatoon)	417	Missouri (Joplin, Springfield)
		418	Quebec, Canada (Quebec City)
307	Wyoming	419	Ohio (Sandusky, Toledo)
308	Nebraska (Grand Island, North Platte)	423	Tennessee (Chattanooga, Knoxville)
309	Illinois (Moline, Peoria)	441	Bermuda
310	California (Long Beach, Santa Monica)	501	Arkansas
312	Illinois (Chicago)	502	Kentucky (Bowling Green, Louisville)
313	Michigan (Ann Arbor, Detroit)	503	Oregon (Portland, Salem)
314	Missouri (Jefferson City, St. Louis)	504	Louisiana (Baton Rouge, New Orleans)
315	New York (Syracuse, Utica)	505	New Mexico
		506	New Brunswick, Canada

Area Code	Location
507	Minnesota (Rochester, Winona)
508	Massachusetts (Cape Cod, Lowell, New Bedford)
509	Washington (Spokane, Yakima)
510	California (Berkeley, Oakland)
512	Texas (Austin, Corpus Christi)
513	Ohio (Cincinnati, Dayton)
514	Quebec, Canada (Montreal)
515	Iowa (Ames, Des Moines)
516	New York (Hempstead, Long Island)
517	Michigan (Lansing, Midland, Saginaw)
518	New York (Albany, Schenectady)
519	Ontario, Canada (London)
520	Arizona (except Phoenix)
540	Virginia (western)
541	Oregon (Corvallis, Eugene, Ashland, and eastern cities)
562	California (southern—cellular, pagers)
601	Mississippi
602	Arizona (Phoenix)
603	New Hampshire
604	British Columbia, Canada (except Alaskan Highway)
605	South Dakota
606	Kentucky (Covington, Lexington)
607	New York (Binghamton, Elmira)
608	Wisconsin (La Crosse, Madison)
609	New Jersey (Atlantic City, Trenton)
610	Pennsylvania (Allentown, Easton, Reading)
612	Minnesota (Minneapolis/St. Paul, Minnetonka)
613	Ontario, Canada (Ottawa)
614	Ohio (Columbus, Steubenville)
615	Tennessee (Nashville)
616	Michigan (Grand Rapids, Kalamazoo)
617	Massachusetts (Boston)
618	Illinois (Centralia, East St. Louis)
619	California (San Diego)
630	Illinois (Chicago—cellular, pagers)
701	North Dakota
702	Nevada
703	Virginia (northern)
704	North Carolina (Asheville, Charlotte)
705	Ontario, Canada (North Bay)
706	Georgia (Augusta, Columbus)
707	California (Eureka, Santa Rosa)
708	Illinois (Evanston, Glencoe)
709	Newfoundland, Canada
712	Iowa (Council Bluffs, Sioux City)
713	Texas (Houston)
714	California (Anaheim, Orange County)
715	Wisconsin (Eau Claire, Wausau)
716	New York (Buffalo, Rochester)
717	Pennsylvania (Harrisburg, Scranton)
718	New York City (Bronx, Brooklyn, Queens, Staten Island)
719	Colorado (Colorado Springs, Pueblo)
770	Georgia (Smyrna, Marietta, Stone Mountain)
801	Utah
802	Vermont
803	South Carolina (except northwestern)
804	Virginia (Richmond, Virginia Beach)
805	California (Bakersfield)
806	Texas (Amarillo)
807	Ontario, Canada (Ft. William, Thunder Bay)
808	Hawaii
809	Caribbean Islands (including Bahamas, Puerto Rico, and Virgin Islands)
810	Michigan (Flint, Pontiac)
812	Indiana (Evansville, Terre Haute)
813	Florida (Tampa)
814	Pennsylvania (Altoona, Erie)
815	Illinois (Joliet, Rockford)
816	Missouri (Kansas City, St. Joseph)
817	Texas (Arlington, Ft. Worth)
818	California (Burbank, Pasadena)
819	Quebec, Canada (Sherbrooke)

Area Code	Location	Area Code	Location
847	Illinois (suburban Chicago)	910	North Carolina (Greensboro, Winston-Salem)
860	Connecticut (except Fairfield and New Haven Counties)	912	Georgia (Macon, Savannah)
864	South Carolina (northwestern)	913	Kansas (Kansas City, Salina, Topeka)
901	Tennessee (Jackson, Memphis)	914	New York (White Plains, Yonkers)
902	Nova Scotia and Prince Edward Island, Canada	915	Texas (Abilene, El Paso)
903	Texas (Paris, Texarkana)	916	California (Sacramento)
904	Florida (Jacksonville, Tallahassee)	917	New York (cellular, pagers)
905	Ontario, Canada (Hamilton)	918	Oklahoma (Muskogee, Tulsa)
906	Michigan (Escanaba, Upper Peninsula)	919	North Carolina (Durham, Raleigh)
907	Alaska	941	Florida (southwest)
908	New Jersey (Elizabeth, Woodbridge)	954	Florida (Fort Lauderdale)
909	California (Riverside, San Bernardino)	970	Colorado (northern)
		972	Texas (Dallas—new numbers)

Note: See note at the end of the next table for information about pending area code changes.

Special Area Codes

Certain area codes have special uses:

- 800 numbers are toll-free long-distance numbers. The owner of the number, usually a business, pays the long-distance charge for each call received, in addition to a monthly service charge. A new toll-free area code, 888, is being added.

- 900 numbers are not toll-free; you will be charged by the minute for any calls you make to a 900 number.

- 500 numbers are on the horizon; they will allow your calls to be forwarded to up to 3 locations. In the meantime, AT&T is offering 700 numbers, with which you can forward calls to 1 number.

Area Codes of the United States, Canada, and Caribbean Islands, by State, Territory, and Province

Location	Area Code
United States	
Alabama (northern)	205
Alabama (southern)	334
Alaska	907
Arizona (except Phoenix)	520
Arizona (Phoenix)	602
Arkansas	501
California (Fresno)	209
California (Los Angeles)	213
California (Long Beach, Santa Monica)	310
California (San Jose)	408
California (San Francisco)	415
California (Berkeley, Oakland)	510
California (southern—cellular, pagers)	562
California (San Diego)	619
California (Eureka, Santa Rosa)	707
California (Anaheim, Orange County)	714
California (Bakersfield)	805
California (Burbank, Pasadena)	818
California (Riverside, San Bernardino)	909
California (Sacramento)	916
Colorado (Denver)	303
Colorado (Colorado Springs, Pueblo)	719
Colorado (northern)	970
Connecticut (Fairfield and New Haven Counties)	203
Connecticut (except Fairfield and New Haven Counties)	860
Delaware	302
District of Columbia	202
Florida (Miami)	305
Florida (Gainesville)	352
Florida (Melbourne, Orlando)	407
Florida (Tampa)	813
Florida (Jacksonville, Tallahassee)	904
Florida (southwest)	941
Florida (Fort Lauderdale)	954
Georgia (Atlanta)	404
Georgia (Augusta, Columbus)	706
Georgia (Smyrna, Marietta, Stone Mountain)	770
Georgia (Macon, Savannah)	912
Hawaii	808
Idaho	208
Illinois (Decatur, Springfield)	217
Illinois (Moline, Peoria)	309
Illinois (Chicago)	312
Illinois (Centralia, East St. Louis)	618
Illinois (Chicago—cellular, pagers)	630
Illinois (Evanston, Glencoe)	708
Illinois (Joliet, Rockford)	815
Illinois (suburban Chicago)	847
Indiana (Ft. Wayne, Gary)	219
Indiana (Indianapolis, Muncie)	317
Indiana (Evansville, Terre Haute)	812
Iowa (Cedar Rapids, Dubuque)	319
Iowa (Ames, Des Moines)	515
Iowa (Council Bluffs, Sioux City)	712
Kansas (Emporia, Wichita)	316
Kansas (Kansas City, Salina, Topeka)	913
Kentucky (Bowling Green, Louisville)	502
Kentucky (Covington, Lexington)	606
Louisiana (Lafayette, Shreveport)	318
Louisiana (Baton Rouge, New Orleans)	504
Maine	207
Maryland (Bethesda, Frederick, Rockville)	301
Maryland (Annapolis, Baltimore)	410
Massachusetts (Amherst, Springfield)	413
Massachusetts (Cape Cod, Lowell, New Bedford)	508
Massachusetts (Boston)	617
Michigan (Ann Arbor, Detroit)	313
Michigan (Lansing, Midland, Saginaw)	517
Michigan (Grand Rapids, Kalamazoo)	616

Location	Area Code
Michigan (Flint, Pontiac)	810
Michigan (Escanaba, Upper Peninsula)	906
Minnesota (Duluth)	218
Minnesota (Rochester, Winona)	507
Minnesota (Minneapolis/St. Paul, Minnetonka)	612
Mississippi	601
Missouri (Jefferson City, St. Louis)	314
Missouri (Joplin, Springfield)	417
Missouri (Kansas City, St. Joseph)	816
Montana	406
Nebraska (Grand Island, North Platte)	308
Nebraska (Lincoln, Omaha)	402
Nevada	702
New Hampshire	603
New Jersey (Jersey City, Newark)	201
New Jersey (Atlantic City, Trenton)	609
New Jersey (Elizabeth, Woodbridge)	908
New Mexico	505
New York City (Manhattan)	212
New York (Syracuse, Utica)	315
New York (Hempstead, Long Island)	516
New York (Albany, Schenectady)	518
New York (Binghamton, Elmira)	607
New York (Buffalo, Rochester)	716
New York City (Bronx, Brooklyn, Queens, Staten Island)	718
New York (White Plains, Yonkers)	914
New York (cellular, pagers)	917
North Carolina (Asheville, Charlotte)	704
North Carolina (Greensboro, Winston-Salem)	910
North Carolina (Durham, Raleigh)	919
North Dakota	701
Ohio (Akron, Cleveland)	216
Ohio (Sandusky, Toledo)	419
Ohio (Cincinnati, Dayton)	513
Ohio (Columbus, Steubenville)	614

Location	Area Code
Oklahoma (Lawton, Oklahoma City)	405
Oklahoma (Muskogee, Tulsa)	918
Oregon (Portland, Salem)	503
Oregon (Corvallis, Eugene, Ashland, and eastern cities)	541
Pennsylvania (Philadelphia)	215
Pennsylvania (Pittsburgh)	412
Pennsylvania (Allentown, Easton, Reading)	610
Pennsylvania (Harrisburg, Scranton)	717
Pennsylvania (Altoona, Erie)	814
Puerto Rico	809
Rhode Island	401
South Carolina (northwestern)	864
South Carolina (except northwestern)	803
South Dakota	605
Tennessee (Chattanooga, Knoxville)	423
Tennessee (Nashville)	615
Tennessee (Jackson, Memphis)	901
Texas (Laredo, San Antonio)	210
Texas (Dallas)	214
Texas (Houston—new numbers)	281
Texas (Beaumont, Galveston)	409
Texas (Austin, Corpus Christi)	512
Texas (Houston)	713
Texas (Amarillo)	806
Texas (Arlington, Ft. Worth)	817
Texas (Paris, Texarkana)	903
Texas (Abilene, El Paso)	915
Texas (Dallas—new numbers)	972
Utah	801
Vermont	802
Virgin Islands	809
Virginia (western)	540
Virginia (northern)	703
Virginia (Richmond, Virginia Beach)	804
Washington (Seattle)	206
Washington (western)	360
Washington (Spokane, Yakima)	509

Location	Area Code	Location	Area Code
West Virginia	304	Ontario (Ottawa)	613
Wisconsin (Green Bay, Milwaukee)	414	Ontario (North Bay)	705
Wisconsin (La Crosse, Madison)	608	Ontario (Ft. William, Thunder Bay)	807
Wisconsin (Eau Claire, Wausau)	715	Ontario (Hamilton)	905
Wyoming	307	Prince Edward Island	902
		Quebec (Quebec City)	418
Canada		Quebec (Montreal)	514
Alberta	403	Quebec (Sherbrooke)	819
British Columbia (Alaskan Highway)	403	Saskatchewan (Creighton)	204
British Columbia (except Alaskan Highway)	604	Saskatchewan (Regina, Saskatoon)	306
Manitoba	204	Saskatchewan (Aberfeldy, Greenstreet, Tangleflag)	403
New Brunswick	506	Yukon	403
Newfoundland	709		
Northwest Territories	403	**Bahamas, Bermuda, and Caribbean Islands**	
Nova Scotia	902	Bahamas	809
Ontario (Toronto)	416	Bermuda	441
Ontario (London)	519	Caribbean Islands	809

Note: New area codes are being added yearly; this list was current as of late 1995. Area code changes are pending for 1996 and beyond for suburban Chicago (area code 708 to split with new area codes 630 and 847); Minnesota (area code 612 to split with new area code 320); Missouri (area code 314 to split with new area code 573); Ohio (area code 216 to split with new area code 330); British Columbia (area code 604 to split with new area code 250); Southern California (area code 310 to split with new area code 562); and the Bahamas (area code 809 to split with new area code 242). Dial "00" for area code assistance.

The Demise of the Fictional 555 Phone Exchange

For more than 30 years, characters on television and in the movies have dialed phone numbers beginning with 555 to avoid using real phone numbers; the only assigned use for the 555 prefix was to dial long-distance information (1 + (area code) + 555-1212). But in 1995, Bellcore, the research division of the Baby Bell phone companies, began assigning the 555 exchange to businesses across the country. One concession was made, however: 100 numbers—from 555-0100 to 555-0199—have been set aside as fictitious and will never be assigned to a real business.

U.S. Towns and Cities: ZIP Codes, Area Codes, and 1990 Population (for places of 5,000 or more population)

Area codes were current as of late 1995. New area codes are being added yearly; if you encounter problems in dialing to a particular location, dial "00" for operator assistance. The asterisks before ZIP codes indicate the general delivery ZIP code for those towns and cities with more than one ZIP code.

Place	ZIP Code	Area Code	1990 Population	Place	ZIP Code	Area Code	1990 Population
Alabama				Guntersville	35976	205	7,038
Alabaster	35007	205	14,619	Hamilton	35570	205	5,787
Albertville	35950	205	14,507	Hartselle	35640	205	10,867
Alexander City	35010	205	14,917	Homewood	35209	205	23,644
Andalusia	36420	334	9,269	Hoover	35236	205	40,000
Anniston	*36201	205	26,638	Hueytown	35023	205	15,280
Arab	35016	205	6,321	Huntsville	*35801	205	159,880
Athens	35611	205	16,901	Irondale	35210	205	9,458
Atmore	*36502	334	8,046	Jackson	36545	334	5,819
Attalla	35954	205	6,859	Jacksonville	36265	205	10,283
Auburn	*36830	334	33,830	Jasper	*35501	205	13,553
Bay Minette	36507	334	7,168	Lanett	36863	205	8,985
Bessemer	*35020	205	33,581	Leeds	35094	205	10,009
Birmingham	*35203	205	265,347	Madison	35758	205	14,792
Boaz	35957	205	6,928	Midfield	35228	205	5,559
Brewton	*36426	334	5,885	Millbrook	36054	205	6,046
Center Point	35215	205	22,658	Mobile	*36601	334	196,263
Chickasaw	36611	205	6,649	Monroeville	*36460	334	6,993
Clanton	35045	205	7,669	Montgomery	*36104	334	187,543
Cullman	*35055	205	13,367	Mountain Brook	35223	205	19,810
Daleville	36322	334	5,117	Muscle Shoals	35667	205	9,611
Daphne	36526	205	11,291	Northport	35476	205	17,297
Decatur	*35601	205	48,778	Opelika	*36801	334	22,122
Demopolis	36732	334	7,512	Opp	36467	334	7,011
Dothan	*36301	334	53,721	Oxford	36203	205	9,537
Enterprise	*36330	334	20,119	Ozark	*36360	334	13,030
Eufaula	*36027	334	13,220	Pelham	35124	205	9,356
Fairfield	35064	205	12,200	Pell City	35125	205	7,945
Fairhope	36532	334	8,400	Phenix City	36867	334	25,311
Florence	*35630	205	36,426	Piedmont	36272	205	5,347
Forestdale	35214	205	10,395	Pinson-Clay-Chalkville	35126	205	10,987
Fort Payne	35967	205	11,838				
Fort Rucker	36362	205	7,593	Pleasant Grove	35127	205	8,458
Fultondale	35068	205	6,400	Prattville	*36067	334	19,816
Gadsden	*35901	205	42,523	Prichard	36610	205	34,320
Gardendale	35071	205	9,251	Rainbow City	35906	205	7,667
Greenville	36037	334	7,494	Roanoke	36274	334	6,362

Place	ZIP Code	Area Code	1990 Population
Alabama (continued)			
Russellville	35653	205	7,812
Saks	36201	205	11,138
Saraland	36571	205	11,760
Satsuma	36572	205	5,194
Scottsboro	35768	205	13,786
Selma	*36701	334	23,755
Sheffield	35660	205	10,380
Southside	35901	205	5,580
Sylacauga	35150	205	12,520
Talladega	35160	205	18,175
Tallassee	36045	334	5,112
Tarrant City	35217	205	8,046
Theodore	*36582	205	6,509
Tillman's Corner	36619	205	17,988
Troy	36081	334	13,051
Trussville	35173	205	8,283
Tuscaloosa	*35401	205	77,866
Tuscumbia	35674	205	8,413
Tuskegee	36083	334	12,257
Valley	*36854	205	8,215
Vestavia Hills	35216	205	19,550
Alaska			
Anchorage	*99501	907	26,338
College	*99708	907	11,249
Eielson AFB	99702	907	5,251
Fairbanks	*99701	907	30,843
Juneau	*99801	907	26,751
Kenai	99611	907	6,327
Ketchikan	*99901	907	8,263
Kodiak	*99615	907	6,365
Ninilchik	99639	907	10,523
Sitka	99835	907	8,588
Arizona			
Apache Junction	*85220	602	18,092
Avondale	85323	602	16,182
Bisbee	85603	520	6,288
Buckeye	85326	602	4,436
Bullhead City	*86430	520	21,951
Camp Verde	86322	520	6,243
Casa Grande	*85222	520	19,076
Chandler	*85225	602	89,862
Chinle	86503	520	5,059
Coolidge	85228	520	6,934
Cottonwood	86326	520	5,918
Douglas	*85607	520	13,137

Place	ZIP Code	Area Code	1990 Population
El Mirage	85335	602	5,001
Eloy	85231	520	7,211
Flagstaff	*86004	520	45,857
Florence	85232	520	7,321
Flowing Wells	85726	520	14,013
Fountain Hills	*85266	602	10,030
Gilbert	*85234	602	29,122
Glendale	*85301	602	147,864
Globe	*85501	520	6,062
Goodyear	85338	602	6,258
Green Valley	*85622	520	13,231
Guadalupe	85283	602	5,458
Kingman	86401	520	12,722
Lake Havasu City	*86403	520	24,363
Mesa	*85201	602	288,104
Mohave Valley	86440	520	6,962
Nogales	*85621	520	19,489
Oro Valley	85737	520	6,670
Page	86040	520	6,589
Paradise Valley	85253	602	11,773
Payson	*85541	520	8,377
Peoria	*85345	602	50,675
Phoenix	*85026	602	983,403
Prescott	*86301	520	26,592
Prescott Valley	*86314	520	8,904
Safford	*85546	520	7,359
Scottsdale	*85251	602	130,075
Sedona	*86336	520	7,720
Show Low	85901	520	5,020
Sierra Vista	*85635	520	32,983
Sierra Vista Southeast	85635	520	9,237
Somerton	85350	520	5,282
South Tucson	85713	520	5,171
Sun City	*85351	602	38,126
Sun City West	85375	602	15,997
Sun Lakes	85248	602	6,578
Surprise	85374	602	7,122
Tempe	*85282	602	141,993
Tuba City	86045	520	7,323
Tucson	*85726	520	405,323
Winslow	86047	520	9,279
Yuma	*85364	520	54,923
Arkansas			
Arkadelphia	71923	501	10,014
Ashdown	71822	501	5,150
Batesville	*72501	501	9,187

Place	ZIP Code	Area Code	1990 Population	Place	ZIP Code	Area Code	1990 Population
Arkansas (continued)				Texarkana	75502	501	22,631
Bella Vista	72714	501	9,083	Trumann	72472	501	6,346
Benton	*72015	501	18,177	Van Buren	72956	501	14,899
Bentonville	72712	501	11,257	Warren	71671	501	6,455
Blytheville	72315	501	22,523	West Helena	72390	501	10,137
Bryant	72022	501	5,269	West Memphis	*72301	501	28,259
Cabot	72023	501	8,319	Wynne	72396	501	8,817
Camden	71701	501	14,701	**California**			
Clarksville	72830	501	5,833	Agoura Hills	91301	818	11,399
Conway	*72032	501	26,481	Alameda	*94501	510	73,979
Crossett	71635	501	6,282	Alamo	94507	510	12,277
Dumas	71639	501	5,520	Albany	94706	510	16,327
El Dorado	*71730	501	23,146	Alhambra	*91802	818	82,087
Fayetteville	*72701	501	42,247	Alondra Park	90249	310	12,215
Forrest City	72335	501	13,364	Alpine	*91901	619	9,695
Fort Smith	*72901	501	72,798	Altadena	*91001	818	42,658
Harrison	*72601	501	9,936	American Canyon	94589	707	7,706
Heber Springs	72543	501	5,628	Anaheim	*92803	714	266,406
Helena	72342	501	7,491	Anderson	96007	916	8,299
Hope	71801	501	9,768	Antioch	*94509	510	62,195
Hot Springs	*71901	501	32,462	Apple Valley	*92307	619	46,079
Hot Springs Village	71909	501	6,361	Aptos	*95003	408	9,061
Jacksonville	*72076	501	29,101	Arcadia	*91006	818	48,284
Jonesboro	72401	501	46,535	Arcata	95521	707	15,211
Little Rock	*72201	501	175,727	Arden-Arcade	95825	916	92,040
Magnolia	71753	501	11,151	Arroyo Grande	*93420	805	14,432
Malvern	72104	501	9,236	Artesia	*90701	310	15,464
Marianna	72360	501	6,033	Arvin	93203	805	9,286
Maumelle	72113	501	6,714	Ashland	94577	510	16,590
Mena	71953	501	5,475	Atascadero	93422	805	23,138
Monticello	71655	501	8,119	Atherton	94025	415	7,163
Morrilton	72110	501	6,551	Atwater	95301	209	22,282
Mountain Home	72653	501	9,027	Auburn	*95603	916	10,653
Newport	72112	501	7,459	August	95205	714	6,376
North Little Rock	*72114	501	61,829	Avocado Heights	91746	818	14,232
Osceola	72370	501	8,930	Azusa	91702	818	41,203
Paragould	72450	501	18,540	Bakersfield	*93302	805	174,978
Pine Bluff	*71601	501	57,140	Baldwin Park	91706	818	69,330
Pocahontas	72455	501	6,151	Banning	92220	714	20,572
Rogers	*72756	501	24,692	Barstow	*92312	619	21,472
Russellville	72801	501	21,260	Baywood-Los Osos	93402	805	14,377
Searcy	72143	501	15,180	Beale AFB	95903	916	6,912
Sherwood	72120	501	18,878	Beaumont	92223	714	9,685
Siloam Springs	72761	501	8,151	Bell	90201	213	34,365
Springdale	*72764	501	29,945	Bellflower	*90706	310	61,815
Stuttgart	72160	501	10,420	Bell Gardens	90201	213	42,315

Place	ZIP Code	Area Code	1990 Population	Place	ZIP Code	Area Code	1990 Population
California (continued)				Coachella	92236	619	16,896
Belmont	94002	415	24,165	Coalinga	93210	209	8,212
Benicia	94510	707	24,437	Colton	92324	714	40,213
Ben Lomond	95005	408	7,884	Commerce	90022	310	12,135
Berkeley	*94704	510	102,724	Compton	*90221	310	90,454
Beverly Hills	*90210	310	31,971	Concord	*94520	510	111,308
Big Bear Lake	92314	714	5,351	Corcoran	93212	209	13,360
Bloomington	92316	714	15,116	Corning	96021	916	5,870
Blythe	*92226	619	8,448	Corona	91718	714	75,943
Bonita	91902	619	12,542	Coronado	*92118	619	26,540
Boulder Creek	95006	408	6,725	Corte Madera	*94925	415	8,272
Brawley	92227	619	18,923	Costa Mesa	*92628	714	96,357
Brea	92622	714	32,873	Cotati	94631	707	5,714
Brentwood	94513	510	7,563	Country Club	94556	209	9,325
Buena Park	*90622	714	68,784	Covina	*91722	818	43,332
Burbank	*91505	818	93,649	Crestine	92325	714	8,594
Burlingame	*94010	415	26,666	Cudahy	90201	213	22,817
Calexico	*92232	619	18,633	Culver City	*90230	310	38,793
Camarillo	*93010	805	52,297	Cupertino	*95014	408	39,967
Cameron Park	95682	916	11,897	Cypress	90630	714	42,655
Campbell	95008	408	36,048	Daly City	*94015	415	92,088
Camp Pendleton North	92055	714	10,373	Dana Point	92629	714	31,896
Camp Pendleton South	92055	714	11,299	Danville	*94526	510	31,306
				Davis	95616	916	46,322
Capitola	95010	408	10,171	Del Aire	90250	310	8,040
Carlsbad	*92008	619	63,292	Delano	*93215	805	22,762
Carmichael	95608	916	48,702	Desert Hot Springs	*92240	619	11,668
Carpinteria	93013	805	13,747	Diamond Bar	*91765	714	53,672
Carson	90745	310	83,995	Dinuba	93618	209	12,743
Casa de Oro-Mt. Helix	92077	619	30,727	Dixon	95620	916	10,417
Castro Valley	*94546	510	48,619	Downey	*90241	310	91,444
Castroville	95012	408	5,272	Duarte	*91010	818	20,716
Cathedral City	*92235	619	30,085	Dublin	94568	510	23,229
Ceres	95307	209	26,413	Earlimart	93219	805	5,881
Cerritos	90703	310	53,244	East Compton	90220	310	7,967
Charter Oak	91724	818	8,858	East Hemet	92343	714	17,611
Cherryland	94541	415	11,088	East La Mirada	90638	310	9,367
Cherry Valley	92223	714	5,945	East Los Angeles	90022	310	126,379
Chico	*95926	916	39,970	East Palo Alto	94303	415	23,451
Chino	*91708	714	59,682	East Pasadena	91117	209	5,910
Chowchilla	93610	209	5,930	East Porterville	93257	805	5,790
Chula Vista	*91910	619	135,160	Edwards AFB	*93523	805	7,423
Citrus	95610	916	9,481	El Cajon	*92020	619	88,693
Citrus Heights	*95621	916	107,439	El Centro	92244	619	31,405
Claremont	91711	714	32,610	El Cerrito	94530	510	22,869
Clovis	*93612	209	50,323	Elk Grove	95624	916	17,483

Place	ZIP Code	Area Code	1990 Population	Place	ZIP Code	Area Code	1990 Population
California (continued)				Grover City	93433	805	11,602
El Monte	*91734	818	106,162	Hacienda Heights	91745	818	52,354
El Paso de Robles	*93446	310	18,583	Half Moon Bay	94019	415	8,886
El Rio	93030	805	6,419	Hanford	93230	209	30,463
El Segundo	90245	310	15,223	Hawaiian Gardens	90716	213	13,639
El Sobrante	94802	510	9,852	Hawthorne	*90250	310	71,349
El Toro	92630	714	62,685	Hayward	*94544	510	111,343
El Toro Station	92709	714	6,869	Healdsburg	95448	707	9,469
Emeryville	*94608	510	5,740	Hemet	92546	714	36,094
Encinitas	*92024	619	55,406	Hercules	94547	415	16,829
Escondido	*92025	619	108,648	Hermosa Beach	90254	310	18,219
Eureka	*95501	707	27,025	Hesperia	*92340	619	50,418
Exeter	93221	209	7,276	Highland	92346	714	34,439
Fairfax	*94930	415	6,931	Hillsborough	94010	415	10,667
Fairfield	94533	707	78,650	Hollister	*95023	408	19,318
Fair Oaks	95628	916	26,867	Home Gardens	91720	714	7,780
Fallbrook	*92028	619	22,095	Huntington Beach	*92647	714	181,519
Farmersville	93223	209	6,235	Huntington Park	90255	213	56,129
Felton	95018	408	5,350	Imperial Beach	*91932	619	26,512
Fillmore	*93015	805	11,992	Indio	*92202	619	36,850
Florence-Graham	90001	213	57,147	Inglewood	*90301	310	109,602
Florin	95828	916	24,330	Irvine	*92716	714	110,330
Folsom	95630	916	29,802	Isla Vista	93117	805	20,395
Fontana	*92335	714	87,535	Kentfield	94904	415	6,030
Foothill Farms	95841	916	17,135	Kerman	93630	209	5,448
Fort Bragg	95437	707	6,078	King City	93930	408	7,634
Fortuna	95540	707	8,788	Kingsburg	93631	209	7,245
Foster City	94404	415	28,176	La Canada Flintridge	*91011	818	19,378
Fountain Valley	92728	714	53,691	La Crescenta-Montrose	91214	818	16,968
Freedom	95019	408	8,361	Ladera Heights	90045	310	6,316
Fremont	*94537	510	173,339	Lafayette	94549	510	23,366
Fresno	*93706	209	354,091	Laguna Beach	*92607	714	23,170
Fullerton	*92634	714	114,144	Laguna Hills	*92607	714	46,731
Galt	95632	209	8,889	Laguna Niguel	92607	714	44,723
Gardena	*90247	310	49,841	La Habra	*90631	310	51,263
Garden Acres	95205	213	8,547	Lake Arrowhead	92352	714	6,539
Garden Grove	*92642	714	143,965	Lake Elsinore	*92331	714	18,316
George AFB	92394	619	5,085	Lakeside	92040	619	39,412
Gilroy	95020	408	31,487	Lakewood	*90714	310	73,553
Glen Avon	92509	714	12,663	La Mesa	*91941	619	52,911
Glendale	*91205	818	180,038	La Mirada	*90638	714	40,452
Glendora	91740	818	47,832	Lamont	93241	805	11,517
Grand Terrace	92324	714	10,946	Lancaster	93534	805	97,300
Grass Valley	*95945	916	9,048	La Palma	90623	714	15,392
Greenacres	93308	805	7,379	La Puente	91747	818	36,955
Greenfield	93927	805	7,464	Larkspur	*94939	415	11,068

Place	ZIP Code	Area Code	1990 Population	Place	ZIP Code	Area Code	1990 Population
California (continued)				Modesto	*95350	209	164,746
La Verne	91750	714	30,843	Monrovia	*91016	818	35,733
Lawndale	90260	310	27,331	Montclair	91763	714	28,434
Lemon Grove	91945	619	23,984	Montebello	90640	213	59,564
Lemoore	93245	209	13,622	Monterey	93940	408	31,954
Lennox	90304	310	22,757	Monterey Park	91754	818	60,738
Lincoln	95648	916	7,248	Moorpark	*93021	805	25,494
Linda	95901	916	13,033	Moraga	94556	510	15,987
Lindsay	93247	209	8,338	Moreno Valley	*92552	714	118,779
Live Oak	95953	916	15,212	Morgan Hill	*95037	408	23,928
Livermore	*94550	510	56,741	Morro Bay	*93442	805	9,664
Livingston	95334	209	7,317	Mountain View	*94041	415	67,365
Lodi	*95240	209	51,874	Muscoy	92405	714	7,541
Loma Linda	92354	714	18,470	Napa	*94558	707	61,865
Lomita	90717	213	19,442	National City	*91950	619	54,249
Lompoc	*93436	805	37,649	Needles	92363	619	5,191
Long Beach	*90801	310	429,321	Newark	94560	510	37,861
Los Alamitos	*90720	310	11,788	Newport Beach	*92658	714	66,643
Los Altos	*94022	415	26,599	Nipomo	93444	805	7,109
Los Altos Hills	94022	415	7,514	Norco	91760	714	23,302
Los Angeles	*90086	213	3,485,557	North Auburn	95603	916	10,301
Los Banos	93635	209	14,519	North Fair Oaks	94025	415	13,912
Los Gatos	*95030	408	27,357	North Highlands	95660	916	42,105
Lucas Valley-Marinwood	94903	415	5,982	Norwalk	90650	310	94,279
				Novato	*94947	415	47,585
Lynwood	90262	310	61,945	Oakdale	95361	209	11,978
McFarland	93250	805	7,005	Oakland	*94617	510	372,242
McKinleyville	95521	707	10,749	Oceanside	*92054	619	128,090
Madera	*93638	209	29,282	Oildale	93308	805	26,553
Manhattan Beach	90266	310	32,063	Ojai	93023	805	7,613
Manteca	95336	209	40,773	Olivehurst	95961	916	9,738
March AFB	92518	714	5,523	Ontario	*91761	714	133,179
Marina	93933	408	26,512	Opal Cliffs	95060	408	5,940
Marina Del Rey	*90292	310	7,431	Orange	*92613	714	110,658
Martinez	94553	510	31,808	Orange Cove	93646	209	5,604
Marysville	95901	916	12,324	Orangevale	95662	916	26,266
Maywood	90270	213	27,893	Orinda	94563	510	16,642
Mendota	93640	209	6,821	Orland	95963	916	5,052
Menlo Park	*94025	415	28,403	Oroville	95965	916	11,885
Mentone	92359	714	5,675	Oxnard	*93030	805	142,560
Merced	*95340	209	56,155	Pacifica	94044	415	37,670
Millbrae	94030	415	20,414	Pacific Grove	93950	408	16,117
Mill Valley	*94941	415	13,038	Palmdale	96550	805	68,946
Milpitas	*95035	408	50,690	Palm Desert	92261	619	23,252
Mira Loma	91752	714	15,786	Palm Springs	92263	619	40,144
Mission Viejo	*92690	714	72,820	Palo Alto	*94303	415	55,900

Place	ZIP Code	Area Code	1990 Population	Place	ZIP Code	Area Code	1990 Population
California (continued)				Rohnert Park	*94928	707	36,326
Palos Verdes Estates	90274	310	13,512	Rolling Hills Estates	90274	310	7,789
Paradise	*95969	916	25,401	Rosamond	93560	805	7,430
Paramount	90723	310	47,669	Roseland	95401	707	8,779
Parkway-So. Sacramento	95823	916	31,903	Rosemead	91770	818	51,638
				Rosemont	95826	916	22,851
Parlier	93648	209	7,938	Roseville	*95678	916	44,685
Pasadena	*91109	818	131,586	Rossmoor	90720	310	9,893
Patterson	95363	209	8,626	Rowland Heights	91748	818	42,647
Perris	*92572	714	21,500	Rubidoux	92509	714	24,367
Petaluma	*94952	707	43,166	Sacramento	*95814	916	369,365
Pico Rivera	*90660	310	59,177	Salinas	*93907	408	108,777
Piedmont	*94611	510	10,602	San Anselmo	*94960	415	11,735
Pinole	94564	510	17,460	San Bernardino	*92401	909	164,676
Pismo Beach	*93449	805	7,669	San Bruno	94066	415	38,961
Pittsburg	94565	510	47,607	San Buenaventura (Ventura)	*93001	805	92,557
Placentia	92670	714	41,259				
Placerville	95667	916	8,286	San Carlos	94070	415	26,382
Pleasant Hill	94523	510	31,583	San Clemente	92674	714	41,100
Pleasanton	*94566	510	50,570	San Diego	*92138	619	1,110,554
Pomona	*91769	714	131,700	San Dimas	91773	714	32,398
Porterville	93257	209	29,521	San Fernando	*91340	818	22,580
Port Hueneme	93044	805	20,322	San Francisco	*94142	415	723,959
Poway	92064	619	43,396	San Gabriel	*91778	818	37,120
Quartz Hill	*93551	805	9,626	Sanger	93657	209	16,839
Ramona	92065	619	13,040	San Jacinto	*92581	714	16,210
Rancho Cordova	95670	916	48,731	San Jose	*95113	408	782,248
Rancho Cucamonga	91730	714	101,409	San Juan Capistrano	*92690	714	26,183
Rancho Mirage	92270	619	9,778	San Leandro	*94577	510	68,223
Rancho Palos Verdes	*90274	310	41,667	San Lorenzo	94580	510	19,987
Red Bluff	96080	916	12,363	San Luis Obispo	*93401	805	41,958
Redding	*96049	916	66,462	San Marcos	*92069	619	38,974
Redlands	*92373	714	60,395	San Marino	*91108	818	12,959
Redondo Beach	*90277	310	60,167	San Mateo	*94402	415	85,610
Redwood City	*94063	415	66,072	San Pablo	94806	510	25,158
Reedley	93654	209	15,791	San Rafael	*94915	415	48,410
Rialto	*92377	714	72,395	San Ramon	94583	510	35,303
Richmond	*94802	510	86,019	Santa Ana	*92711	714	293,827
Ridgecrest	*93555	619	28,295	Santa Barbara	*93102	805	85,571
Rio Del Mar	95003	408	8,919	Santa Clara	*95050	408	93,613
Rio Linda	95673	916	9,481	Santa Clarita	*91354	805	110,690
Ripon	95366	209	7,455	Santa Cruz	*95060	408	49,711
Riverbank	95367	209	8,591	Santa Fe Springs	90670	310	15,520
Riverside	*92502	909	226,546	Santa Maria	93454	805	61,552
Rocklin	*95677	916	18,806	Santa Monica	*90401	310	86,905
Rodeo	94572	415	7,589	Santa Paula	93060	805	25,062

Place	ZIP Code	Area Code	1990 Population	Place	ZIP Code	Area Code	1990 Population
California (continued)				Tulare	*93274	209	33,249
Santa Rosa	*95402	707	113,261	Turlock	*95380	209	42,224
Santee	92071	619	52,902	Tustin	92681	714	50,689
Saratoga	95070	408	28,061	Tustin Foothills	92705	714	24,358
Sausalito	94965	415	7,152	Twentynine Palms	*92277	619	11,821
Scotts Valley	95066	408	8,667	Twentynine Palms Base	92278	619	10,606
Seal Beach	90740	310	25,098	Ukiah	95482	707	14,632
Seaside	93955	408	38,826	Union City	94587	510	53,762
Sebastopol	*95472	707	7,008	Upland	*91785	714	63,374
Selma	93662	209	14,757	Vacaville	*95687	707	71,476
Shafter	93263	805	8,409	Valinda	91744	818	18,735
Sierra Madre	*91024	818	10,762	Vallejo	*94590	707	109,199
Signal Hill	90806	310	8,371	Valle Vista	92343	714	8,751
Simi Valley	*93065	805	100,218	Vandenberg AFB	93437	805	9,846
Solana Beach	92075	619	12,956	Vandenberg Village	93436	805	5,971
Soledad	93960	408	7,161	Victorville	*92393	619	40,674
Sonoma	95476	707	8,168	View Park-Windsor Hills	90043	310	11,769
Soquel	95073	408	9,188	Villa Park	92667	714	6,299
South El Monte	91733	213	20,850	Visalia	93277	209	75,659
South Gate	90280	213	86,284	Vista	*92083	619	71,865
South Lake Tahoe	*91030	916	21,586	Walnut	*91788	714	29,105
South Oroville	95965	916	7,463	Walnut Creek	*94596	510	60,569
South Pasadena	91030	818	23,936	Walnut Park	90255	310	14,722
South San Francisco	94080	415	54,312	Wasco	93280	805	12,412
South San Gabriel	91770	213	7,700	Watsonville	95076	408	31,099
South San Jose Hills	91744	408	17,814	West Athens	90044	310	8,859
South Whittier	90605	310	49,514	West Carson	90502	213	20,143
South Yuba	95991	916	8,816	West Compton	90247	310	5,451
Spring Valley	*91979	619	55,331	West Covina	*91790	818	96,226
Stanford	94305	415	18,097	West Hollywood	90069	310	36,118
Stanton	90680	714	30,491	Westminster	92684	714	78,293
Stockton	*95208	209	210,943	Westmont	90047	213	31,044
Suisun City	94585	707	22,704	West Pittsburg	94565	510	17,453
Sun City	*92386	714	14,930	West Puente Valley	91746	818	20,254
Sunnyvale	*94086	408	117,324	West Sacramento	95691	916	28,898
Susanville	96130	916	7,279	West Whittier-Los Nietos	*90606	310	24,164
Taft	93268	805	5,902	Whittier	*90605	310	77,671
Tamalpais-Homestead Valley	94941	415	9,601	Willits	95490	707	5,027
Tehachapi	*93561	805	6,182	Willowbrook	90222	213	32,772
Temecula	*92589	909	27,177	Willows	95988	916	5,988
Temple City	91780	818	31,153	Woodlake	93286	209	5,678
Thousand Oaks	*91359	805	104,381	Woodland	*95695	916	40,230
Tiburon	94920	415	7,554	Woodside	94062	415	5,034
Torrance	*90503	310	133,107	Yorba Linda	*92686	714	52,422
Tracy	95396	209	33,558	Yreka City	96097	916	6,948

Place	ZIP Code	Area Code	1990 Population	Place	ZIP Code	Area Code	1990 Population
California (continued)				Longmont	*80501	303	51,529
Yuba City	*95991	916	27,385	Louisville	80027	303	12,363
Yucaipa	92399	714	32,824	Loveland	*80538	303	37,357
Yucca Valley	*92284	619	13,701	Montrose	*81401	970	8,854
Colorado				Northglenn	80233	303	27,195
Air Force Academy	*80840	719	9,062	Parker	80134	303	5,450
Alamosa	81101	719	7,579	Pueblo	*81003	719	98,640
Applewood	90401	970	11,069	Security-Widefield	80911	719	23,822
Arvada	*80004	303	89,218	Sherrelwood	80221	303	16,636
Aspen	*81611	970	5,049	Southglenn	80122	303	43,087
Aurora	*80017	303	222,103	Steamboat Springs	*80477	970	6,695
Boulder	*80302	303	83,295	Sterling	80751	970	10,362
Brighton	80601	303	14,203	Stratmoor	80906	719	5,854
Broomfield	*80020	303	24,638	Thornton	80229	303	55,031
Canon City	*81212	719	12,687	Trinidad	81082	719	8,580
Castle Rock	80104	303	8,710	Welby	80229	303	10,218
Cherry Hills Village	80110	303	5,245	Westminster	80030	303	74,619
Cimarron Hills	81220	719	11,160	Westminster East	80221	303	5,197
Clifton	81520	970	12,671	Wheat Ridge	*80033	303	29,419
Colorado Springs	*80903	719	280,430	Windsor	80550	970	5,062
Columbine	80120	970	23,969	**Connecticut**			
Commerce City	*80022	303	16,466	Ansonia	06401	203	18,403
Cortez	81321	303	7,284	Avon	06001	860	13,937
Craig	*81625	303	8,091	Beacon Falls	06403	203	5,083
Denver	*80202	303	467,610	Berlin	06037	860	16,787
Durango	*81301	970	12,439	Bethel	06801	203	17,541
Englewood	*80110	303	29,396	Bloomfield	06002	860	19,483
Evans	80620	970	5,876	Branford	06405	203	27,603
Evergreen	80439	303	7,582	Bridgeport	*06602	203	141,686
Federal Heights	80221	303	9,342	Bristol	*06010	860	60,640
Fort Carson	80913	719	11,309	Brookfield	06804	203	14,113
Fort Collins	*80525	970	87,491	Brooklyn	06234	203	6,681
Fort Lupton	80621	970	5,159	Burlington	06013	860	7,026
Fort Morgan	80701	970	9,068	Canton	06019	860	8,268
Fountain	80817	719	10,175	Central Manchester	06040	860	30,934
Gateway	81522	970	7,510	Cheshire	06410	203	25,684
Glenwood Springs	81601	970	6,561	Clinton	06413	860	12,767
Golden	*80401	303	13,127	Colchester	06415	860	10,980
Grand Junction	*81501	970	29,255	Coventry	06238	860	10,063
Greeley	*80631	970	60,454	Cromwell	06416	860	12,286
Greenwood Village	*80111	303	7,589	Danbury	*06810	203	65,585
Lafayette	80026	303	14,708	Darien	06820	203	18,130
La Junta	81050	719	7,678	Derby	06418	203	12,199
Lakewood	80215	303	126,475	East Haddam	06423	860	6,676
Lamar	81052	719	8,343	East Hampton	06424	860	10,428
Littleton	*80120	303	33,711	East Hartford	*06101	860	50,452

Place	ZIP Code	Area Code	1990 Population	Place	ZIP Code	Area Code	1990 Population
Connecticut (continued)				North Haven	06473	203	22,249
East Haven	06512	203	26,144	Norwalk	*06856	203	78,331
East Lyme	06333	860	15,340	Norwich	06360	860	37,391
East Windsor	06016	860	10,081	Oakville	06779	203	8,741
Ellington	06029	860	11,197	Old Lyme	06371	860	6,535
Enfield	*06082	860	45,532	Old Saybrook	06475	860	9,552
Essex	06426	860	5,904	Orange	06477	203	12,830
Fairfield	06430	203	53,418	Oxford	06483	203	8,685
Farmington	*06032	860	20,608	Pawcatuck	02891	860	5,289
Glastonbury Center	06033	860	7,082	Plainfield	06374	860	14,363
Granby	06035	860	9,369	Plainville	06062	860	17,392
Greenwich	*06830	203	58,441	Plymouth	06782	860	11,822
Groton	06340	860	45,144	Portland	06480	860	8,418
Groton Borough	06340	860	9,837	Preston	06360	860	5,006
Guilford	06437	203	19,848	Prospect	06712	203	7,775
Haddam	06438	860	6,769	Putnam	06260	860	6,835
Hamden	*06514	203	52,434	Redding	06875	203	7,927
Hartford	*06101	860	139,739	Ridgefield Center	06877	203	6,363
Harwinton	06791	860	5,228	Ridgefield	06877	203	20,919
Hazardville	06082	860	5,179	Rocky Hill	06067	860	16,554
Hebron	06248	860	7,079	Seymour	06483	203	14,288
Kensington	06037	860	8,306	Shelton	06484	203	35,418
Lebanon	06249	860	6,041	Simsbury	06070	860	22,023
Ledyard	06339	860	14,913	Somers	06071	860	9,108
Litchfield	06759	860	8,365	Southbury	06488	203	15,818
Madison	06443	203	15,485	Southington	06489	860	38,518
Manchester	06040	860	51,618	South Windsor	06074	860	22,090
Mansfield	06250	860	21,103	Stafford	06075	860	11,091
Marlborough	06447	860	5,535	Stamford	*06904	203	108,056
Meriden	*06450	203	59,479	Stonington	06378	860	16,919
Middlebury	06762	203	6,145	Storrs	06268	860	12,198
Middletown	06457	860	42,762	Stratford	06497	203	49,389
Milford	06460	203	49,938	Suffield	06078	860	11,427
Monroe	06468	203	16,896	Terryville	06786	860	5,426
Montville	06353	860	16,673	Thomaston	06787	860	6,947
Naugatuck	06770	203	30,625	Thompson	06277	860	8,668
New Britain	*06050	860	75,491	Thompsonville	06082	860	8,458
New Canaan	06840	203	17,864	Tolland	06084	860	11,001
New Fairfield	06810	203	12,911	Torrington	06790	860	33,687
New Hartford	06057	860	5,769	Trumbull	06611	203	32,000
New Haven	*06510	203	130,474	Vernon	06066	860	29,841
Newington	*06111	860	29,208	Wallingford	06492	203	40,822
New London	06320	860	28,540	Waterbury	*06701	203	108,961
New Milford	06776	203	23,629	Waterford	06385	860	17,930
Newtown	06470	203	20,779	Watertown	06795	860	20,456
North Branford	06471	203	12,996	West Hartford	06107	860	60,110

Place	ZIP Code	Area Code	1990 Population	Place	ZIP Code	Area Code	1990 Population
Connecticut (continued)				Bayonet Point	34667	941	21,860
West Haven	06516	203	54,021	Bayshore Gardens	33505	813	17,062
Westbrook	06498	860	5,414	Beacon Square	33589	813	6,265
Weston	06883	203	8,648	Bellair-Meadowbrook Terrace	32073	813	15,606
Westport	*06880	203	24,407	Belle Glade	33430	407	16,177
Wethersfield	06109	860	25,651	Belleview	32526	904	19,386
Willimantic	06226	860	14,746	Beverly Hills	34464	904	6,163
Willington	06279	860	5,979	Boca Raton	*33431	407	61,486
Wilton	06897	203	15,989	Bonita Springs	*33923	813	13,600
Winchester	06094	860	11,524	Boynton Beach	*33436	407	46,284
Windham	06280	860	22,039	Bradenton	*34206	813	43,769
Windsor	06095	860	27,817	Brandon	*33509	941	57,985
Windsor Locks	06096	860	12,358	Brent	32503	904	21,624
Winsted	06098	860	8,254	Broadview Park	33317	305	6,109
Wolcott	06716	860	13,700	Broadview-Pompano Park	33313	305	5,230
Woodbridge	06525	203	7,924	Brooksville	*34601	904	7,589
Woodbury	06798	860	8,131	Brownsville	33142	813	15,607
Woodstock	06281	860	6,008	Callaway	32404	904	12,253
Delaware				Cape Canaveral	32920	407	8,014
Brookside	19713	302	15,307	Cape Coral	*33990	813	74,991
Claymont	19703	302	9,800	Carol City	33055	305	53,331
Dover	*19901	302	27,630	Casselberry	*32707	407	18,849
Edgemoor	19809	302	5,853	Century Village	33401	305	8,363
Elsmere	19805	302	5,935	Clearwater	*34618	941	98,746
Milford	19963	302	6,032	Clermont	*32711	904	6,910
Newark	*19711	302	26,463	Clewiston	33440	813	6,085
Pike Creek	19800	302	10,163	Cocoa	32923	407	17,710
Seaford	19973	302	5,689	Cocoa Beach	*32931	407	12,123
Smyrna	19977	302	5,231	Cocoa West	32922	407	6,160
Stanton	19804	302	5,028	Coconut Creek	33063	305	27,269
Talleyville	19803	302	6,346	Collier Manor-Cresthaven	33064	305	7,322
Wilmington	*19899	302	71,529	Conway	32809	407	13,159
Wilmington Manor	19720	302	8,568	Cooper City	33328	305	21,335
District of Columbia				Coral Gables	33114	305	40,091
Washington	*20090	202	606,900	Coral Springs	33077	954	78,864
Florida				Crestview	32536	904	9,886
Altamonte Springs	*32714	407	35,167	Crystal Lake	33803	813	5,300
Apollo Beach	33572	941	6,025	Cutler	33157	305	16,201
Apopka	*32712	407	13,611	Cutler Ridge	33157	305	21,268
Arcadia	33821	813	6,488	Cypress Gardens	33884	813	9,188
Atlantic Beach	32233	904	11,636	Dade City	*33525	904	5,633
Auburndale	33823	813	8,846	Dania	33004	954	13,183
Aventura	33280	305	14,914	Davie	33314	954	47,143
Avon Park	33825	813	8,078	Daytona Beach	*32114	904	61,991
Azalea Park	32857	407	8,926				
Bartow	33830	813	14,716				

Place	ZIP Code	Area Code	1990 Population	Place	ZIP Code	Area Code	1990 Population
Florida (continued)				Homestead AFB	33039	305	5,153
De Bary	32713	407	7,176	Homosassa Springs	34447	904	6,271
Deerfield Beach	*33441	954	46,997	Hudson	*34667	813	7,344
DeFuniak Springs	32433	904	5,200	Immokalee	33934	813	14,120
De Land	*32720	904	16,622	Indian Harbour Beach	32937	407	6,933
Delray Beach	*33444	407	47,184	Inverness	*34450	904	5,797
Del Rio	33617	813	8,248	Jacksonville Beach	32250	904	17,839
Deltona	*32725	407	50,828	Jacksonville	*32203	904	635,230
Destin	*32541	904	8,090	Jensen Beach	*34957	407	9,884
Dunedin	*34698	941	34,427	Jupiter	*33458	407	24,907
East Lake-Orient Park	33610	813	6,171	Kendale Lakes	33183	305	48,524
East Naples	33940	813	22,951	Kendall	33256	305	87,271
Edgewater	*32132	904	15,351	Key Biscayne	33149	305	8,854
Eglin AFB	32542	904	8,347	Key Largo	33037	305	11,336
Elfers	34680	941	12,356	Key West	*33040	305	24,832
Englewood	*34223	813	15,025	Kissimmee	*34744	407	30,337
Ensley	32534	904	16,362	Lake City	*32055	904	9,626
Eustis	*32726	904	12,856	Lakeland	*33804	813	70,576
Fernandina Beach	*32034	904	8,765	Lakeland Highlands	33801	813	9,972
Fern Park	32730	407	8,294	Lake Mary	*32746	407	5,929
Ferry Pass	32514	904	26,301	Lake Park	33403	407	6,704
Florida City	33034	305	5,978	Lake Wales	*33853	813	9,670
Forest City	32714	407	10,638	Lake Worth	*33461	407	28,564
Fort Lauderdale	*33310	954	149,238	Land O'Lakes	34639	941	7,892
Fort Myers	*33902	813	44,947	Lantana	33462	407	8,392
Fort Myers Beach	*33931	813	9,284	Largo	*34640	941	65,910
Fort Pierce	*34981	407	36,830	Lauderdale Lakes	33313	954	27,341
Fort Pierce North	33452	407	5,833	Lauderhill	33313	954	49,015
Fort Walton Beach	*32548	904	21,407	Laurel	34272	813	8,245
Gainesville	*32602	352	85,075	Lealman	33714	813	21,748
Gibsonton	33534	941	7,706	Leesburg	*34748	904	14,783
Gifford	32960	407	6,278	Lehigh Acres	*33936	813	13,611
Goldenrod	32733	407	12,362	Leisure City	33033	305	19,379
Gonzalez	32560	904	7,669	Lighthouse Point	33074	954	10,378
Greenacres City	33463	407	18,683	Live Oak	32060	904	6,332
Gulf Breeze	*32561	904	5,530	Lockhart	32860	407	11,636
Gulfport	33707	941	11,709	Longboat Key	34228	813	5,937
Haines City	33844	813	11,683	Longwood	*32750	407	13,316
Hallandale	*33009	305	30,997	Lutz	33549	941	10,552
Hialeah	*33010	305	188,008	Lynn Haven	32444	904	9,270
Hialeah Gardens	33016	305	7,727	Maitland	*32751	407	8,932
Hobe Sound	*33455	407	11,507	Marathon	33050	305	8,857
Holiday	*34690	941	19,360	Margate	33063	954	42,985
Holly Hill	32117	904	11,141	Marianna	*32446	904	6,292
Hollywood	*33022	954	121,720	Melbourne	*32901	407	60,034
Homestead	*33030	305	26,694	Melrose Park	32666	904	6,477

Place	ZIP Code	Area Code	1990 Population	Place	ZIP Code	Area Code	1990 Population
Florida (continued)				Palatka	*32177	904	10,447
Merritt Island	*32953	407	32,886	Palm Bay	*32906	407	62,543
Miami	*33101	305	358,648	Palm Beach	33480	407	9,814
Miami Beach	33109	305	92,639	Palm Beach Gardens	33403	407	22,990
Miami Gardens-Utopia-Carver	33023	305	7,448	Palmetto	*34221	813	9,268
				Palmetto Estates	33157	305	12,293
Miami Lakes	33014	305	12,750	Palm Harbor	*34683	941	50,256
Miami Shores	33153	305	10,084	Palm River-Clair Mel	33619	813	13,691
Miami Springs	33166	305	13,268	Palm Springs	*33460	407	9,763
Milton	32570	904	7,216	Palm Springs North	33012	407	5,300
Mims	32754	407	9,412	Panama City	*32401	904	34,396
Miramar	33023	954	40,663	Pembroke Pines	33029	954	65,566
Mount Dora	32757	904	7,316	Pensacola	*32502	904	59,198
Myrtle Grove	32526	904	17,402	Perrine	33257	305	15,576
Naples	*33940	813	19,505	Perry	32347	904	7,151
Naples Park	33940	813	8,002	Pine Castle	32859	407	8,276
Naranja	33092	305	5,790	Pine Hills	32858	407	35,322
Neptune Beach	32266	904	6,816	Pinellas Park	*34665	941	43,571
New Port Richey	*34653	941	14,044	Plantation	33318	954	66,814
New Port Richey East	33552	941	9,683	Plant City	*33566	813	22,754
New Smyrna Beach	32168	904	16,549	Pompano Beach	*33060	954	72,411
Niceville	*32578	904	10,509	Pompano Beach Highlands	33064	954	17,915
Norland	33269	305	22,109				
North Andrews Gardens	33308	305	9,002	Port Charlotte	33952	813	41,535
North Fort Myers	33918	813	30,027	Port Orange	32124	904	35,399
North Miami	33961	305	50,001	Port St. Lucie	34985	407	55,761
North Miami Beach	33160	305	35,361	Princeton	33032	305	7,073
North Naples	33940	813	13,422	Punta Gorda	*33950	813	10,637
North Palm Beach	33408	407	11,284	Quincy	*32351	904	7,452
North Port	34287	813	11,973	Riverview	33569	941	6,478
Oakland Park	33334	954	26,326	Riviera Beach	33419	407	27,646
Oak Ridge	33860	813	15,388	Rockledge	*32955	407	16,023
Ocala	*34478	904	42,045	Ruskin	*33570	941	6,046
Ocean City	32548	904	5,422	Safety Harbor	34695	813	15,120
Ocoee	32761	407	12,778	Saint Augustine	*32084	904	11,695
Oldsmar	34677	941	8,361	Saint Cloud	*34769	407	12,684
Opa-Locka	*33054	305	15,283	Saint Petersburg	*33733	941	240,318
Opa-Locka North	33054	305	6,568	Saint Petersburg Beach	33736	941	9,200
Orange Park	*32073	904	9,488	Sandalfoot Cove	33432	305	14,214
Orlando	*32802	407	164,674	Sanford	*32771	407	32,387
Orlo Vista	32861	407	5,990	Sanibel	33957	813	5,468
Ormond Beach	*32174	904	29,721	Sarasota	*34236	813	50,897
Ormond By-The-Sea	32074	904	8,157	Sarasota Springs	33577	813	16,088
Oviedo	*32765	407	11,114	Satellite Beach	32937	407	9,889
Pace	32571	904	6,277	Sebastian	*32958	407	10,248
Pahokee	33476	407	6,822	Sebring	*33871	813	8,841

Place	ZIP Code	Area Code	1990 Population	Place	ZIP Code	Area Code	1990 Population
Florida (continued)				Wilton Manors	33305	305	11,804
Seffner	33584	941	5,371	Winter Garden	*34787	407	9,863
Seminole	*34642	813	9,251	Winter Haven	*33880	813	24,725
Siesta Key	33578	813	7,772	Winter Park	*32789	407	22,623
Sky Lake	32809	407	6,202	Winter Springs	*32708	407	22,151
South Apopka	32703	407	6,360	Wright	32547	904	18,945
South Bradenton	33505	813	20,398	Zephyrhills	33540	813	8,220
South Daytona	32121	904	12,488	**Georgia**			
South Miami	33243	305	10,404	Adel	31620	912	5,093
South Miami Heights	33157	305	30,030	Albany	*31706	912	78,804
South Pasadena	33707	813	5,644	Americus	31709	912	16,516
South Patrick Shores	32937	407	10,249	Athens	*30603	706	45,734
South Venice	33595	813	11,951	Atlanta	*30301	404	393,929
Springfield	32401	904	8,719	Augusta	*30903	706	44,707
Spring Hill	*34606	904	31,117	Bainbridge	31717	912	10,803
Starke	32091	904	5,226	Belvedere Park	30032	770	18,089
Stuart	*34994	407	11,936	Blakely	31723	912	5,595
Sun City Center	33573	941	8,326	Brunswick	*31520	912	16,433
Sunrise	*33322	305	65,683	Buford	30518	770	8,771
Sweetwater	33144	305	13,909	Cairo	31728	912	9,035
Tallahassee	*32301	904	124,773	Calhoun	*30701	706	7,135
Tamarac	33320	954	44,822	Camilla	31730	912	5,124
Tamiami	33144	305	33,845	Candler-McAfee	30032	770	29,491
Tampa	*33602	813	280,015	Carrollton	*30117	770	16,029
Tarpon Springs	*34689	941	17,874	Cartersville	30120	770	12,037
Tavares	32778	904	7,383	Cedartown	30125	770	7,976
Temple Terrace	33687	813	16,444	College Park	30337	770	20,645
Titusville	*32780	407	39,394	Columbus	*31908	706	178,681
Town 'n' Country	32505	813	60,946	Conley	30027	770	5,528
Treasure Island	33706	941	7,266	Conyers	*30208	770	7,380
Union Park	32867	407	6,890	Cordele	31015	912	10,833
University West	33620	904	23,760	Dalton	*30720	706	22,218
Valparaiso	32580	904	6,316	Dawson	31742	912	5,295
Venice	*34285	813	17,052	Decatur	*30030	404	17,304
Venice Gardens	33595	813	7,701	Dock Junction	31520	912	7,094
Vero Beach	*32960	407	17,350	Doraville	30362	404	7,626
Vero Beach South	32960	407	16,973	Douglas	31533	912	10,464
Vilas	33901	813	9,898	Douglasville	*30134	770	11,635
Warrington	32507	904	16,040	Druid Hills	30333	770	12,174
Wekiva Springs	32703	407	23,026	Dublin	*31021	912	16,312
Westchester	33155	305	29,883	Dunwoody	30356	770	26,302
West Little River	33138	305	33,575	Eastman	31023	912	5,241
West Melbourne	32904	407	8,398	East Point	30364	404	34,595
West Miami	33144	305	5,727	Elberton	30635	706	4,973
West Palm Beach	*33406	407	67,764	Evans	30809	770	13,713
West Pensacola	32505	904	22,107	Fair Oaks	30060	770	6,996

Place	ZIP Code	Area Code	1990 Population	Place	ZIP Code	Area Code	1990 Population
Georgia (continued)				Scottdale	30079	770	8,636
Fairview	30535	770	6,444	Smyrna	*30080	770	30,981
Fitzgerald	31750	912	8,901	Snellville	30278	770	12,084
Forest Park	*30050	770	16,958	Statesboro	*30458	912	15,854
Fort Benning South	31905	770	14,617	Stone Mountain	*30086	770	6,544
Fort Gordon	30905	770	9,140	Summerville	30747	706	5,025
Fort Oglethorpe	30742	770	5,880	Swainsboro	30401	912	7,361
Fort Stewart	31314	912	13,774	Sylvester	31791	912	6,023
Fort Valley	31030	912	8,198	Thomaston	30286	706	9,127
Gainesville	*30501	770	17,885	Thomasville	*31792	912	17,554
Garden City	31408	912	7,410	Thomson	30824	706	6,862
Georgetown	31754	912	5,554	Tifton	*31794	912	14,215
Gresham Park	30316	770	9,000	Toccoa	30577	706	8,720
Griffin	*30223	770	21,325	Tucker	*30084	770	25,781
Hapeville	30354	770	5,483	Union City	30291	770	8,887
Hinesville	31313	912	21,596	Valdosta	*31603	912	40,038
Jesup	31545	912	8,958	Vidalia	30474	912	11,118
Kennesaw	30144	770	8,936	Warner Robins	*31088	912	43,861
Kingsland	31548	912	5,474	Waycross	*31501	912	16,410
La Fayette	30728	706	6,313	Waynesboro	30830	706	5,669
La Grange	*30240	706	25,574	West Augusta	30901	706	27,637
Lawrenceville	*30245	770	17,250	Wilmington Island	31410	912	11,230
Lithia Springs	30057	770	11,403	Winder	30680	770	7,373
Mableton	30059	770	25,725	**Hawaii**			
Macon	*31201	912	107,365	Aiea	96706	808	8,906
Marietta	*30060	770	44,129	Ewa Beach	96706	808	14,315
Martinez	30917	770	33,731	Hilo	96720	808	37,808
Milledgeville	31061	912	17,727	Honolulu	*96815	808	365,272
Monroe	30655	770	9,759	Kahului	96732	808	16,889
Moultrie	*31768	912	14,865	Kailua	96863	808	36,818
Mountain Park	30087	770	11,025	Kaneohe	96744	808	35,448
Newnan	*30263	770	12,497	Kapaa	96746	808	8,149
Norcross	*30071	770	5,947	Kihei	96753	808	11,107
North Atlanta	30310	770	27,812	Lahaina	96761	808	9,073
North Decatur	30033	770	13,936	Laie	96762	808	5,577
North Druid Hills	30033	770	14,170	Lihue	96766	808	5,536
Peachtree City	30269	770	19,027	Makawao	96768	808	5,405
Perry	31069	912	9,452	Mililani Town	96789	808	29,359
Quitman	31643	912	5,292	Nanakuli	96792	808	9,575
Riverdale	*30274	404	9,455	Pearl City	96782	808	30,993
Rome	*30161	706	30,325	Pukalani	96788	808	5,879
Roswell	*30077	770	47,986	Wahiawa	96786	808	17,386
Saint Simons Island	31522	912	12,026	Waianae	96792	808	8,758
Sandersville	31082	912	6,290	Wailuku	96793	808	10,688
Sandy Springs	30358	404	67,842	Waimea	96796	808	5,972
Savannah	*31402	912	137,812	Waipahu	96797	808	31,435

Place	ZIP Code	Area Code	1990 Population
Idaho			
Ammon	83401	208	5,002
Blackfoot	83221	208	9,646
Boise City	*83707	208	125,551
Burley	83318	208	8,702
Caldwell	*83605	208	18,400
Chubbuck	83202	208	7,794
Coeur D'Alene	*83814	208	24,561
Garden City	83714	208	6,369
Idaho Falls	*83402	208	43,973
Jerome	83338	208	6,529
Lewiston	83501	208	28,082
Meridian	*83642	208	9,596
Moscow	83843	208	18,398
Mountain Home	83647	208	7,913
Mountain Home AFB	83648	208	5,936
Nampa	*83651	208	28,365
Payette	83661	208	5,672
Pocatello	*83201	208	46,117
Post Falls	83854	208	7,349
Rexburg	83440	208	14,298
Rupert	83350	208	5,455
Sandpoint	83864	208	5,203
Twin Falls	*83301	208	27,634
Illinois			

In 1996, towns and cities in the suburban Chicago area (area code 708) will be assigned new area codes: 847 and 630. Call the local operator for assistance.

Place	ZIP Code	Area Code	1990 Population
Addison	60101	708	32,053
Algonquin	60102	708	11,693
Alsip	60658	708	18,227
Alton	62002	618	33,064
Antioch	60002	708	6,105
Arlington Heights	*60005	708	75,463
Aurora	*60505	708	99,556
Barrington	*60010	708	9,538
Bartlett	60103	708	19,395
Bartonville	61607	309	5,671
Batavia	60510	708	17,076
Beardstown	62618	217	5,270
Belleville	*62220	618	42,806
Bellwood	60104	708	20,241
Belvidere	61008	815	15,962
Bensenville	60106	708	17,767
Benton	62812	618	7,216
Berkeley	60163	708	5,137
Berwyn	60402	708	45,426

Place	ZIP Code	Area Code	1990 Population
Bethalto	62010	618	9,507
Bloomingdale	60108	708	16,614
Bloomington	*61701	309	51,889
Blue Island	60406	708	21,203
Bolingbrook	60440	708	40,843
Boulder Hill	60538	708	8,894
Bourbonnais	60914	815	13,929
Bradley	60915	815	10,918
Bridgeview	60455	708	14,402
Broadview	60153	708	8,538
Brookfield	60513	708	18,876
Buffalo Grove	60089	708	36,417
Burbank	60459	708	27,600
Burr Ridge	60521	708	7,684
Cahokia	62206	618	17,550
Calumet City	60409	708	37,840
Calumet Park	60643	708	8,418
Canton	61520	309	13,959
Carbondale	*62901	618	27,033
Carlinville	62626	217	5,416
Carmi	62821	618	5,626
Carol Stream	*60188	708	31,759
Carpentersville	60110	708	23,049
Cary	60013	708	10,043
Centralia	62801	618	14,274
Centreville	62206	618	7,489
Champaign	*61821	217	63,502
Charleston	61920	217	20,398
Chatham	62629	217	6,074
Chester	62233	618	8,204
Chicago	*60607	312	2,783,726
Chicago Heights	60411	708	32,966
Chicago Ridge	60415	708	13,643
Chillicothe	61523	309	5,959
Cicero	60650	708	67,436
Clarendon Hills	60514	708	6,994
Clinton	61727	217	7,437
Collinsville	62234	618	22,424
Columbia	62236	618	5,524
Country Club Hills	60478	708	15,431
Countryside	60525	708	5,961
Crest Hill	60435	815	10,999
Crestwood	60445	708	10,823
Crete	60417	708	6,773
Creve Coeur	61611	309	5,938
Crystal Lake	*60014	815	24,696

Place	ZIP Code	Area Code	1990 Population	Place	ZIP Code	Area Code	1990 Population
Illinois (continued)				Hanover Park	60103	708	32,918
Danville	*61832	217	33,828	Harrisburg	62946	618	9,318
Darien	60561	708	18,148	Harvard	60033	815	5,975
Decatur	*62525	217	83,900	Harvey	60426	708	29,771
Deerfield	60015	708	17,327	Harwood Heights	60656	312	7,680
De Kalb	60115	815	35,076	Hazel Crest	60429	708	13,334
Des Plaines	*60018	708	53,414	Herrin	62948	618	10,857
Dixon	61021	815	15,134	Hickory Hills	60457	708	13,021
Dolton	60419	708	23,956	Highland	62249	618	7,546
Downers Grove	60515	708	46,845	Highland Park	60035	708	30,575
Du Quoin	62832	618	6,697	Highwood	60040	708	5,331
East Alton	62024	618	7,063	Hillside	60162	708	7,672
East Moline	61244	309	20,147	Hinsdale	*60521	708	16,029
East Peoria	61611	309	21,378	Hoffman Estates	60195	708	46,363
East St. Louis	*62201	618	40,944	Homewood	60430	708	19,278
Edwardsville	62025	618	14,582	Hoopeston	60942	217	5,871
Effingham	62401	217	11,927	Inverness	60067	708	6,516
Elgin	*60120	708	77,010	Itasca	60143	708	6,947
Elk Grove Village	*60009	708	33,429	Jacksonville	*62650	217	19,327
Elmhurst	60126	708	42,029	Jerseyville	62052	618	7,382
Elmwood Park	60635	708	23,206	Joliet	*60436	815	77,217
Evanston	*60201	708	73,233	Justice	60458	708	11,137
Evergreen Park	60642	708	20,874	Kankakee	60901	815	27,541
Fairfield	62837	618	5,442	Kewanee	61443	309	12,969
Fairview Heights	62208	618	14,351	La Grange	60525	708	15,362
Flora	62839	618	5,093	La Grange Park	60525	708	12,861
Flossmoor	60422	708	8,651	Lake Bluff	60044	708	5,486
Forest Park	60130	708	14,918	Lake Forest	60045	708	17,836
Fox Lake	60020	708	7,539	Lake in the Hills	60102	708	5,900
Franklin Park	60131	708	18,485	Lake Zurich	60047	708	14,927
Freeport	61032	815	25,840	Lansing	60438	708	28,131
Gages Lake	60030	708	8,349	La Salle	61301	815	9,717
Galesburg	*61401	309	33,530	Lemont	60439	708	7,359
Geneseo	61254	309	5,990	Libertyville	*60048	708	19,174
Geneva	60134	708	12,625	Lincoln	62656	217	15,418
Glen Carbon	62034	618	7,774	Lincolnwood	60645	708	11,365
Glencoe	60022	708	8,499	Lindenhurst	60046	708	8,044
Glendale Heights	60139	708	27,915	Lisle	60532	708	19,584
Glen Ellyn	*60137	708	24,919	Litchfield	62056	217	6,883
Glenview	60025	708	37,052	Lockport	60441	815	9,401
Glenwood	60425	708	9,289	Lombard	60148	708	39,408
Godfrey	62035	618	5,436	Loves Park	61130	815	15,462
Granite City	62040	618	32,766	Lynwood	60411	708	6,535
Grayslake	60030	708	7,388	Lyons	60534	708	9,828
Greenville	62246	618	5,108	McHenry	*60050	815	16,343
Gurnee	60031	708	13,715	Machesney Park	61115	815	19,042

Numbers

Place	ZIP Code	Area Code	1990 Population	Place	ZIP Code	Area Code	1990 Population
Illinois (continued)				Park Forest	60466	708	24,656
Macomb	61455	309	19,952	Park Ridge	60068	708	36,175
Marion	62959	618	14,545	Pekin	*61554	309	32,254
Markham	60426	708	13,136	Peoria	*61601	309	113,504
Mascoutah	62258	618	5,511	Peoria Heights	61614	309	6,930
Matteson	60443	708	11,378	Peru	61354	815	9,302
Mattoon	61938	217	18,441	Plano	60545	708	5,104
Maywood	60153	708	27,139	Pontiac	61764	815	11,428
Melrose Park	*60160	708	20,859	Princeton	61356	815	7,197
Mendota	61342	815	7,017	Prospect Heights	60070	708	15,236
Metropolis	62960	618	6,734	Quincy	*62301	217	39,682
Midlothian	60445	708	14,372	Rantoul	61866	217	17,212
Milan	61264	309	5,753	Richton Park	60471	708	10,523
Moline	*61265	309	43,080	Riverdale	60627	708	13,671
Monmouth	61462	309	9,489	River Forest	60305	708	11,669
Morris	60450	815	10,274	River Grove	60171	708	9,961
Morton	61550	309	13,799	Riverside	60546	708	8,774
Morton Grove	60053	708	22,373	Robbins	60472	708	7,498
Mount Carmel	62863	618	8,287	Robinson	62454	618	6,740
Mount Prospect	60056	708	53,168	Rochelle	61068	815	8,769
Mount Vernon	62864	618	17,082	Rock Falls	61071	815	9,669
Mundelein	60060	708	21,224	Rockford	*61125	815	139,704
Murphysboro	62966	618	9,176	Rock Island	*61201	309	40,630
Naperville	*60540	708	85,806	Rolling Meadows	60008	708	22,598
New Lenox	60451	815	9,698	Romeoville	60441	815	14,101
Niles	60714	708	28,375	Roselle	60172	708	20,803
Normal	61761	309	40,023	Round Lake Beach	60073	708	16,406
Norridge	60634	708	14,459	St. Charles	*60174	708	22,620
North Aurora	60542	708	6,010	Salem	62881	618	7,470
Northbrook	*60062	708	32,572	Sandwich	60548	815	5,607
North Chicago	60064	708	34,978	Sauk Village	60411	708	9,926
Northlake	60164	708	12,505	Schaumburg	*60194	708	68,586
North Riverside	60546	708	6,180	Schiller Park	60176	708	11,189
Oak Brook	60521	708	9,087	Scott AFB	62225	618	7,245
Oak Forest	60452	708	26,202	Shorewood	60436	815	6,264
Oak Lawn	*60455	708	56,182	Silvis	61282	309	6,926
Oak Park	*60303	708	53,648	Skokie	*60077	708	59,432
O'Fallon	62269	618	16,064	South Elgin	60177	708	7,474
Olney	62450	618	8,661	South Holland	60473	708	22,105
Orland Park	60462	708	35,720	Springfield	*62703	217	105,417
Ottawa	61350	815	17,528	Spring Valley	61362	815	5,246
Palatine	*60067	708	38,894	Steger	60475	708	8,592
Palos Heights	60463	708	11,478	Sterling	61081	815	15,142
Palos Hills	60465	708	17,803	Stickney	60402	708	5,678
Pana	62557	217	5,796	Streamwood	60107	708	31,197
Paris	61944	217	9,016	Streator	61364	815	14,121

Place	ZIP Code	Area Code	1990 Population	Place	ZIP Code	Area Code	1990 Population
Illinois (continued)				Bloomington	*47408	812	60,633
Summit	60501	708	9,971	Bluffton	46714	219	9,104
Swansea	62221	618	8,201	Boonville	47601	812	6,686
Sycamore	60178	815	9,896	Brazil	47834	812	7,640
Taylorville	62568	217	11,133	Brownsburg	46112	317	7,628
Tinley Park	60477	708	37,115	Carmel	*46032	317	25,380
Troy	62294	618	6,019	Cedar Lake	46303	219	8,885
University Park	60466	708	6,204	Charlestown	47111	812	5,889
Urbana	61801	217	36,383	Chesterton	46304	219	9,118
Vandalia	62471	618	6,114	Clarksville	47129	812	19,838
Vernon Hills	60061	708	15,319	Clinton	47842	317	5,040
Villa Park	60181	708	22,279	Columbia City	46725	219	5,700
Warrenville	60555	708	11,389	Columbus	*47201	812	31,802
Washington	61571	309	10,136	Conersville	47331	317	15,550
Washington Park	62204	618	7,431	Crawfordsville	47933	317	13,584
Waterloo	62298	618	5,030	Crown Point	46307	219	17,728
Watseka	60970	815	5,424	Decatur	46733	219	8,642
Wauconda	60084	708	6,294	Dunlap	46514	219	5,705
Waukegan	*60085	708	69,481	Dyer	46311	219	10,923
Westchester	60154	708	17,301	East Chicago	46312	219	33,892
West Chicago	*60185	708	14,808	Elkhart	*46513	219	43,627
Western Springs	60558	708	11,956	Elwood	46036	317	9,494
West Frankfort	62896	618	8,526	Evansville	*47708	812	126,272
Westmont	60559	708	21,402	Fort Wayne	*46802	219	172,971
West Peoria	61604	309	5,314	Frankfort	46041	317	14,754
Wheaton	*60187	708	51,441	Franklin	46131	317	12,932
Wheeling	60090	708	29,911	Garrett	46738	219	5,349
Willowbrook	60514	708	8,701	Gary	*46401	219	116,646
Wilmette	60091	708	26,694	Gas City	46933	317	6,296
Winfield	60190	708	7,096	Goshen	*46526	219	23,794
Winnetka	60093	708	12,210	Granger	46530	219	20,241
Winthrop Harbor	60096	708	6,240	Greencastle	46135	317	8,984
Wonder Lake	60097	815	6,664	Greenfield	46140	317	11,657
Wood Dale	60191	708	12,394	Greensburg	47240	812	9,286
Woodridge	60517	708	26,359	Greenwood	*46142	317	26,507
Wood River	62095	618	11,490	Griffith	46319	219	17,914
Woodstock	60098	815	14,368	Hammond	*46320	219	84,236
Worth	60482	708	11,208	Hartford City	47348	317	6,960
Zion	60099	708	19,783	Highland	46322	219	23,696
Indiana				Hobart	46342	219	21,822
Alexandria	46001	317	5,709	Huntingburg	47542	812	5,236
Anderson	*46011	317	59,459	Huntington	46750	219	16,389
Angola	46703	219	5,851	Indianapolis	*46206	317	731,327
Auburn	46706	219	9,386	Jasper	47546	812	10,030
Bedford	47421	812	13,817	Jeffersonville	*47130	812	21,968
Beech Grove	46107	317	13,383	Kendallville	46755	219	7,773

Place	ZIP Code	Area Code	1990 Population	Place	ZIP Code	Area Code	1990 Population
Indiana (continued)				Speedway	46224	317	13,092
Kokomo	*46902	317	44,996	Tell City	47586	812	8,088
Lafayette	*47901	317	43,758	Terre Haute	*47808	812	55,430
Lake Station	46405	219	13,899	Valparaiso	46383	219	24,414
La Porte	46350	219	21,507	Vincennes	47591	812	19,867
Lawrence	46226	317	26,779	Wabash	46992	219	12,127
Lebanon	46052	317	12,059	Warsaw	*46580	219	10,968
Linton	47441	812	5,814	Washington	47501	812	10,864
Logansport	46947	219	16,865	West Lafayette	47906	317	26,144
Lowell	*46356	219	6,430	Whiting	46394	219	5,155
Madison	47250	812	12,006	Winchester	47394	317	5,095
Marion	*46952	317	32,607	**Iowa**			
Martinsville	46151	317	11,677	Algona	50511	515	6,015
Merrillville	46410	219	27,257	Altoona	50009	515	7,242
Michigan City	46360	219	33,822	Ames	*50010	515	47,198
Mishawaka	*46544	219	42,635	Anamosa	52205	319	5,100
Monticello	47960	219	5,237	Ankeny	50021	515	18,482
Mooresville	46158	317	5,541	Atlantic	50022	712	7,432
Mount Vernon	47620	812	7,217	Bettendorf	52722	319	28,139
Muncie	*47302	317	71,170	Boone	50036	515	12,392
Munster	46321	219	19,949	Burlington	52601	319	27,208
New Albany	*47150	812	36,322	Carroll	51401	712	9,579
New Castle	47362	317	17,753	Cedar Falls	50613	319	34,298
New Haven	46774	219	9,338	Cedar Rapids	*52401	319	108,772
Noblesville	46060	317	17,655	Centerville	52544	515	5,936
North Manchester	46962	219	6,383	Charles City	50616	515	7,878
North Vernon	47265	812	5,129	Cherokee	51012	712	6,026
Oak Park	47130	812	5,630	Clarinda	51632	712	5,104
Peru	46970	317	12,843	Clear Lake	50428	515	8,183
Plainfield	46168	317	10,438	Clinton	*52732	319	29,201
Plymouth	46563	219	8,291	Clive	50053	515	7,462
Portage	46368	219	29,062	Coralville	52241	319	10,347
Portland	47371	219	6,483	Council Bluffs	*51501	712	54,315
Princeton	47670	812	8,127	Creston	50801	515	7,911
Rensselaer	47978	219	5,045	Davenport	*52802	319	95,333
Richmond	*47374	317	38,705	Decorah	52101	319	8,063
Rochester	46975	219	5,969	Denison	51442	712	6,604
Rushville	46173	317	5,533	Des Moines	*50318	515	193,189
Salem	47167	812	5,619	Dubuque	52001	319	57,538
Schererville	46375	219	20,155	Estherville	51334	712	6,720
Scottsburg	47170	812	5,334	Fairfield	52556	515	9,768
Sellersburg	47172	812	5,914	Fort Dodge	50501	515	25,894
Seymour	47274	812	15,579	Fort Madison	52627	319	11,614
Shelbyville	46176	317	15,347	Grinnell	50112	515	8,902
South Bend	*46624	219	105,511	Hartan	51537	712	5,148
South Haven	46383	219	6,112	Independence	50644	319	5,972

Place	ZIP Code	Area Code	1990 Population	Place	ZIP Code	Area Code	1990 Population
Iowa (continued)				Dodge City	67801	316	21,129
Indianola	50125	515	11,340	El Dorado	67042	316	11,495
Iowa City	52240	319	59,735	Emporia	66801	316	25,512
Iowa Falls	50126	515	5,435	Fort Riley North	66442	913	12,848
Keokuk	52632	319	12,451	Fort Scott	66701	316	8,362
Knoxville	50138	515	8,232	Garden City	67846	316	24,097
Le Mars	51031	712	8,454	Great Bend	67530	316	15,427
Manchester	52057	319	5,137	Hays	67601	913	17,814
Maquoketa	52060	319	6,130	Haysville	67060	316	8,364
Marion	52302	319	20,442	Hutchinson	*67501	316	39,308
Marshalltown	50158	515	25,178	Independence	67301	316	10,030
Mason City	50401	515	29,040	Iola	66749	316	6,351
Mount Pleasant	52641	319	7,959	Junction City	66441	913	20,642
Muscatine	52761	319	22,881	Kansas City	*66102	913	149,800
Nevada	50201	515	6,009	Lansing	66043	913	7,120
Newton	50208	515	14,799	Lawrence	*66044	913	65,608
Oelwein	50662	319	6,493	Leavenworth	66048	913	38,495
Oskaloosa	52577	515	10,600	Leawood	66209	913	19,693
Ottumwa	52501	515	24,488	Lenexa	66210	913	34,110
Pella	50219	515	9,270	Liberal	*67901	316	16,573
Perry	50220	515	6,652	McPherson	67460	316	12,422
Red Oak	51566	712	6,264	Manhattan	66502	913	37,737
Shenandoah	51601	712	5,572	Merriam	66202	913	11,819
Sioux Center	51250	712	5,074	Mission	66202	913	9,504
Sioux City	*51101	712	80,505	Newton	67114	316	16,700
Spencer	51301	712	11,066	Olathe	*66061	913	63,402
Storm Lake	50588	712	8,769	Ottawa	66067	913	10,667
Urbandale	50322	515	23,500	Overland Park	66202	913	111,790
Vinton	52349	319	5,103	Parsons	67357	316	11,919
Washington	52353	319	7,074	Pittsburg	66762	316	17,789
Waterloo	*50701	319	66,467	Prairie Village	66202	913	23,186
Waverly	50677	319	8,539	Pratt	67124	316	6,687
Webster City	50595	515	7,894	Roeland Park	66202	913	7,706
West Des Moines	*50265	515	31,702	Salina	*67401	913	42,299
Windsor Heights	50311	515	5,190	Shawnee	66203	913	37,962
Kansas				Topeka	*66601	913	119,883
Abilene	67410	913	6,242	Ulysses	67880	316	5,474
Arkansas City	67005	316	12,762	Wellington	67152	316	8,517
Atchison	66002	913	10,656	Wichita	*67209	316	304,017
Augusta	67010	316	7,848	Winfield	67156	316	11,931
Bonner Springs	66012	913	6,413	**Kentucky**			
Chanute	66720	316	9,488	Ashland	*41101	606	23,622
Coffeyville	67337	316	12,917	Bardstown	40004	502	6,712
Colby	67701	913	5,510	Bellevue	41073	606	6,997
Concordia	66901	913	6,152	Berea	40403	606	9,129
Derby	67037	316	14,691	Bowling Green	*42102	502	41,688

Place	ZIP Code	Area Code	1990 Population	Place	ZIP Code	Area Code	1990 Population
Kentucky (continued)				Newburg	40218	502	21,647
Buechel	40261	502	7,081	Newport	*41071	606	18,871
Campbellsville	*42718	502	9,592	Nicholasville	*40356	606	13,603
Corbin	*40701	606	7,644	Okolona	40259	502	18,902
Covington	*41011	606	43,646	Owensboro	42303	502	53,577
Cynthiana	41031	606	6,497	Paducah	*42003	502	27,256
Danville	*40422	606	12,559	Paris	*40361	606	8,730
Dayton	41074	606	6,576	Pikeville	41501	606	6,324
Edgewood	41017	606	8,143	Pleasure Ridge Park	40268	502	25,131
Elizabethtown	*42701	502	18,167	Princeton	42445	502	6,940
Elsmere	41018	606	6,847	Radcliff	*40160	502	19,778
Erlanger	41018	606	15,979	Richmond	*40475	606	21,183
Fairdale	40118	502	6,563	Russellville	42276	502	7,454
Fern Creek	40291	502	16,406	Saint Matthews	40206	502	15,691
Flatwoods	41139	606	7,799	Shelbyville	*40066	502	6,155
Florence	*41042	606	18,586	Shively	40256	502	15,535
Fort Campbell North	42223	502	18,861	Somerset	*42501	606	10,735
Fort Knox	40121	502	21,495	Valley Station	40272	502	22,840
Fort Mitchell	41017	606	7,438	Versailles	40383	606	7,269
Fort Thomas	41075	606	16,032	Westwoods	41101	606	5,300
Frankfort	*40601	502	26,535	Williamsburg	40769	606	5,493
Franklin	*42134	502	7,607	Winchester	*40391	606	15,799
Georgetown	40324	502	11,414	**Louisiana**			
Glasgow	*42141	502	12,351	Abbeville	*70510	318	11,184
Harrodsburg	40330	606	7,335	Alexandria	*71301	318	49,049
Hazard	*41701	606	5,416	Arabi	70032	504	8,787
Henderson	42420	502	25,945	Avondale	70094	504	5,813
Highview	40228	502	14,814	Baker	*70714	504	13,087
Hillview	40229	502	6,119	Bastrop	*71220	318	13,916
Hopkinsville	*42240	502	29,809	Baton Rouge	*70821	504	219,531
Independence	41051	606	10,444	Bayou Cane	70360	504	15,876
Jeffersontown	40269	502	23,223	Belle Chasse	70037	504	8,512
Lawrenceburg	40342	502	5,911	Bogalusa	*70427	504	14,280
Lebanon	40033	502	5,695	Bossier City	*71111	318	52,721
Lexington	*40507	606	225,366	Breaux Bridge	70517	318	6,694
London	*40741	606	5,757	Brownsville-Bawcomville	71291	318	7,397
Louisville	*40232	502	269,555				
Madisonville	42431	502	16,203	Bunkie	71322	318	5,044
Mayfield	42066	502	9,935	Chalmette	*70043	504	31,860
Maysville	41056	606	7,169	Claiborne	71291	318	8,300
Middlesboro	40965	606	11,328	Covington	*70433	504	7,691
Monticello	42633	606	5,357	Crowley	*70526	318	13,983
Morehead	40351	606	8,357	Cut Off	70345	504	5,325
Mount Sterling	40353	606	5,362	Denham Springs	*70726	504	8,381
Mount Washington	40047	502	5,256	De Ridder	70634	318	9,868
Murray	42071	502	14,442	Destrehan	70047	504	8,031

Place	ZIP Code	Area Code	1990 Population	Place	ZIP Code	Area Code	1990 Population
Louisiana (continued)				Reserve	70084	504	8,847
Donaldsonville	70346	504	7,949	River Ridge	70123	504	14,800
Estelle	70072	504	14,091	Ruston	*71270	318	20,071
Eunice	70535	318	11,162	Saint Martinville	70582	318	7,226
Fort Polk South	71459	318	10,911	Saint Rose	70087	504	6,259
Franklin	70538	318	9,004	Shreveport	*71103	318	198,518
Gonzales	*70737	504	7,208	Slidell	*70458	504	24,124
Gretna	*70053	504	17,208	Springhill	71075	318	5,668
Hammond	*70401	504	15,871	Sulphur	*70663	318	20,125
Harahan	70123	504	9,927	Tallulah	*71282	318	8,526
Harvey	*70058	504	21,222	Terrytown	70056	504	23,787
Houma	*70360	504	30,495	Thibodaux	*70301	504	14,125
Jeanerette	70544	318	6,205	Timberlane	70053	504	12,614
Jefferson	70502	504	14,521	Ville Platte	70586	318	9,037
Jennings	70546	318	11,305	Violet	70092	504	8,574
Kenner	*70062	504	72,033	Waggaman	70094	504	9,405
Lacombe	70445	504	6,523	Westlake	70669	318	5,007
Lafayette	*70501	318	94,438	West Monroe	71291	318	14,096
Lake Charles	*70601	318	70,580	Westwego	*70094	504	11,218
Lake Providence	71254	318	5,380	Winnfield	71483	318	6,138
La Place	*70068	504	24,194	Winnsboro	71295	318	5,755
Larose	70373	504	5,772	Zachary	70791	504	9,036
Leesville	*71446	318	7,638	**Maine**			
Mandeville	*70448	504	7,474	Auburn	*04210	207	24,309
Mansfield	71052	318	5,389	Augusta	*04330	207	21,325
Marksville	71351	318	5,526	Bangor	*04401	207	33,181
Marrero	*70072	504	36,671	Bath	04530	207	9,799
Meraux	70075	504	8,849	Belfast	04915	207	6,355
Metairie	*70009	504	149,428	Berwick	03901	207	5,995
Minden	*71055	318	13,661	Biddeford	*04005	207	20,710
Monroe	*71203	318	54,909	Brewer	04412	207	9,021
Morgan City	*70380	504	14,531	Brunswick Center	04011	207	14,683
Moss Bluff	70612	318	8,039	Brunswick	04011	207	20,906
Natchitoches	*71457	318	16,609	Buxton	04093	207	6,494
New Iberia	*70560	318	31,838	Camden	04843	207	5,060
New Orleans	*70140	504	496,938	Cape Elizabeth	04107	207	8,854
Oakdale	71463	318	6,837	Caribou	04736	207	9,415
Opelousas	*70570	318	19,091	Cumberland	04021	207	5,836
Patterson	70392	504	5,166	Eliot	03903	207	5,329
Pineville	*71360	318	12,255	Ellsworth	04605	207	5,975
Plaquemine	*70764	504	7,101	Fairfield	04937	207	6,718
Ponchatoula	70454	504	5,425	Falmouth	04105	207	7,610
Port Allen	70767	504	6,277	Farmington	04938	207	7,436
Prien	70601	318	6,448	Freeport	04032	207	6,905
Raceland	70394	504	5,564	Gardiner	04345	207	6,746
Rayne	70578	318	8,502	Gorham	04038	207	11,856

Place	ZIP Code	Area Code	1990 Population
Maine (continued)			
Gray	04039	207	5,904
Hampden	04444	207	5,974
Houlton Center	04730	207	5,627
Houlton	04730	207	6,613
Jay	04239	207	5,080
Kennebunk	04043	207	8,004
Kittery Center	03904	207	5,151
Kittery	03904	207	9,372
Lewiston	*04240	207	39,757
Limestone	04750	207	9,922
Lincoln	04457	207	· 5,587
Lisbon	04250	207	9,457
Loring AFB	04750	207	7,829
Millinocket Center	04462	207	6,922
Millinocket	04462	207	6,956
Oakland	04963	207	5,595
Old Orchard Beach	04064	207	7,789
Old Town	04468	207	8,317
Orono Center	04473	207	9,789
Orono	04473	207	10,573
Portland	*04101	207	64,358
Presque Isle	04769	207	10,550
Rockland	04841	207	7,972
Rumford Compact	04276	207	5,419
Rumford	04276	207	7,078
Saco	04072	207	15,181
Sanford Center	04073	207	10,296
Sanford	04073	207	20,463
Scarborough	*04074	207	12,518
Skowhegan Center	04976	207	6,990
Skowhegan	04976	207	8,725
South Berwick	03908	207	5,877
South Portland	04106	207	23,163
Standish	04084	207	7,678
Topsham	04086	207	8,746
Waterville	*04901	207	17,173
Wells	04090	207	7,778
Westbrook	*04092	207	16,121
Windham	04082	207	13,020
Winslow Center	04901	207	5,436
Winslow	04901	207	7,997
Winthrop	04364	207	5,986
Yarmouth	04096	207	7,862
York	03909	207	9,818

Place	ZIP Code	Area Code	1990 Population
Maryland			
Aberdeen	21001	410	13,087
Aberdeen Proving Ground	21005	410	5,267
Adelphi	20783	301	13,524
Andrews AFB	20331	410	10,228
Annapolis	*21401	410	33,195
Arbutus	21227	410	19,750
Arnold	21012	410	20,261
Aspen Hill	20916	301	45,494
Baltimore	*21203	410	736,014
Bel Air	21014	410	8,942
Bel Air North	21050	410	14,880
Bel Air South	21014	410	26,421
Beltsville	*20705	301	14,476
Bethesda	*20814	301	62,936
Bladensburg	20710	301	8,064
Bowie	*20715	301	37,642
Brooklyn Park	21225	410	10,987
California	20619	410	7,626
Cambridge	21613	410	11,514
Camp Springs	20748	301	16,392
Cape St. Clair	21401	410	7,878
Carney	21234	410	25,578
Catonsville	21228	410	35,233
Cheverly	20784	301	6,023
Chevy Chase	*20825	301	8,559
Chillum	20783	301	31,309
Clinton	20735	301	19,987
Cloverly	20904	301	7,904
Cockeysville	21030	410	18,668
Colesville	20914	301	18,819
College Park	*20740	301	23,714
Columbia	*21045	301	75,883
Coral Hills	20743	410	11,032
Crofton	21114	410	12,781
Cumberland	*21502	301	23,712
Damascus	20872	301	9,817
District Heights	20747	301	6,711
Dundalk	21222	410	65,800
Easton	21601	410	9,372
East Riverdale	20737	301	14,187
Edgemere	21219	410	9,226
Edgewood	21040	410	23,903
Elkton	*21921	410	9,073
Ellicott City	*21043	410	41,396

Place	ZIP Code	Area Code	1990 Population	Place	ZIP Code	Area Code	1990 Population
Maryland (continued)				North Kensington	20895	301	8,607
Essex	21221	410	40,872	North Laurel	20707	301	15,008
Fairland	20904	301	19,828	North Potomac	20878	301	18,456
Fallston	21047	410	5,730	Ocean City	21842	410	5,146
Ferndale	21061	410	16,355	Odenton	21113	410	12,833
Forestville	20747	301	16,731	Olney	*20832	301	23,019
Fort Meade	20755	301	12,509	Overlea	21206	410	12,137
Fort Washington	20744	301	24,032	Owings Mills	21117	410	9,474
Frederick	*21701	301	40,186	Oxon Hill-Glassmanor	20745	301	35,794
Frostburg	21532	301	8,069	Palmer Park	20785	301	7,019
Gaithersburg	*20877	301	39,676	Parkville	21234	410	31,617
Germantown	*20874	301	41,145	Parole	21401	410	10,054
Glen Burnie	*21061	410	37,305	Pasadena	21122	410	10,012
Glenn Dale	20769	301	9,689	Perry Hall	21128	410	22,723
Greenbelt	*20770	301	20,561	Pikesville	21208	410	24,815
Green Haven	21122	410	14,416	Potomac	*20854	301	45,634
Hagerstown	*21740	301	35,306	Randallstown	21133	301	26,277
Halfway	21740	301	8,873	Reisterstown	21136	410	19,314
Havre de Grace	21078	410	8,952	Riverdale	*20737	301	4,843
Hillcrest Heights	20748	301	17,136	Riviera Beach	21122	410	11,376
Hyattsville	*20780	301	13,864	Rockville	*20850	301	44,830
Jessup	20794	410	6,537	Rosedale	21237	410	18,703
Joppatowne	21085	410	11,084	Rossville	21221	410	9,492
Kentland	20785	301	7,967	St. Charles	20602	301	28,717
Kettering	20772	301	9,901	Salisbury	*21801	410	20,592
Lake Shore	21122	410	13,269	Savage-Guilford	20763	410	9,669
Landover	20785	301	5,052	Seat Pleasant	20743	301	5,359
Langley Park	20787	301	17,474	Severn	21144	410	24,499
Lanham-Seabrook	20706	301	16,692	Severna Park	21146	410	25,879
Lansdowne-Baltimore Highlands	21227	410	15,509	Silver Spring	*20907	301	76,046
La Plata	20646	301	5,841	South Gate	21061	410	27,564
Largo	20772	301	9,475	South Kensington	20895	301	8,777
Laurel	*20707	301	19,086	South Laurel	20707	301	18,591
Lexington Park	20653	410	9,943	Suitland-Silver Hills	20746	301	35,111
Linthicum	21090	410	7,547	Takoma Park	*20912	301	16,724
Lochearn	21207	410	25,240	Temple Hills	*20748	301	6,865
Londontowne	21037	410	6,992	Towson	21285	410	49,445
Lutherville-Timonium	*21093	410	16,442	Upper Marlboro	20772	301	11,528
Marlow Heights	20748	301	5,885	Waldorf	*20602	301	15,058
Middle River	21220	410	24,616	Walker Mill	20743	301	10,920
Montgomery Village	20886	301	32,315	Westminster	*21157	410	13,060
Mount Rainier	20712	301	7,954	Wheaton Glenmont	20902	301	53,720
Naval Academy	21402	410	5,420	White Marsh	21162	301	8,183
New Carrollton	20784	301	12,002	White Oak	20903	301	18,671
North Bethesda	20815	301	29,656	Woodlawn	21207	410	32,907

Place	ZIP Code	Area Code	1990 Population	Place	ZIP Code	Area Code	1990 Population
Massachusetts				Cochituate	01778	508	6,046
Abington	02351	617	13,817	Cohasset	02025	617	7,075
Acton	01720	508	17,872	Concord	01742	508	17,076
Acushnet	02743	508	9,554	Dalton	*01226	413	7,155
Adams Center	01220	413	6,356	Danvers	01923	508	24,174
Adams	01220	413	9,445	Dartmouth	02714	508	27,244
Agawam	01001	413	27,323	Dedham	02026	617	23,782
Amesbury Center	01913	508	12,109	Deerfield	01342	413	5,018
Amesbury	01913	508	14,997	Dennis	02638	508	13,864
Amherst Center	01002	413	17,824	Dighton	02715	508	5,631
Amherst	*01002	413	35,228	Dracut	01826	508	25,594
Arlington	02174	617	44,630	Dudley	01570	508	9,540
Ashburnham	01430	508	5,433	Duxbury	*02332	617	13,895
Ashland	01721	508	12,066	East Bridgewater	02333	508	11,104
Athol Center	01331	508	8,732	East Falmouth	02536	508	5,577
Athol	01331	508	11,451	Easthampton	01027	413	15,537
Attleboro	02703	508	38,383	East Longmeadow	01028	413	13,367
Auburn	01501	508	15,005	Easton	02334	508	19,807
Ayer	*01432	508	6,871	Everett	02149	617	35,701
Barnstable	02630	508	40,949	Fairhaven	02719	508	16,132
Bedford	01730	617	12,996	Fall River	*02722	508	92,703
Belchertown	01007	413	10,579	Falmouth	*02540	508	27,960
Bellingham	02019	508	14,877	Fitchburg	04120	508	41,194
Belmont	02178	617	24,720	Fort Devens	01433	508	8,973
Beverly	01915	508	38,195	Foxborough	02035	508	5,706
Billerica	*01821	508	37,609	Framingham	01701	508	64,989
Blackstone	01504	508	8,023	Franklin Center	02038	508	9,965
Boston	*02109	617	574,283	Franklin	02038	508	22,095
Bourne	02532	508	16,064	Freetown	02702	508	8,522
Boxford	01921	508	6,266	Gardner	01440	508	20,125
Braintree	*02184	617	33,836	Georgetown	01833	508	6,384
Brewster	02631	508	8,440	Gloucester	*01930	508	28,716
Bridgewater	02324	508	21,249	Grafton	01519	508	13,035
Brockton	*02403	508	92,788	Granby	01033	413	5,565
Brookline	02146	617	54,718	Great Barrington	01230	413	7,725
Burlington	01803	617	23,302	Greenfield Center	*01301	413	14,016
Cambridge	*02139	617	95,802	Greenfield	01302	413	18,666
Canton	02021	617	18,530	Groton	01450	508	7,511
Carver	02330	508	10,590	Groveland	01834	508	5,214
Centerville	02632	508	9,190	Halifax	02338	617	6,526
Charlton	01507	508	9,576	Hamilton	01936	508	7,280
Chatham	02633	508	6,579	Hanover	02339	617	11,912
Chelmsford	01824	508	32,383	Hanson	02341	617	9,028
Chelsea	02150	617	28,710	Harvard	01451	508	12,329
Chicopee	*01020	413	56,632	Harwich	02645	508	10,275
Clinton	01510	508	13,222	Haverhill	*01830	508	51,418

Place	ZIP Code	Area Code	1990 Population	Place	ZIP Code	Area Code	1990 Population
Massachusetts (continued)				Middleborough	02346	617	17,867
Hingham	02043	617	19,821	Milford Center	01757	508	23,339
Holbrook	02343	617	11,041	Milford	01757	508	25,355
Holden	01520	508	14,628	Millbury	01527	508	12,228
Holliston	01746	508	12,926	Millis	02054	508	7,613
Holyoke	*01040	413	43,704	Milton	02186	617	25,725
Hopedale	01747	508	5,666	Monson	01057	413	7,776
Hopkinton	01748	508	9,191	Montague	01351	413	8,316
Hudson Center	01749	508	14,267	Nantucket	02554	508	6,012
Hudson	01749	508	17,233	Natick	01760	508	30,510
Hull	02045	617	10,466	Needham	*02192	617	27,557
Hyannis	02601	508	14,120	New Bedford	*02740	508	99,992
Ipswich	01938	508	11,873	Newbury	01951	508	5,623
Kingston	02364	617	9,045	Newburyport	01950	508	16,317
Lakeville	02346	617	7,785	Newton	*02205	617	82,585
Lancaster	01523	508	6,661	Norfolk	02056	508	9,270
Lawrence	*01842	508	70,207	North Adams	01247	413	16,797
Lee	01238	413	5,849	North Amherst	01002	413	6,239
Leicester	01524	508	10,191	Northampton	01060	413	29,289
Lenox	01240	413	5,069	North Andover	01845	508	22,792
Leominster	01453	508	38,145	North Attleborough	*02760	508	25,038
Lexington	02173	617	28,974	Northborough	01532	508	11,929
Lincoln	01773	617	7,666	Northbridge	01534	508	13,371
Littleton	01460	508	7,051	North Reading	01864	508	12,002
Longmeadow	01106	413	15,467	Norton	02766	508	14,265
Lowell	*01853	508	103,439	Norwell	02061	617	9,279
Ludlow	01056	413	18,820	Norwood	02062	617	28,700
Lunenburg	01462	508	9,117	Orange	01364	508	7,312
Lynn	*01901	617	81,245	Orleans	02653	508	5,838
Lynnfield	01940	617	11,274	Oxford Center	01540	508	5,969
Malden	02148	617	53,884	Oxford	01540	508	12,588
Manchester	01944	508	5,286	Palmer	01069	413	12,054
Mansfield	02048	508	16,568	Peabody	01960	508	47,264
Marblehead	01945	617	19,971	Pembroke	02359	617	14,544
Marlborough	01752	508	31,813	Pepperell	01463	508	10,098
Marshfield	02050	617	21,531	Pinehurst	01866	508	6,614
Mashpee	02649	508	7,884	Pittsfield	01201	413	48,622
Mattapoisett	02739	508	5,850	Plainville	02762	508	6,871
Maynard	01754	508	10,325	Plymouth Center	*02360	508	7,258
Medfield	02052	508	10,531	Plymouth	02360	508	45,608
Medford	02155	617	57,407	Quincy	02169	617	84,985
Medway	02053	508	9,931	Randolph	02368	617	30,093
Melrose	02176	617	28,150	Raynham	02767	508	9,867
Merrimac	01860	508	5,166	Reading	01867	617	22,539
Methuen	01844	508	39,990	Rehoboth	02769	508	8,656
Middleborough Center	02346	617	6,837	Revere	02151	617	42,786

Place	ZIP Code	Area Code	1990 Population	Place	ZIP Code	Area Code	1990 Population
Massachusetts (continued)				Watertown	02205	617	33,284
Rockland	02370	617	16,123	Wayland	01778	508	11,874
Rockport	01966	508	7,482	Webster Center	01570	508	11,849
Salem	01970	508	38,091	Webster	01570	508	16,196
Salisbury	01950	508	6,882	Wellesley	02181	617	26,615
Sandwich	02563	508	15,489	Westborough	01581	508	14,133
Saugus	01906	617	25,549	West Boylston	01583	508	6,611
Scituate	02066	617	16,786	West Bridgewater	02379	508	6,389
Seekonk	02771	508	13,046	West Concord	01742	508	5,761
Sharon	02067	617	15,517	Westfield	01085	413	38,372
Shirley	01464	508	6,118	Westford	01886	508	16,392
Shrewsbury	01545	508	24,146	Westminster	01473	508	6,191
Somerset	*02725	508	17,655	Weston	02193	617	10,200
Somerville	*02205	617	76,210	Westport	02790	508	13,852
South Amherst	01002	413	5,053	West Springfield	01089	413	27,537
Southborough	01772	508	6,628	Westwood	02090	617	12,557
Southbridge Center	01550	508	13,631	West Yarmouth	02673	508	5,409
Southbridge	01550	508	17,816	Weymouth	*02205	617	54,063
South Hadley	01075	413	16,685	Whitinsville	01588	508	5,639
Southwick	01077	413	7,667	Whitman	02382	617	13,240
South Yarmouth	02664	508	10,358	Wilbraham	01095	413	12,635
Spencer Center	01562	508	6,306	Williamstown	01267	413	8,220
Spencer	01562	508	11,645	Wilmington	01887	508	17,651
Springfield	*01101	413	156,983	Winchendon	01475	508	8,805
Sterling	01564	508	6,481	Winchester	01890	617	20,267
Stoneham	02180	617	22,203	Winthrop	02152	617	18,127
Stoughton	02072	617	26,777	Woburn	01801	617	35,943
Stow	01775	508	5,328	Worcester	*01613	508	169,759
Sturbridge	01566	508	7,775	Wrentham	02093	508	9,006
Sudbury	01776	508	14,358	Yarmouth	02675	508	21,174
Sutton	01527	508	6,824	**Michigan**			
Swampscott	01907	617	13,650	Adrian	49221	517	22,097
Swansea	02777	508	15,411	Albion	49224	517	10,066
Taunton	02780	508	49,832	Allen Park	48101	313	31,092
Templeton	01468	508	6,438	Alma	48801	517	9,034
Tewksbury	01876	508	27,266	Alpena	49707	517	11,354
Topsfield	01983	508	5,754	Ann Arbor	*48107	313	109,608
Townsend	01469	508	8,496	Auburn Hills	48321	313	17,076
Tyngsborough	01879	508	8,642	Battle Creek	*49016	616	53,516
Uxbridge	01569	508	10,415	Bay City	*48707	517	38,936
Wakefield	01880	617	24,825	Beecher	48505	313	14,465
Walpole	02081	508	20,223	Belding	48809	616	5,969
Waltham	02154	617	57,878	Benton Harbor	*49022	616	12,818
Ware Center	01082	413	6,533	Benton Heights	49022	616	5,465
Ware	01082	413	9,808	Berkley	48072	313	16,960
Wareham	02571	508	19,232	Beverly Hills	48025	313	10,610

Place	ZIP Code	Area Code	1990 Population	Place	ZIP Code	Area Code	1990 Population
Michigan (continued)				Grosse Pointe Farms	48230	313	10,092
Big Rapids	49307	616	12,603	Grosse Pointe Park	48230	313	12,857
Birmingham	*48012	810	19,997	Grosse Pointe Woods	48230	313	17,715
Bloomfield	48301	313	42,137	Hamtramck	48212	313	18,372
Bridgeport	48722	517	8,569	Harper Woods	48225	313	14,903
Brighton	48116	810	5,686	Harrison	48625	517	24,685
Burton	*48509	810	27,437	Haslett	48840	517	10,230
Cadillac	49601	616	10,104	Hastings	49058	616	6,549
Canton	48187	313	57,047	Hazel Park	48030	810	20,051
Carrollton	48724	517	6,521	Highland Park	48203	313	20,121
Center Line	48015	810	9,026	Hillsdale	49242	517	8,175
Charlotte	48813	517	8,083	Holland	*49423	616	30,745
Clawson	48017	313	13,874	Holly	48442	810	5,595
Clinton	49236	517	85,866	Holt	48842	517	11,744
Coldwater	49036	517	9,607	Houghton	49931	906	7,498
Comstock Park	49321	616	6,530	Howell	*48843	517	8,147
Cutlerville	49508	616	11,228	Huntington Woods	48070	313	6,419
Davison	48423	810	5,693	Inkster	48141	313	30,772
Dearborn	*48120	313	89,286	Ionia	48846	616	5,990
Dearborn Heights	48127	313	60,838	Iron Mountain	49801	906	8,525
Detroit	*48231	313	1,027,974	Ironwood	49938	906	6,849
Dowagiac	49047	616	6,418	Ishpeming	49849	906	7,200
East Grand Rapids	49506	616	10,807	Jackson	*49204	517	37,425
East Lansing	48826	517	50,677	Jenison	*49428	616	17,882
Eastpointe	48021	810	35,283	Kalamazoo	*49001	616	80,277
Eastwood	49001	616	6,340	Kentwood	49518	616	37,826
Ecorse	48229	313	12,180	Kingsford	49801	906	5,480
Escanaba	49829	906	13,659	K.I. Sawyer AFB	49843	906	6,577
Fair Plain	49022	616	8,051	Lambertville	48144	313	7,860
Farmington	*48333	810	10,170	Lansing	*48901	517	127,321
Farmington Hills	48333	810	74,614	Lapeer	48446	810	7,759
Fenton	48430	313	8,434	Lincoln Park	48146	313	41,832
Ferndale	48220	313	25,084	Livonia	*48150	313	100,850
Flat Rock	48134	313	7,290	Ludington	49431	616	8,507
Flint	*48501	810	140,925	Madison Heights	48071	313	32,196
Flushing	48433	810	8,542	Manistee	49660	616	6,734
Fraser	48026	810	13,899	Marquette	49855	906	21,977
Garden City	*48135	313	31,846	Marshall	49068	616	6,941
Grand Blanc	48439	810	7,760	Marysville	48040	810	8,515
Grand Haven	49417	616	11,951	Mason	48854	517	6,768
Grand Ledge	48837	517	7,562	Melvindale	48122	313	11,216
Grand Rapids	*49501	616	189,126	Menominee	49858	906	9,398
Grandville	*49418	616	15,624	Midland	*48640	517	38,053
Greenville	48838	616	8,101	Milford	48381	810	5,500
Grosse Ile	48138	313	9,781	Monroe	48161	313	22,902
Grosse Pointe	48236	313	5,681	Mount Clemens	48046	810	18,405

Place	ZIP Code	Area Code	1990 Population
Michigan (continued)			
Mount Pleasant	*48804	517	23,299
Muskegon	*49440	616	39,809
Muskegon Heights	49444	616	13,176
New Baltimore	48047	810	5,798
Niles	49120	616	12,458
Northville	48167	313	6,226
Norton Shores	49441	616	21,755
Novi	48376	810	32,998
Oak Park	48237	313	30,468
Okemos	48805	517	20,216
Owosso	48867	517	16,322
Petoskey	49770	616	6,056
Plymouth	48170	313	9,560
Plymouth Township	48170	313	23,646
Pontiac	*48343	810	71,136
Portage	49081	616	41,042
Port Huron	48061	810	33,694
Redford	48239	313	54,387
River Rouge	48218	313	11,314
Riverview	48192	313	13,894
Rochester	*48308	810	7,130
Rochester Hills	48306	313	61,766
Romulus	48174	313	22,897
Roseville	48066	313	51,412
Royal Oak	*48068	810	65,410
Saginaw	*48605	517	69,512
Saginaw Township North	48604	517	23,018
Saginaw Township South	48603	517	13,987
Saint Clair	48079	810	5,116
Saint Clair Shores	*48080	810	68,107
Saint Johns	48879	517	7,392
Saint Joseph	49085	616	9,214
Saline	48176	313	6,663
Sault Sainte Marie	49783	906	14,689
Shelby	49455	616	48,655
Southfield	*48037	810	75,727
Southgate	48195	313	30,771
South Haven	49090	616	5,563
South Lyon	48178	810	6,479
Springfield	49015	616	5,582
Sterling Heights	*48311	313	117,810
Sturgis	49091	616	10,130
Taylor	48180	313	70,811
Tecumseh	49286	517	7,462
Temperance	48182	313	6,542
Three Rivers	49093	616	7,464
Traverse City	*49684	616	15,155
Trenton	48183	313	20,586
Troy	*48099	810	72,884
Utica	48318	810	5,081
Walker	49504	616	17,279
Warren	*48390	313	144,864
Waterford	48329	313	66,692
Wayne	48184	313	19,899
West Bloomfield	*48325	313	54,843
Westland	48185	313	84,724
Westwood	49019	616	8,957
Wixom	48393	313	8,550
Woodhaven	48183	313	11,631
Wurtsmith AFB	48753	517	5,080
Wyandotte	*48192	313	30,938
Wyoming	49509	616	63,891
Ypsilanti	48197	313	24,846
Zeeland	49464	616	5,417
Minnesota			

Minnesota

Area code 612 will be split with a new area code, 320, beginning in March 1996. Call the local operator for assistance.

Place	ZIP Code	Area Code	1990 Population
Albert Lea	56007	507	18,310
Alexandria	56308	612	8,029
Andover	55303	612	15,216
Anoka	55303	612	17,192
Apple Valley	55124	612	34,598
Arden Hills	55112	612	9,199
Austin	55912	507	21,926
Bemidji	*56601	218	11,165
Blaine	55449	612	38,975
Bloomington	*55420	612	86,335
Brainerd	56401	218	12,353
Brooklyn Center	55429	612	28,887
Brooklyn Park	55443	612	56,381
Buffalo	55313	612	6,856
Burnsville	55337	612	51,288
Champlin	55316	612	16,849
Chanhassen	55317	612	11,732
Chaska	55318	612	11,339
Chisholm	55719	218	5,290
Cloquet	55720	218	10,885
Columbia Heights	55421	612	18,910
Coon Rapids	55433	612	52,978
Cottage Grove	55016	612	22,935

Place	ZIP Code	Area Code	1990 Population
Minnesota (continued)			
Crookston	56716	218	8,119
Crystal	55428	612	23,788
Detroit Lakes	*56501	218	6,635
Duluth	*55806	218	85,493
Eagan	55121	612	47,409
East Bethel	55005	612	8,050
East Grand Forks	56721	218	8,658
Eden Prairie	*55344	612	39,311
Edina	55424	612	46,075
Elk River	55330	612	11,143
Fairmont	56031	507	11,265
Falcon Heights	55113	612	5,380
Faribault	55021	507	17,085
Farmington	55024	612	5,940
Fergus Falls	*56537	218	12,362
Forest Lake	55025	612	5,833
Fridley	55432	612	28,335
Golden Valley	55427	612	20,971
Grand Rapids	*55744	218	7,976
Ham Lake	*55304	612	8,924
Hastings	55033	612	15,478
Hermantown	55810	218	6,761
Hibbing	*55746	218	18,046
Hopkins	55343	612	16,529
Hutchinson	55350	612	11,459
International Falls	56649	218	8,301
Inver Grove Heights	55076	612	22,477
Lake Elmo	55042	612	5,900
Lakeville	55044	612	24,854
Litchfield	55355	612	6,041
Little Canada	55117	612	8,971
Little Falls	56345	612	7,371
Mankato	*56001	507	31,405
Maple Grove	55311	612	38,736
Maplewood	55109	612	30,954
Marshall	56258	507	12,023
Mendota Heights	55118	612	9,388
Minneapolis	*55440	612	368,383
Minnetonka	55345	612	48,370
Montevideo	56265	612	5,499
Monticello	*55362	612	5,045
Moorhead	*56560	218	32,295
Morris	56267	612	5,613
Mound	55364	612	9,634
Mounds View	55112	612	12,541
New Brighton	55112	612	22,207
New Hope	54427	612	21,853
New Ulm	56073	507	13,132
Northfield	55057	507	14,684
North Mankato	56001	507	10,662
North Saint Paul	55109	612	12,376
Oakdale	55119	612	18,377
Orono	55323	612	7,285
Owatonna	55060	507	19,386
Plymouth	55421	612	50,889
Prior Lake	55372	612	11,482
Ramsey	55303	612	12,408
Red Wing	55066	612	15,134
Richfield	55423	612	35,710
Robbinsdale	55422	612	14,396
Rochester	*55901	507	70,729
Rosemount	55068	612	8,622
Roseville	55113	612	33,485
Saint Anthony	55418	612	7,727
Saint Cloud	*56301	612	48,812
Saint Louis Park	55426	612	43,787
Saint Paul	*55101	612	272,235
Saint Peter	56082	507	9,481
Sartell	56377	612	5,409
Sauk Rapids	56379	612	7,823
Savage	56378	612	9,906
Shakopee	55379	612	11,739
Shoreview	55126	612	24,587
South Saint Paul	55075	612	20,197
Spring Lake Park	55432	612	6,532
Stillwater	*55082	612	13,882
Thief River Falls	56701	218	8,010
Vadnais Heights	55127	612	11,041
Virginia	*55792	218	9,410
Waseca	56093	507	8,385
West Saint Paul	55118	612	19,248
White Bear Lake	55110	612	24,622
Willmar	56201	612	17,531
Winona	55987	507	25,435
Woodbury	55125	612	20,075
Worthington	56187	507	9,977
Mississippi			
Aberdeen	39730	601	6,837
Amory	38821	601	7,093
Batesville	38606	601	6,403
Bay Saint Louis	*39520	601	8,063

Place	ZIP Code	Area Code	1990 Population
Mississippi (continued)			
Biloxi	*39530	601	46,319
Booneville	38829	601	7,955
Brandon	*39042	601	11,077
Brookhaven	39601	601	10,243
Canton	39046	601	10,062
Clarksdale	38614	601	19,717
Cleveland	38732	601	15,384
Clinton	*39056	601	21,847
Columbia	39429	601	6,815
Columbus	*39701	601	23,799
Corinth	38834	601	11,820
Crystal Springs	39059	601	5,643
D'Iberville	39532	601	6,566
Forest	39074	601	5,062
Gautier	39553	601	10,088
Greenville	*38701	601	45,226
Greenwood	*38930	601	18,906
Grenada	*38901	601	10,864
Gulfport	*39501	601	40,775
Hattiesburg	*39401	601	41,906
Holly Springs	*38635	601	7,261
Horn Lake	38637	601	9,069
Indianola	38751	601	11,809
Jackson	*39205	601	196,637
Kosciusko	39090	601	6,986
Laurel	*39440	601	18,827
Leland	38756	601	6,366
Long Beach	39560	601	15,804
Louisville	39339	601	7,165
McComb	39648	601	11,797
Madison	*39110	601	7,471
Meridian	*39302	601	41,036
Moss Point	*39563	601	17,837
Natchez	*39120	601	19,460
New Albany	38652	601	6,775
Ocean Springs	*39564	601	14,673
Orange Grove	39567	601	15,676
Oxford	38655	601	10,026
Pascagoula	*39567	601	25,899
Pass Christian	39571	601	5,557
Pearl	39288	601	19,588
Petal	39465	601	7,883
Philadelphia	39350	601	6,758
Picayune	39466	601	10,633
Ridgeland	*39157	601	11,714

Place	ZIP Code	Area Code	1990 Population
Ripley	38663	601	5,371
Southaven	38671	601	17,949
Starkville	39759	601	18,458
Tupelo	*38801	601	30,685
Vicksburg	*39180	601	20,909
Waveland	39576	601	5,369
Waynesboro	39367	601	5,143
West Point	39773	601	8,489
Winona	38967	601	5,724
Yazoo City	39194	601	12,427
Missouri			

Area code 314 will be split into a new area code, 573, effective January 1996. Call the local operator for assistance.

Place	ZIP Code	Area Code	1990 Population
Affton	63123	314	21,106
Arnold	63010	314	18,828
Aurora	65605	417	6,459
Ballwin	*63011	314	21,406
Bellefontaine Neighbors	63137	314	10,918
Belton	64012	816	18,145
Berkeley	63134	314	12,250
Black Jack	63031	314	6,131
Blue Springs	*64015	816	40,103
Bolivar	65613	417	6,845
Boonville	65233	816	7,095
Breckenridge Hills	63114	314	5,181
Brentwood	63144	314	8,150
Bridgeton	63044	314	17,732
Cape Girardeau	*63701	314	34,475
Carthage	64836	417	10,747
Caruthersville	63830	314	7,389
Charleston	63834	314	5,085
Chesterfield	63017	314	38,630
Chillicothe	64601	816	8,799
Clayton	63105	314	13,926
Clinton	64735	816	8,703
Columbia	*65201	314	69,133
Concord	63128	314	19,859
Crestwood	63126	314	11,229
Creve Coeur	63141	314	12,289
Dellwood	63136	314	5,245
De Soto	63020	314	5,993
Des Peres	63131	314	8,388
Dexter	63841	314	7,506
Ellisville	63011	314	7,183
Excelsior Springs	64024	816	10,373
Farmington	63640	314	11,596

Place	ZIP Code	Area Code	1990 Population	Place	ZIP Code	Area Code	1990 Population
Missouri (continued)				Pine Lawn	63120	314	5,083
Ferguson	63135	314	22,290	Poplar Bluff	63901	314	16,841
Festus	63028	314	8,105	Raymore	64083	816	5,592
Florissant	*63033	314	51,038	Raytown	64133	816	30,601
Fort Leonard Wood	65473	314	15,863	Richmond	64085	816	5,738
Fulton	65251	314	10,033	Richmond Heights	63117	314	10,448
Gladstone	64118	816	26,243	Rock Hill	63124	314	5,217
Glasgow Village	65254	314	5,199	Rolla	65401	314	14,090
Glendale	63122	314	5,945	Saint Ann	63074	314	14,449
Grandview	64030	816	24,973	Saint Charles	*63301	314	50,634
Hannibal	63401	314	18,004	Saint John	63114	314	7,502
Harrisonville	64701	816	7,696	Saint Joseph	*64501	816	71,852
Hazelwood	*63042	314	15,512	Saint Louis	*63166	314	396,685
Independence	*64050	816	112,301	Saint Peters	63376	314	40,660
Jackson	63755	314	9,256	Sappington	63126	314	10,917
Jefferson City	*65101	314	35,517	Sedalia	*65301	816	19,800
Jennings	63136	314	15,841	Shrewsbury	63119	314	6,416
Joplin	*64801	417	40,866	Sikeston	63801	314	17,641
Kansas City	*64108	816	434,829	Spanish Lake	63138	314	20,322
Kennett	63857	314	10,941	Springfield	*65801	417	140,494
Kirksville	63501	816	17,152	Sullivan	63080	314	5,661
Kirkwood	63122	314	27,291	Trenton	64683	816	6,129
Ladue	63124	314	8,795	Union	63084	314	6,048
Lebanon	65536	417	9,983	University City	63130	314	40,087
Lee's Summit	*64063	816	46,418	Warrensburg	64093	816	15,244
Lemay	63125	314	18,005	Washington	63090	314	10,704
Liberty	64068	816	20,459	Webb City	64870	417	7,449
Macon	63552	816	5,571	Webster Groves	63119	314	22,992
Malden	63863	314	5,123	Wentzville	63385	314	4,640
Manchester	63011	314	6,537	West Plains	65775	417	8,913
Maplewood	63143	314	9,962	**Montana**			
Marshall	65340	816	12,711	Anaconda	59711	406	10,356
Maryland Heights	63043	314	25,440	Billings	*59106	406	81,125
Maryville	64468	816	10,663	Bozeman	*59715	406	22,660
Mexico	65265	314	11,290	Butte	*59701	406	33,336
Moberly	65270	816	12,839	Great Falls	*59401	406	55,125
Monett	65708	417	6,529	Havre	59501	406	10,201
Murphy	63026	314	9,342	Helena	*59601	406	24,609
Neosho	64850	417	9,254	Kalispell	*59901	406	11,917
Nevada	64772	417	8,597	Laurel	59044	406	5,686
Normandy	63121	314	5,063	Lewistown	59457	406	6,097
Northwoods	63121	314	5,106	Livingston	59047	406	6,701
O'Fallon	63366	314	17,427	Malmstrom AFB	59402	406	5,938
Olivette	63132	314	7,573	Miles City	59301	406	8,461
Overland	63114	314	17,987	Missoula	*59801	406	42,918
Perryville	63775	314	6,933	Orchard Homes	59801	406	10,317

Place	ZIP Code	Area Code	1990 Population	Place	ZIP Code	Area Code	1990 Population
Montana (continued)				Nellis AFB	89191	702	8,377
Sidney	59270	406	5,217	North Las Vegas	*89030	702	47,849
Nebraska				Pahrump	89041	702	7,424
Alliance	69301	308	9,765	Paradise	89109	702	124,682
Beatrice	68310	402	12,352	Reno	*89501	702	133,850
Bellevue	68005	402	30,948	Sparks	*89431	702	53,367
Blair	68008	402	6,860	Sunrise Manor	89110	702	95,362
Chadron	69337	308	5,588	Sun Valley	89433	702	11,391
Columbus	*68601	402	19,480	Winchester	89101	702	23,365
Fremont	68025	402	23,680	Winnemucca	*89445	702	6,102
Gering	69341	308	7,946	**New Hampshire**			
Grand Island	*68802	308	39,487	Amherst	03031	603	9,068
Hastings	*68901	402	22,837	Atkinson	03811	603	5,188
Holdrege	68949	308	5,671	Barrington	03825	603	6,164
Kearney	*68847	308	24,396	Bedford	03102	603	12,563
La Vista	68128	402	9,840	Belmont	03220	603	5,796
Lexington	68850	308	6,600	Berlin	03570	603	11,824
Lincoln	*68501	402	191,972	Bow	03304	603	5,500
McCook	69001	308	8,112	Claremont	03743	603	13,902
Nebraska City	68410	402	6,547	Concord	*03301	603	36,006
Norfolk	*68701	402	21,476	Conway	03818	603	7,940
North Platte	69101	308	22,605	Derry Compact	03038	603	20,446
Offutt AFB West	68113	402	10,883	Derry	03038	603	29,603
Ogallala	69153	308	5,095	Dover	03820	603	25,042
Omaha	*68108	402	335,719	Durham Compact	03824	603	9,236
Papillion	68046	402	10,378	Durham	03824	603	11,818
Plattsmouth	68048	402	6,415	Epping	03042	603	5,162
Ralston	68127	402	6,236	Exeter Compact	03833	603	9,556
Scottsbluff	*69361	308	13,711	Exeter	03833	603	12,481
Seward	68434	402	5,641	Farmington	03835	603	5,739
Sidney	69162	308	5,959	Franklin	03235	603	8,304
South Sioux City	68776	402	9,677	Gilford	03246	603	5,867
Wayne	68787	402	5,142	Goffstown	03045	603	14,621
York	68467	402	7,940	Hampstead	03841	603	6,732
Nevada				Hampton Compact	03842	603	7,989
Boulder City	*89005	702	12,567	Hampton	03842	603	12,278
Carson City	*89701	702	40,443	Hanover Compact	03755	603	6,538
East Las Vegas	89112	702	11,087	Hanover	03756	603	9,212
Elko	*89801	702	14,836	Hollis	03049	603	5,705
Fallon	*89406	702	6,430	Hooksett	03106	603	9,002
Fernley	89408	702	5,164	Hudson	03051	603	19,530
Gardnerville Ranchos	89410	702	7,455	Jaffrey	03452	603	5,361
Henderson	*89015	702	64,948	Keene	03431	603	22,430
Incline Village-Crystal Bay	*89450	702	7,119	Kingston	03848	603	5,591
				Laconia	*03246	603	15,743
Las Vegas	*89125	702	258,204	Lebanon	*03766	603	12,183

Place	ZIP Code	Area Code	1990 Population	Place	ZIP Code	Area Code	1990 Population
New Hampshire (continued)				Boonton	07005	201	8,343
Littleton	03561	603	5,827	Bound Brook	08805	908	9,487
Londonderry Compact	03053	603	10,114	Brick Twp.	*08724	201	66,473
Londonderry	03053	603	19,781	Bridgeton	08302	609	18,942
Manchester	*03103	603	99,332	Bridgewater Twp.	08807	908	32,509
Merrimack	03054	603	22,156	Brigantine	08203	609	11,354
Milford Compact	03055	603	8,015	Browns Mills	08015	609	11,429
Milford	03055	603	11,795	Budd Lake	07828	201	7,272
Nashua	*03060	603	79,662	Burlington	08016	609	9,835
Newmarket	03857	603	7,157	Butler	07405	201	7,392
Newport	03773	603	6,110	Caldwell	*07006	201	7,549
Pelham	03076	603	9,408	Camden	*08101	609	87,492
Pembroke	03275	603	6,561	Carlstadt	07072	201	5,510
Peterborough	03458	603	5,239	Carney's Point Twp.	08069	609	8,443
Plaistow	03865	603	7,316	Carteret	07008	908	19,025
Plymouth	03264	603	5,811	Cedar Grove Twp.	07009	201	12,053
Portsmouth	*03801	603	25,925	Chatham	07928	201	8,007
Raymond	03077	603	8,713	Cherry Hill Twp.	*08034	609	69,319
Rochester	*03867	603	26,630	Cinnaminson Twp.	08077	609	14,583
Salem	03079	603	25,746	Clark Twp.	07066	908	14,629
Seabrook	03874	603	6,503	Clayton	08312	609	6,155
Somersworth	03878	603	11,249	Clementon	08021	609	5,601
Suncook	03275	603	5,214	Cliffside Park	07010	201	20,393
Weare	03281	603	6,193	Clifton	*07015	201	71,984
Windham	03087	603	9,000	Closter	07624	201	8,094
New Jersey				Collingswood	08108	609	15,289
Absecon	08201	609	7,298	Colonia	07067	908	18,238
Allendale	07401	201	5,900	Cranford Twp.	07016	908	22,633
Asbury Park	07712	908	16,799	Cresskill	07626	201	7,558
Atlantic City	*08401	609	37,986	Dover	*07801	201	15,115
Audubon	08106	609	9,205	Dumont	07628	201	17,187
Avenel	07001	908	15,504	Dunellen	08812	908	6,528
Barrington	08007	609	6,792	East Brunswick Twp.	08816	908	43,548
Bayonne	07002	201	61,464	East Hanover Twp.	07936	201	9,926
Beachwood	08722	908	9,324	East Orange	*07019	201	73,552
Belleville	07109	201	34,213	East Rutherford	07073	201	7,902
Bellmawr	*08031	609	12,603	Eatontown	07724	908	13,800
Belmar	07719	908	5,877	Edgewater	07020	201	5,001
Bergenfield	07621	201	24,458	Edgewater Park Twp.	08010	609	8,388
Berkeley Heights Twp.	07922	908	11,980	Edison Twp.	*08818	908	88,680
Berlin	08009	609	5,672	Elizabeth	*07207	908	110,002
Bernardsville	07924	908	6,597	Elmwood Park	07407	201	17,623
Blackwood	08012	609	5,120	Emerson	07630	201	6,930
Bloomfield	07003	201	45,061	Englewood	*07631	201	24,850
Bloomingdale	07403	201	7,530	Englewood Cliffs	07632	201	5,634
Bogota	07603	201	7,824	Ewing Twp.	08618	609	34,185

399

Place	ZIP Code	Area Code	1990 Population	Place	ZIP Code	Area Code	1990 Population
New Jersey (continued)				Keansburg	07734	908	11,069
Fairfield	07006	201	7,615	Kearny	07032	201	34,874
Fair Haven	07701	908	5,270	Kendall Park	08824	908	7,127
Fair Lawn	07410	201	30,548	Kenilworth	07033	908	7,574
Fairview	07022	201	10,733	Keyport	07735	908	7,586
Fanwood	07023	908	7,115	Kinnelon	07405	201	8,470
Florence-Roebling	08518	609	8,564	Lake Mohawk	07871	201	8,930
Florham Park	07932	201	8,521	Lakewood	08701	908	26,095
Fords	08863	908	14,392	Laurence Harbor	08879	908	6,361
Fort Dix	08640	609	10,205	Leonia	07605	201	8,365
Fort Lee	07024	201	31,997	Lincoln Park	07035	201	10,978
Franklin Lakes	07417	201	9,873	Linden	07036	908	36,701
Freehold	07728	908	10,742	Lindenwold	08021	609	18,734
Garfield	07026	201	26,727	Linwood	08221	609	6,866
Gilford Park	08753	908	8,668	Little Falls Twp.	07424	201	11,294
Glassboro	08028	609	15,614	Little Ferry	07643	201	9,989
Glendora	08029	609	5,201	Little Silver	07739	908	5,721
Glen Ridge	07028	201	7,076	Livingston Twp.	07039	201	26,609
Glen Rock	07452	201	10,883	Lodi	07644	201	22,355
Gloucester City	08030	609	12,649	Long Branch	07740	908	28,658
Guttenberg	07093	201	8,268	Long Hill Twp.	*07946	908	7,826
Hackensack	*07602	201	37,049	Lyndhurst Twp.	07071	201	18,262
Hackettstown	07840	908	8,120	McGuire AFB	08641	609	7,580
Haddonfield	08033	609	11,633	Madison	07940	201	15,850
Haddon Heights	08035	609	7,860	Madison Park	08859	201	7,490
Haledon	*07508	201	6,951	Mahwah Twp.	*07430	201	17,905
Hamilton Twp. (Mercer)	*08609	609	86,553	Manasquan	08736	908	5,369
				Manville	08835	908	10,567
Hammonton	08037	609	12,208	Maple Shade Twp.	08052	609	19,211
Hanover Twp.	07981	201	11,538	Maplewood Twp.	07040	201	21,756
Harrison	07029	201	13,425	Margate City	08402	609	8,431
Hasbrouck Heights	07604	201	11,488	Marlboro Twp.	07746	908	27,974
Hawthorne	*07506	201	17,084	Marlton	08053	609	10,228
Hazlet Twp.	07730	908	21,976	Matawan	07747	908	9,239
Highland Park	08904	201	13,279	Maywood	07607	201	9,536
Hightstown	08520	609	5,126	Mercerville-Hamilton Sq.	08619	609	26,873
Hillsdale	07642	201	9,750				
Hillside Twp.	07205	908	21,044	Metuchen	08840	908	12,804
Hoboken	07030	201	33,397	Middlesex	08846	908	13,055
Holiday City-Berkeley	08753	908	14,293	Middletown Twp.	07748	908	68,183
Hopatcong	07843	201	15,586	Midland Park	07432	201	7,047
Hopewell Twp. (Mercer)	*08560	609	10,893	Millburn Twp.	07041	201	18,630
				Milltown	08850	908	6,968
Irvington	07111	201	59,774	Millville	08332	609	25,992
Iselin	08830	908	16,141	Monroe Twp. (Gloucester)	08094	609	26,703
Jackson Twp.	*08527	908	33,283				
Jersey City	*07303	201	228,517	Montclair	*07042	201	37,729

Place	ZIP Code	Area Code	1990 Population
New Jersey (continued)			
Montvale	07645	201	6,946
Montville Twp.	07045	201	15,600
Moorestown-Lenola	08057	609	13,242
Morris Plains	07950	201	5,219
Morristown	07960	201	16,189
Mountainside	07092	908	6,657
Mount Holly Twp.	08060	609	10,639
Neptune Twp.	*07753	908	28,148
Newark	*07102	201	275,221
New Brunswick	*08901	908	41,711
New Milford	07646	201	15,990
New Providence	07974	908	11,439
Newton	07860	201	7,521
North Arlington	07032	201	13,790
North Bergen Twp.	07047	201	48,414
North Brunswick Twp.	08902	908	31,287
North Caldwell	07006	201	6,706
Northfield	08225	609	7,305
North Haledon	07508	201	7,987
North Plainfield	07060	908	18,820
Nutley	07110	201	27,099
Oakland	07436	201	11,997
Ocean Twp. (Ocean)	*08758	908	5,416
Ocean City	08226	609	15,512
Oceanport	07757	908	6,146
Old Bridge	08857	908	22,151
Old Bridge Twp.	08857	908	56,493
Oradell	07649	201	8,024
Orange	*07050	201	29,925
Palisades Park	07650	201	14,536
Palmyra	08065	609	7,056
Paramus	07652	201	25,004
Park Ridge	07656	201	8,102
Parsippany-Troy Hills Twp.	07054	201	48,478
Passaic	*07055	201	58,041
Paterson	*07510	201	140,891
Paulsboro	08066	609	6,577
Pennsauken Twp.	08110	609	34,738
Penns Grove	08069	609	5,228
Pennsville Center	08070	609	12,218
Pequannock Twp.	07440	201	12,844
Perth Amboy	*08861	908	41,967
Phillipsburg	08865	908	15,757
Pine Hill	08021	609	9,854

Place	ZIP Code	Area Code	1990 Population
Piscataway Twp.	*08854	908	47,089
Pitman	08071	609	9,365
Plainfield	*07061	908	46,577
Pleasantville	08232	609	16,027
Point Pleasant	08742	908	18,177
Point Pleasant Beach	08742	908	5,112
Pompton Lakes	07442	201	10,539
Princeton	*08540	609	12,016
Prospect Park	07508	201	5,053
Rahway	07065	908	25,325
Ramblewood	08057	609	6,181
Ramsey	07446	201	13,228
Randolph Twp.	07869	201	19,974
Raritan	08869	908	5,798
Red Bank	07701	908	10,636
Ridgefield	07657	201	9,996
Ridgefield Park	07660	201	12,454
Ridgewood	*07451	201	24,152
Ringwood	07456	201	12,623
River Edge	07661	201	10,603
Riverside Twp.	08075	609	7,974
River Vale	07675	201	9,410
Robertsville	07726	908	9,841
Rochelle Park Twp.	07662	201	5,587
Rockaway	07866	201	6,243
Roselle	07203	908	20,314
Roselle Park	07204	201	12,805
Rumson	07760	908	6,701
Runnemede	08078	609	9,042
Rutherford	*07070	201	17,790
Saddle Brook Twp.	07662	201	13,296
Salem	08079	609	6,883
Sayreville	08872	908	34,998
Scotch Plains Twp.	07076	908	21,150
Secaucus	07094	201	14,061
Silverton	08753	908	9,175
Somerdale	08083	609	5,440
Somerset	*08873	908	22,070
Somers Point	08244	609	11,216
Somerville	08876	908	11,632
South Amboy	08879	908	7,851
South Orange Twp.	07079	201	16,390
South Plainfield	07080	908	20,489
South River	08882	908	13,692
Sparta Twp.	07871	201	15,157
Spotswood	08884	908	7,983

Place	ZIP Code	Area Code	1990 Population	Place	ZIP Code	Area Code	1990 Population
New Jersey (continued)				Woodbridge	07095	908	17,434
Springfield Twp.	07081	201	13,420	Woodbridge Twp.	*07095	908	93,092
Spring Lake Heights	07762	908	5,341	Woodbury	08096	609	10,904
Stratford	08084	609	7,614	Woodcliff Lake	07675	201	5,303
Strathmore	07747	201	7,060	Wood-Ridge	07075	201	7,506
Succasunna-Kenvil	07876	201	11,781	Wyckoff Twp.	07481	201	15,372
Summit	07901	908	19,757	Yardville-Groveville	08620	609	9,248
Teaneck Twp.	07666	201	37,825	**New Mexico**			
Tenafly	07670	201	13,326	Alamogordo	*88310	505	27,596
Tinton Falls	07724	908	12,361	Albuquerque	*87101	505	384,619
Toms River	*08753	908	7,524	Anthony	88021	505	5,160
Totowa	07512	201	10,177	Artesia	88210	505	10,610
Trenton	*08650	609	88,675	Aztec	87410	505	5,480
Twin Rivers	08520	609	7,715	Belen	87002	505	6,547
Union Twp. (Union)	07083	908	50,024	Bernalillo	87004	505	5,960
Union Beach	07735	908	6,156	Bloomfield	87413	505	5,214
Union City	07087	201	58,012	Carlsbad	*88220	505	24,952
Upper Saddle River	07458	201	7,198	Clovis	*88101	505	30,954
Ventnor City	08406	609	11,005	Corrales	87048	505	5,453
Verona	07044	201	13,597	Deming	*88030	505	10,970
Villas	08251	609	8,136	Espanola	87532	505	8,389
Vineland	08360	609	54,780	Farmington	*87401	505	33,997
Waldwick	07463	201	9,757	Gallup	*87301	505	19,157
Wallington	07057	201	10,828	Grants	87020	505	8,626
Wanaque	07465	201	9,711	Hobbs	*88240	505	29,121
Washington	07882	908	6,474	Holloman AFB	88330	505	5,891
Washington Twp. (Bergen)	07675	201	9,245	Las Cruces	*88001	505	62,360
				Las Vegas	87701	505	14,753
Watchung	07060	908	5,110	Los Alamos	87544	505	11,455
Wayne Twp.	*07470	201	47,025	Los Lunas	87031	505	6,013
Weehawken Twp.	07087	201	12,385	Lovington	88260	505	9,322
West Caldwell	07006	201	10,422	North Valley	87107	505	12,507
Westfield	*07091	908	28,870	Paradise Hills	87114	505	5,513
West Freehold	07728	908	11,166	Portales	88130	505	10,690
West Long Branch	07764	908	7,690	Raton	87740	505	7,372
West Milford Twp.	07480	201	25,430	Rio Rancho	87124	505	32,512
West New York	07093	201	38,125	Roswell	*88201	505	44,260
West Orange	07052	201	39,103	Sandia	87115	505	6,742
West Paterson	07424	201	10,982	Santa Fe	*87501	505	56,537
Westwood	07675	201	10,446	Shiprock	87420	505	7,687
Wharton	07885	201	5,405	Silver City	*88061	505	10,683
White Horse	08610	609	9,397	Socorro	87801	505	8,159
White Meadow Lake	07886	201	8,002	South Valley	87105	505	35,701
Williamstown	08094	609	10,891	Sunland Park	88063	505	8,179
Willingboro Twp.	08046	609	36,291	Truth or Consequences	87901	505	6,221
Winslow Twp.	08095	609	30,087				

Daily Life

Place	ZIP Code	Area Code	1990 Population
New Mexico (continued)			
Tucumcari	88401	505	6,827
White Rock	87544	505	6,192
Zuni Pueblo	87327	505	5,857
New York			
Albany	*12201	518	100,031
Albertson	11507	516	5,166
Albion	14411	716	5,863
Amityville	11701	516	9,286
Amsterdam	12010	518	20,714
Arlington	12603	914	11,948
Auburn	13021	315	31,258
Babylon	*11702	516	12,249
Baldwin	11510	516	22,719
Baldwinsville	13027	315	6,591
Ballston Spa	12020	518	5,194
Batavia	14020	716	16,310
Bath	14810	607	5,801
Bayport	11705	516	7,702
Bay Shore	11706	516	21,279
Bayville	11709	516	7,193
Beacon	12508	914	13,243
Bellmore	11710	516	16,438
Bethpage	11714	516	15,761
Binghamton	*13902	607	53,008
Bohemia	11716	516	9,556
Brentwood	11717	516	45,218
Briarcliff Manor	10510	914	7,070
Brighton	14610	716	34,455
Brockport	14420	716	8,749
Bronxville	10708	914	6,028
Buffalo	*14240	716	328,175
Canandaigua	*14424	716	10,725
Canton	13617	315	6,379
Carle Place	11514	516	5,107
Cedarhurst	11516	516	5,716
Centereach	11720	516	26,720
Center Moriches	11934	516	5,987
Centerport	11721	516	5,333
Central Islip	11722	516	26,028
Cheektowaga	*14225	716	84,387
Cobleskill	12043	518	5,268
Cohoes	12047	518	16,825
Colonie	12205	518	8,019
Commack	11725	516	36,124
Congers	10920	914	8,003
Copiague	11726	516	20,769
Coram	11727	516	30,111
Corning	14830	607	11,938
Cortland	13045	607	19,801
Croton-on-Hudson	10520	914	7,018
Dansville	14437	716	5,002
Deer Park	11729	516	28,840
Delmar	12054	518	8,360
Depew	14043	716	17,673
DeWitt	13214	315	8,244
Dix Hills	11746	516	25,849
Dobbs Ferry	10522	914	9,940
Dunkirk	14048	716	13,989
East Aurora	14052	716	6,647
Eastchester	10709	914	18,537
East Glenville	12302	518	6,518
East Hills	11576	516	6,746
East Islip	11730	516	14,325
East Massapequa	11758	516	19,550
East Meadow	11554	516	36,909
East Northport	11731	516	20,411
East Patchogue	11772	516	20,195
East Rochester	14445	716	6,932
East Rockaway	11518	516	10,152
East Shoreham	11786	516	5,461
Elmira	*14901	607	33,724
Elmont	11003	516	28,612
Elwood	11731	516	10,916
Endicott	*13760	607	13,531
Endwell	13760	607	12,602
Fairmount	13219	315	12,266
Fairport	14450	716	5,943
Farmingdale	11735	516	8,022
Farmingville	11738	516	14,842
Floral Park	*11001	516	15,947
Fort Drum	13603	315	11,578
Fort Salonga	11768	516	9,176
Franklin Square	11010	516	28,205
Fredonia	14063	716	10,436
Freeport	11520	516	39,894
Fulton	13069	315	12,929
Garden City	11530	516	21,675
Garden City Park	11040	516	7,437
Gates-North Gates	14624	716	14,995
Geneseo	14454	716	7,187
Geneva	14456	315	14,143

403

Place	ZIP Code	Area Code	1990 Population
New York (continued)			
Glen Cove	11542	516	24,149
Glens Falls	12801	518	15,023
Glens Falls North	12801	518	7,978
Gloversville	12078	518	16,656
Goshen	10924	914	5,255
Great Neck	*11021	516	8,745
Great Neck Plaza	11020	516	5,897
Greece	14616	716	15,632
Greenlawn	11740	516	13,208
Greenville	*12771	914	9,528
Hamburg	14075	716	10,442
Hampton Bays	11946	516	7,893
Harrison	10528	914	23,308
Hartsdale	10530	914	9,587
Hastings-on-Hudson	10706	914	8,000
Hauppauge	11787	516	19,750
Haverstraw	10927	914	9,438
Hempstead	*11551	516	47,982
Herkimer	13350	315	7,945
Hewlett	11557	516	6,620
Hicksville	*11802	516	40,174
Hillcrest	10977	914	6,447
Holbrook	11741	516	25,273
Holtsville	11742	516	14,972
Hornell	14843	607	9,877
Horseheads	*14845	607	6,802
Hudson	12534	518	8,034
Hudson Falls	12839	518	7,651
Huntington	11743	516	18,243
Huntington Station	11746	516	28,247
Ilion	13357	315	8,888
Inwood	11696	516	7,767
Irondequoit	14617	716	52,322
Irvington	10533	914	6,348
Islip	11751	516	18,924
Islip Terrace	11752	516	5,530
Ithaca	*14850	607	29,541
Jamestown	*14702	716	34,681
Jefferson Valley-Yorktown	10535	914	14,118
Jericho	11753	516	13,141
Johnson City	13790	607	16,578
Johnstown	12095	518	9,058
Kenmore	14217	716	17,180
Kings Park	11754	516	17,773

Place	ZIP Code	Area Code	1990 Population
Kingston	12401	914	23,095
Lackawanna	14218	716	20,585
Lake Carmel	10512	914	8,489
Lake Grove	11755	516	9,612
Lake Ronkonkoma	11779	516	18,997
Lakeview	11552	516	5,476
Lancaster	14086	716	11,940
Larchmont	10538	914	6,181
Latham	12110	518	10,131
Lawrence	11559	516	6,513
Levittown	11756	516	53,286
Lindenhurst	11757	516	26,879
Little Falls	13365	315	5,829
Lockport	14094	716	24,426
Long Beach	11561	516	33,510
Loudonville	12211	518	10,822
Lynbrook	11563	516	19,208
Mahopac	10541	914	7,755
Malone	12953	518	6,777
Malverne	11565	516	9,054
Mamaroneck	10543	914	17,325
Manhasset	11030	516	7,718
Manorhaven	11050	516	5,672
Manorville	11949	516	6,198
Massapequa	11758	516	22,018
Massapequa Park	11762	516	18,044
Massena	13662	315	11,716
Mastic	11950	516	13,778
Mastic Beach	11951	516	10,293
Mattydale	13211	315	6,418
Mechanicville	12118	518	5,249
Medford	11763	516	21,274
Medina	14103	716	6,686
Melville	11746	516	12,586
Merrick	11566	516	23,042
Middle Island	11953	516	7,848
Middletown	10940	914	24,160
Miller Place	11764	516	9,315
Mineola	11501	516	19,005
Monroe	10950	914	6,672
Monsey	10952	914	13,986
Monticello	12701	914	6,597
Mount Kisco	10549	914	9,108
Mount Sinai	11766	516	8,023
Mount Vernon	*10551	914	67,153
Myers Corner	12590	914	5,599

Place	ZIP Code	Area Code	1990 Population	Place	ZIP Code	Area Code	1990 Population
New York (continued)				Pelham	10803	914	6,413
Nanuet	10954	914	14,065	Pelham Manor	10803	914	5,443
Nesconset	11767	516	10,712	Penn Yan	14527	315	5,257
Newark	14513	315	9,849	Plainedge	11714	516	8,739
Newburgh	*12550	914	26,454	Plainview	11803	516	26,207
New Cassel	11590	516	10,257	Plattsburgh	12901	518	21,255
New City	10956	914	33,673	Plattsburgh AFB	12903	518	5,483
New Hyde Park	11040	516	9,728	Pleasantville	10570	914	6,592
New Paltz	12561	914	5,470	Port Chester	10573	914	24,728
New Rochelle	*10802	914	67,265	Port Jefferson	11777	516	7,455
New Windsor Center	*12550	914	8,898	Port Jefferson Station	11776	516	7,232
New York	*10001	212/ 718	7,322,564	Port Jervis	12771	914	9,060
				Port Washington	11050	516	15,387
Niagara Falls	*14302	716	61,840	Potsdam	13676	315	10,251
North Amityville	11701	516	13,849	Poughkeepsie	*12601	914	28,844
North Babylon	11703	516	18,081	Rensselaer	12144	518	8,255
North Bay Shore	11706	516	12,799	Ridge	11961	516	11,734
North Bellmore	11710	516	19,707	Riverhead	11901	516	8,814
North Bellport	11713	516	8,182	Rochester	*14692	716	230,356
North Lindenhurst	11757	516	10,563	Rockville Centre	*11571	516	24,727
North Massapequa	11758	516	19,365	Rocky Point	11778	516	8,596
North Merrick	11566	516	12,113	Roessleville	12205	518	10,753
North New Hyde Park	11040	516	14,359	Rome	13440	315	44,350
North Patchogue	11772	516	7,374	Ronkonkoma	11779	516	20,391
Northport	11768	516	7,572	Roosevelt	11575	516	15,030
North Syracuse	13212	315	7,363	Roslyn Heights	11577	516	6,405
North Tarrytown	10591	914	8,152	Rotterdam	12303	518	21,228
North Tonawanda	14120	716	34,989	Rye	10580	914	14,936
North Valley Stream	11580	516	14,574	Saint James	11780	516	12,703
North Wantagh	11793	516	12,276	Salamanca	14779	716	6,566
Norwich	13815	607	7,613	Saranac Lake	12983	518	5,377
Nyack	10960	914	6,558	Saratoga Springs	12866	518	25,001
Oakdale	11769	516	7,875	Sayville	11782	516	16,550
Oceanside	11572	516	32,423	Scarsdale	10583	914	16,987
Ogdensburg	13669	315	13,521	Schenectady	*12301	518	65,566
Old Bethpage	11804	516	5,610	Scotchtown	10940	914	8,765
Olean	14760	716	16,946	Scotia	12302	518	7,359
Oneida	13421	315	10,850	Sea Cliff	11579	516	5,054
Oneonta	13820	607	13,954	Seaford	11783	516	15,597
Orange Lake	12550	914	5,196	Selden	11784	516	20,608
Ossining	10562	914	22,582	Seneca Falls	13148	315	7,370
Oswego	13126	315	19,195	Setauket-East Setauket	11733	516	13,634
Oyster Bay	11771	516	6,687				
Patchogue	11772	516	11,060	Shirley	11967	516	22,936
Pearl River	10965	914	15,314	Smithtown	11787	516	25,638
Peekskill	10566	914	19,536	Solvay	13209	315	6,717

Place	ZIP Code	Area Code	1990 Population	Place	ZIP Code	Area Code	1990 Population
New York (continued)				Williston Park	11596	516	7,516
Sound Beach	11789	516	9,102	Woodbury	11797	516	8,008
South Farmingdale	11735	516	15,377	Woodmere	11598	516	15,578
South Hill	14850	607	5,423	Wyandach	11792	516	8,950
South Huntington	11746	516	9,624	Yonkers	*10702	914	188,082
Southold	11971	516	5,192	Yorktown Heights	10598	914	7,690
Southport	14904	607	7,753	**North Carolina**			
South Valley Stream	11581	516	5,328	Albemarle	*28001	704	14,940
Spring Valley	10977	914	21,802	Archdale	27263	919	6,975
Stony Brook	11790	516	13,726	Asheboro	*27203	910	16,362
Stony Point	10980	315	10,587	Asheville	*28801	704	61,855
Suffern	10901	914	11,055	Belmont	28012	704	8,434
Syosset	11791	516	18,967	Boone	28607	704	12,949
Syracuse	*13220	315	163,860	Brevard	28712	704	5,388
Tappan	10983	914	6,867	Burlington	*27215	910	39,498
Tarrytown	10591	914	10,739	Camp Le Jeune	28547	919	36,716
Thornwood	10594	914	7,025	Carrboro	27510	919	12,134
Tonawanda	14151	716	65,284	Cary	*27511	919	44,397
Troy	*12180	518	54,269	Chapel Hill	*27514	919	38,711
Tuckahoe	10707	914	6,302	Charlotte	*28204	704	395,925
Uniondale	11553	516	20,328	Clemmons	27012	919	5,982
Utica	*13504	315	68,637	Clinton	28328	910	8,385
Valley Cottage	10989	914	9,007	Concord	*28025	704	27,601
Valley Stream	*11580	516	33,946	Conover	28613	704	5,311
Wading River	11792	516	5,317	Dunn	*28334	910	8,556
Walden	12586	914	5,836	Durham	*27701	919	136,612
Wantagh	11793	516	18,567	Eden	*27288	910	15,238
Waterloo	13165	315	5,116	Edenton	27932	919	5,268
Watertown	*13601	315	29,429	Elizabeth City	*27909	919	14,292
Watervliet	12189	518	11,061	Fayetteville	*28302	910	75,850
Webster	14580	716	5,464	Forest City	28043	704	7,475
Wellsville	14895	716	5,241	Fort Bragg	28307	919	34,744
West Babylon	11704	516	42,410	Garner	27529	919	14,716
Westbury	11590	516	13,060	Gastonia	*28052	704	54,725
West Elmira	14905	607	5,218	Goldsboro	*27530	919	40,709
West Glens Falls	12801	518	5,964	Graham	27253	919	10,368
West Haverstraw	10993	914	9,183	Greensboro	*27420	910	183,894
West Hempstead	11552	516	17,689	Greenville	*27834	919	46,305
West Hills	11743	516	5,849	Hamlet	28345	919	6,324
West Islip	11795	516	28,419	Havelock	28532	919	20,300
Westmere	12203	518	6,750	Henderson	27536	919	15,655
West Point	10996	914	8,024	Hendersonville	*28739	704	7,284
West Seneca	14224	716	47,866	Hickory	28603	704	28,474
Westvale	13219	315	5,952	High Point	*27260	910	69,428
White Plains	*10602	914	48,718	Hope Mills	28348	919	8,272
Williamsville	14221	716	5,583	Jacksonville	*28540	910	30,398

Place	ZIP Code	Area Code	1990 Population	Place	ZIP Code	Area Code	1990 Population
North Carolina (continued)				Whiteville	28472	910	5,078
Kannapolis	*28081	704	29,709	Williamston	27892	919	5,503
Kernersville	*27284	910	10,899	Wilmington	*28402	910	55,530
Kings Mountain	28086	704	8,768	Wilson	*27893	919	36,930
Kinston	*28501	919	25,295	Winston-Salem	*27102	910	143,532
Laurenburg	28352	919	11,643	**North Dakota**			
Lenoir	28645	704	14,223	Bismarck	*58501	701	49,272
Lexington	*27292	704	16,583	Devils Lake	58301	701	7,782
Lincolnton	*28092	704	6,955	Dickinson	*58601	701	16,097
Lumberton	*28358	910	18,656	Fargo	*58102	701	74,084
Matthews	*28105	704	13,651	Grand Forks	*58201	701	49,417
Mint Hill	28212	704	11,615	Grand Forks AFB	58204	701	9,343
Monroe	*28110	704	16,385	Jamestown	*58401	701	15,571
Mooresville	28115	704	9,317	Mandan	58554	701	15,177
Morehead	28557	919	6,046	Minot	*58701	701	34,544
Morganton	*28655	704	15,085	Minot AFB	*58704	701	9,095
Mount Airy	27030	910	7,156	Valley City	58072	701	7,163
Mount Holly	28120	704	7,710	Wahpeton	*58075	701	8,751
New Bern	*28562	919	17,363	West Fargo	58078	701	12,287
New Hope (Wake)	27604	704	5,694	Williston	*58801	701	13,136
New River Station	28540	919	9,732	**Ohio**			
Newton	28658	704	9,077	Area code 216 will be split into new area code, 330, in 1996. Call the local operator for assistance.			
Oxford	27565	919	7,965	Ada	45810	419	5,428
Pinehurst	28374	919	5,091	Akron	*44309	216	223,019
Piney Green	28399	919	8,999	Alliance	44601	216	23,376
Raleigh	*27611	919	212,092	Amherst	44001	216	10,332
Reidsville	*27320	910	12,183	Ashland	44805	419	20,079
Roanoke Rapids	27870	919	15,722	Ashtabula	44004	216	21,633
Rockingham	28379	910	9,399	Athens	45701	614	21,265
Rocky Mount	*27801	919	49,438	Aurora	44202	216	9,192
Roxboro	27573	910	7,332	Austintown	44515	216	32,371
Saint Stephens	28601	704	8,734	Avon	44011	216	7,337
Salisbury	*28144	704	23,626	Avon Lake	44012	216	15,066
Sanford	*27330	919	14,755	Barberton	44203	216	27,623
Shelby	*28150	704	14,669	Bay Village	44140	216	17,000
Smithfield	27577	919	7,540	Beachwood	44122	216	10,644
Southern Pines	*28387	910	9,213	Beavercreek	45434	513	33,626
Spring Lake	28390	919	7,552	Bedford	44146	216	14,822
Statesville	*28677	704	17,567	Bedford Heights	44146	216	12,131
Tarboro	27886	919	11,037	Bellaire	43906	614	6,028
Thomasville	*27360	910	15,915	Bellbrook	45305	513	6,511
Trinity	27370	919	5,469	Bellefontaine	43311	513	12,126
Wake Forest	*27587	919	5,832	Bellevue	44811	419	8,157
Washington	27889	919	9,160	Belpre	45714	614	6,796
Waynesville	28786	704	6,760	Berea	44017	216	19,051

Place	ZIP Code	Area Code	1990 Population	Place	ZIP Code	Area Code	1990 Population
Ohio (continued)				Englewood	45322	513	11,402
Bexley	43209	614	13,088	Euclid	44117	216	54,875
Blacklick Estates	43004	614	10,080	Fairborn	45324	513	31,300
Blue Ash	45242	513	11,923	Fairfield	45014	513	39,709
Boardman	44513	216	38,596	Fairlawn	44313	216	5,779
Bowling Green	43402	419	28,303	Fairview Park	44126	216	18,028
Brecksville	44141	216	11,818	Findlay	45839	419	35,703
Bridgetown North	45211	513	11,748	Forest Park	45405	513	18,621
Broadview Heights	44147	216	12,219	Fort McKinley	45426	513	9,740
Brooklyn	44144	216	11,706	Fostoria	44830	419	14,971
Brookpark	44142	216	22,865	Franklin	45005	513	11,026
Brunswick	44212	216	28,218	Fremont	43420	419	17,619
Bryan	43506	419	8,348	Gahanna	43230	614	23,898
Bucyrus	44820	419	13,496	Galion	44833	419	11,859
Cambridge	43725	614	11,748	Garfield Heights	44125	216	31,739
Campbell	44405	216	10,038	Geneva	44041	216	6,597
Canfield	44406	216	5,409	Girard	44420	216	11,304
Canton	*44711	216	84,161	Grandview Heights	43212	614	7,010
Celina	45822	419	9,923	Greenfield	45123	513	5,172
Centerville	45459	513	21,082	Greenville	45331	513	12,863
Cheviot	45211	513	9,616	Groesbeck	45239	513	6,684
Chillicothe	45601	614	21,923	Grove City	43123	614	19,661
Cincinnati	*45202	513	364,114	Hamilton	*45011	513	61,436
Circleville	43113	614	11,666	Harrison	45030	513	7,520
Cleveland	*44101	216	505,616	Heath	43056	614	7,231
Cleveland Heights	44118	216	54,052	Highland Heights	44124	216	6,249
Clyde	43410	419	5,776	Hilliard	43026	614	11,794
Columbus	*43216	614	632,945	Hillsboro	45133	513	6,235
Conneaut	44030	216	13,241	Howland Center	44484	216	6,732
Cortland	44410	216	5,652	Hubbard	44425	216	8,248
Coshocton	43812	614	12,193	Huber Heights	45424	513	38,696
Covedale	45238	513	6,669	Huber Ridge	43081	614	5,255
Cuyahoga Falls	*44222	216	48,950	Hudson	44236	216	5,159
Dayton	*45401	513	182,005	Huron	44839	419	7,067
Deer Park	45236	513	6,181	Independence	44131	216	6,500
Defiance	43512	419	16,787	Ironton	45638	614	12,751
Delaware	43015	614	19,966	Jackson	45640	614	6,167
Delphos	45833	419	7,093	Kent	44240	216	28,835
Dover	44622	216	11,329	Kenton	43326	419	8,356
Dublin	43016	614	16,366	Kenwood	43606	513	7,469
East Cleveland	44112	216	33,096	Kettering	45429	513	60,569
Eastlake	44094	216	21,161	Kirtland	44094	216	5,881
East Liverpool	43920	216	13,654	Lakewood	44107	216	59,718
East Palestine	44413	216	5,168	Lancaster	43130	614	34,507
Eaton	45320	513	7,396	Lebanon	45036	513	10,461
Elyria	*44035	216	56,746	Lima	*45802	419	45,553

Place	ZIP Code	Area Code	1990 Population	Place	ZIP Code	Area Code	1990 Population
Ohio (continued)				North Royalton	44133	216	23,197
Lincoln Village	43228	614	9,958	Northwood	43619	419	5,506
Logan	43138	614	6,725	Norton	44203	216	11,477
London	43140	614	7,807	Norwalk	44857	419	14,731
Lorain	*44052	216	71,245	Norwood	45212	513	23,674
Louisville	44641	216	8,087	Oakwood	45873	419	8,957
Loveland	45140	513	10,122	Oberlin	44074	216	8,191
Lyndhurst	44124	216	15,982	Olmsted Falls	44138	216	6,741
Macedonia	44056	216	7,509	Oregon	43616	419	18,334
Madeira	45243	513	9,141	Orrville	44667	216	7,712
Mansfield	*44901	419	50,627	Overlook-Page Manor	45431	513	13,242
Maple Heights	44137	216	27,089	Oxford	45056	513	18,937
Marietta	45750	614	15,026	Painesville	44077	216	15,769
Marion	*43302	614	34,075	Parma	44129	216	87,876
Martins Ferry	43935	614	8,003	Parma Heights	44130	216	21,448
Marysville	43040	513	9,656	Pepper Pike	44124	216	6,185
Mason	45040	513	11,450	Perry Heights	44646	216	9,055
Massillon	*44646	216	30,969	Perrysburg	*43551	419	12,551
Maumee	43537	419	15,561	Piqua	45356	513	20,612
Mayfield Heights	44124	216	19,847	Portage Lakes	44319	216	13,373
Medina	44256	216	19,231	Port Clinton	43452	419	7,106
Mentor	*44060	216	47,491	Portsmouth	45662	614	22,676
Mentor-on-the-Lake	44060	216	8,271	Ravenna	44266	216	12,069
Miamisburg	*45342	513	17,834	Reading	45215	513	12,038
Middleburg Heights	44130	216	14,702	Reynoldsburg	43068	614	25,748
Middletown	*45042	513	46,022	Richmond Heights	44143	216	9,611
Milford	45150	513	5,660	Rittman	44270	216	6,147
Montgomery	45242	513	9,733	Rocky River	44116	216	20,410
Moraine	45439	513	5,989	Rossford	43460	419	5,861
Mount Healthy	45231	513	7,580	Saint Bernard	45217	513	5,344
Mount Vernon	43050	614	14,550	Saint Clairsville	43950	614	5,136
Munroe Falls	44262	216	5,359	Saint Marys	45885	419	8,441
Napoleon	43545	419	8,884	Salem	44460	216	12,233
Newark	43055	614	44,396	Sandusky	44870	419	29,764
New Carlisle	45344	513	6,049	Sandusky South	44870	419	6,336
New Lexington	43764	614	5,117	Seven Hills	44131	216	12,339
New Philadelphia	44663	216	15,698	Shaker Heights	44120	216	30,867
Niles	44446	216	21,128	Sharonville	45241	513	13,121
Northbrook	45239	513	11,471	Sheffield Lake	44054	216	9,825
North Canton	44720	216	14,904	Shelby	44875	419	9,610
North College Hill	45239	513	11,002	Shiloh	44878	419	11,607
North Madison	44057	216	8,699	Sidney	45365	513	18,710
North Olmsted	44070	216	34,204	Silverton	45236	513	5,859
Northridge	45502	513	5,939	Solon	44139	216	18,548
Northridge	45414	513	9,448	South Euclid	44121	216	23,866
North Ridgeville	44039	216	21,564	Springboro	45066	513	6,574

Place	ZIP Code	Area Code	1990 Population	Place	ZIP Code	Area Code	1990 Population
Ohio (continued)				Wilmington	45177	513	11,199
Springdale	45246	216	10,621	Woodbourne-Hyde Park	45459	513	7,837
Springfield	*45501	513	70,487				
Steubenville	43952	614	22,125	Wooster	44691	216	22,427
Stow	44224	216	27,998	Worthington	43085	614	14,869
Streetsboro	44241	216	9,932	Wright-Patterson AFB	45433	513	8,579
Strongsville	44136	216	35,308	Wyoming	45215	513	8,128
Struthers	44471	216	12,284	Xenia	45385	513	24,836
Sylvania	43560	419	17,489	Youngstown	*44501	216	95,732
Tallmadge	44278	216	14,870	Zanesville	43701	614	26,778
The Village of Indian Hill	45243	513	5,383	**Oklahoma**			
				Ada	*74820	405	15,765
Tiffin	44883	419	18,604	Altus	*73521	405	21,910
Tipp City	45371	513	6,027	Alva	73717	405	5,495
Toledo	*43601	419	332,943	Anadarko	73005	405	6,586
Toronto	43964	614	6,127	Ardmore	*73401	405	23,079
Trenton	45067	513	6,189	Bartlesville	*74003	918	34,256
Trotwood	45426	513	8,816	Bethany	73008	405	20,075
Troy	45373	513	19,478	Bixby	74008	918	9,502
Twinsburg	44087	216	9,606	Blackwell	74631	405	7,538
Uhrichsville	44683	614	5,604	Broken Arrow	*74012	918	58,082
Union	45322	513	5,531	Chickasha	73018	405	14,988
University Heights	44118	216	14,787	Choctaw	*73020	405	8,545
Upper Arlington	43221	614	34,128	Claremore	*74017	918	13,280
Upper Sandusky	43351	419	5,906	Clinton	73601	405	9,298
Urbana	43078	513	11,353	Cushing	74023	918	7,218
Vandalia	45377	513	13,872	Del City	73115	405	23,928
Van Wert	45891	419	10,922	Duncan	*73533	405	21,732
Vermilion	44089	216	11,127	Durant	*74701	405	12,929
Wadsworth	44281	216	15,718	Edmond	*73034	405	52,310
Wapakoneta	45895	419	9,214	Elk City	*73644	405	10,428
Warren	*44481	216	50,793	El Reno	73036	405	15,414
Warrensville Heights	44122	216	15,745	Enid	*73701	405	45,309
Washington C.H.	43160	614	13,080	Fort Sill	73503	405	12,107
Wauseon	43567	419	6,322	Frederick	73542	405	5,221
Wellston	45692	614	6,049	Glenpool	74033	918	6,688
West Carrollton City	45449	513	14,403	Guthrie	73044	405	10,440
Westerville	43081	614	30,269	Guymon	73942	405	7,803
Westlake	44145	216	27,018	Henryetta	74437	918	5,872
Whitehall	43213	614	20,572	Hugo	74743	405	5,978
White Oak	45239	513	12,430	Idabel	74745	405	6,957
Wickliffe	44092	216	14,558	Jenks	74037	918	7,474
Willard	44890	419	6,210	Lawton	*73501	405	80,561
Willoughby	*44094	216	20,510	McAlester	*74501	918	16,370
Willoughby Hills	44094	216	8,427	Miami	*74354	918	13,142
Willowick	44095	216	15,269	Midwest City	73140	405	52,267

Place	ZIP Code	Area Code	1990 Population	Place	ZIP Code	Area Code	1990 Population
Oklahoma (continued)				Gresham	97030	503	68,249
Moore	73153	405	40,318	Hermiston	97838	541	10,047
Muskogee	*74401	918	37,708	Hillsboro	97123	503	37,598
Mustang	73064	405	10,434	Klamath Falls	*97601	541	17,737
Norman	*73069	405	80,071	La Grande	97850	541	11,766
Oklahoma City	*73125	405	444,724	Lake Oswego	*97034	503	30,576
Okmulgee	74447	918	13,441	Lebanon	97355	541	10,950
Owasso	74055	918	11,151	Lincoln City	97367	541	5,903
Pauls Valley	73075	405	6,150	McMinnville	97128	503	17,894
Ponca City	74601	405	26,359	Medford	*97501	541	47,021
Poteau	74953	918	7,210	Milton-Freewater	97862	541	5,533
Pryor Creek	74361	918	8,327	Milwaukie	97222	503	18,670
Sallisaw	74955	918	7,122	Newberg	97132	503	13,086
Sand Springs	74063	918	15,339	Newport	97365	541	8,437
Sapulpa	*74066	918	18,074	North Bend	97459	541	9,614
Seminole	*74868	405	7,071	Ontario	97914	541	9,394
Shawnee	*74801	405	26,017	Oregon City	97045	503	14,698
Stillwater	*74074	405	36,676	Pendleton	97801	541	15,142
Tahlequah	*74464	918	10,586	Portland	*97208	503	438,802
Tecumseh	74873	405	5,570	Prineville	97754	541	5,355
Tulsa	*74103	918	367,302	Redmond	97756	541	7,165
The Village	73156	405	10,353	Roseburg	97470	541	17,069
Vinita	74301	918	5,804	Roseburg North	97470	541	6,031
Wagoner	*74467	918	6,894	Saint Helens	97051	503	7,535
Warr Acres	73132	405	9,288	Salem	*97301	503	107,793
Weatherford	73096	405	10,124	Seaside	97138	503	5,359
Woodward	*73801	405	12,340	Silverton	97381	503	5,635
Yukon	*73099	405	20,935	Springfield	*97477	541	44,664
Oregon				Stayton	97383	503	5,011
Albany	97321	541	29,540	Sutherlin	97479	541	5,020
Aloha	97006	503	34,284	Sweet Home	97386	541	6,850
Ashland	97520	541	16,252	The Dalles	97058	541	11,021
Astoria	97103	503	10,069	Tigard	97223	503	29,435
Baker City	97814	541	9,140	Tualatin	97062	503	14,664
Beaverton	*97005	503	53,307	West Linn	97068	503	16,389
Bend	*97701	541	20,447	White City	97503	541	5,891
Canby	97013	503	8,990	Wilsonville	97070	503	7,106
Central Point	97502	541	7,512	Woodburn	97071	503	13,404
Coos Bay	97420	541	15,076	**Pennsylvania**			
Corvallis	*97333	541	44,757	** These towns are split between area codes 215 and 610; call the local operator for assistance.			
Cottage Grove	97424	541	7,403				
Dallas	97338	503	9,422	Aliquippa	15001	412	13,374
Eugene	*97401	541	112,773	Allentown	*18105	610	105,301
Florence	97439	541	5,171	Altoona	*16603	814	51,881
Forest Grove	97116	503	13,559	Ambler	19002	**215	6,609
Grants Pass	*97526	541	17,503	Ambridge	15003	412	8,133

Place	ZIP Code	Area Code	1990 Population	Place	ZIP Code	Area Code	1990 Population
Pennsylvania (continued)				Conshohocken	19428	**215	8,064
Archbald	18403	717	6,291	Coraopolis	15108	412	6,747
Ardmore	19003	610	12,646	Corry	16407	814	7,216
Arnold	15068	412	6,113	Crafton	15205	412	7,188
Avalon	15202	412	5,784	Croydon	19020	215	9,967
Baden	15005	412	5,074	Danville	17821	717	5,165
Baldwin	15234	412	21,923	Darby	19023	610	11,140
Bangor	18013	610	5,383	Darby Twp.	19036	610	10,955
Beaver	15009	412	5,028	Devon-Berwyn	19333	610	5,019
Beaver Falls	15010	412	10,687	Dickson City	18519	717	6,276
Bellefonte	16823	814	6,358	Donora	15033	412	5,928
Bellevue	15202	412	9,126	Dormont	15216	412	9,772
Berwick	18603	717	10,976	Downingtown	19335	610	7,749
Bethel Park	15102	412	33,823	Doylestown	18901	215	8,575
Bethlehem	*18016	610	71,427	Drexel Hill	19026	610	29,744
Blakely	18447	717	7,222	Du Bois	15801	814	8,286
Bloomsburg	17815	717	12,439	Dunmore	18512	717	15,403
Bradford	16701	814	9,625	Duquesne	15110	412	8,525
Brentwood	15227	412	10,823	East Norriton	19401	**215	13,324
Bridgeville	15017	412	5,445	Easton	*18042	610	26,276
Bristol	19007	215	10,405	East Stroudsburg	18301	717	8,781
Brookhaven	19015	610	8,567	East York	17405	717	8,487
Broomall	19008	610	10,930	Economy	15005	412	9,305
Butler	*16001	412	15,714	Edinboro	16412	814	7,736
California	15419	412	5,748	Edwardsville	18704	717	5,399
Camp Hill	*17011	717	7,831	Elizabethtown	17022	717	9,952
Canonsburg	15317	412	9,200	Ellwood City	16117	412	8,894
Carbondale	18407	717	10,664	Emmaus	18049	610	11,157
Carlisle	17013	717	18,419	Ephrata	17522	717	12,133
Carnegie	15106	412	9,278	Erie	*16501	814	108,718
Carnot-Moon	15108	412	10,187	Exeter	18643	717	5,691
Castle Shannon	15234	412	9,135	Fairless Hills	19030	215	9,026
Catasauqua	18032	610	6,662	Farrell	16121	412	6,835
Chambersburg	17201	717	16,647	Feasterville-Trevose	19047	215	6,696
Charleroi	15022	412	5,014	Folcroft	19032	610	7,506
Chester	*19013	610	41,856	Forest Hills	15221	412	8,173
Chester Twp.	19013	610	5,399	Forty Fort	18704	717	5,049
Clairton	15025	412	9,656	Fox Chapel	15238	412	5,319
Clarion	16214	814	6,457	Franklin	16323	814	7,329
Clarks Summit	18411	717	5,433	Franklin Park	15143	412	10,109
Clearfield	16830	814	6,633	Fullerton	18052	610	13,127
Clifton Heights	19018	610	7,111	Gettysburg	17325	717	7,025
Coatesville	19320	610	11,038	Glassport	15045	412	5,582
Collingdale	19023	610	9,175	Glenolden	19036	610	7,260
Columbia	17512	717	10,701	Glenside	19038	215	8,704
Connellsville	15425	412	9,229	Greensburg	15601	412	16,318

Place	ZIP Code	Area Code	1990 Population
Pennsylvania (continued)			
Greenville	16125	412	6,734
Grove City	16127	412	8,240
Hanover	17331	717	14,399
Harleysville	19438	**215	7,405
Harrisburg	*17105	717	52,376
Hatboro	19040	215	7,382
Hazleton	18201	717	24,730
Hellertown	18055	610	5,662
Hermitage	16148	412	15,260
Hershey	17033	717	11,860
Hollidaysburg	16648	814	5,624
Homeacre-Lyndora	16001	412	7,511
Horsham	19044	215	15,051
Huntingdon	16652	814	6,843
Indiana	15701	412	15,174
Jeannette	15644	412	11,221
Jefferson	15344	412	9,533
Jim Thorpe	18229	717	5,048
Johnstown	*15907	814	28,124
Kennedy Twp.	15108	412	7,152
Kennett Square	19348	610	5,218
King of Prussia	19406	**215	18,406
Kingston	18704	717	14,507
Kittanning	16201	412	5,120
Lancaster	*17604	717	55,551
Lansdale	19446	215	16,362
Lansdowne	19050	610	11,712
Latrobe	15650	412	9,265
Lebanon	17042	717	24,800
Lehighton	18235	610	5,914
Levittown	19055	215	55,362
Lewisburg	17837	717	5,785
Lewistown	17044	717	9,341
Lionville-Marchwood	19353	610	6,468
Lititz	17543	717	8,280
Lock Haven	17745	717	9,230
Lower Burrell	15068	412	12,251
McCandless Twp.	15237	412	28,781
McKeesport	*15134	412	26,016
McKees Rocks	15136	412	7,691
Mahanoy City	17948	717	5,209
Manheim	17545	717	5,011
Meadville	16335	814	14,318
Mechanicsburg	17055	717	9,452
Media	*19063	610	5,957

Place	ZIP Code	Area Code	1990 Population
Middletown (Dauphin)	17057	717	9,254
Middletown (Northampton)	18017	610	6,866
Millersville	17551	717	8,099
Milton	17847	717	6,746
Monaca	15061	412	6,739
Monessen	15062	412	9,901
Montgomeryville	18936	215	9,114
Moosic	18507	717	5,397
Morrisville	19067	215	9,765
Mount Carmel	17851	717	7,196
Mount Joy	17552	717	6,398
Mount Lebanon	15228	412	33,362
Munhall	15120	412	13,158
Municipality of Monroeville	15146	412	29,169
Municipality of Murrysville	15668	412	17,240
Nanticoke	18634	717	12,267
Nazareth	18064	610	5,713
New Brighton	15066	412	6,854
New Castle	*16108	412	28,334
New Cumberland	17070	717	7,665
New Kensington	15068	412	15,894
Norristown	*19401	610	30,754
Northampton	18067	610	8,717
North Braddock	15104	412	7,036
North Versailles	15137	412	12,302
Northwest Harborcreek	16421	814	6,662
Norwood	19074	610	6,162
Oakmont	15139	412	6,961
Oil City	16301	814	11,949
Old Forge	18518	717	8,834
Olyphant	18447	717	5,222
Palmerton	18071	610	5,394
Palmyra	17078	717	6,910
Paoli	19301	610	5,603
Parkville	17331	717	6,014
Penn Hills	15235	717	51,430
Perkasie	18944	215	7,787
Philadelphia	*19104	215	1,585,577
Phoenixville	19460	610	15,066
Pittsburgh	*15233	412	369,879
Pittston	*18640	717	9,389
Pleasant Hills	15236	412	8,884
Plum	15239	412	25,609

Place	ZIP Code	Area Code	1990 Population
Pennsylvania (continued)			
Plymouth	18651	717	7,134
Plymouth Meeting	19462	**215	6,241
Pottstown	19464	610	21,831
Pottsville	17901	717	16,603
Prospect Park	19076	610	6,764
Punxsutawney	15767	814	6,782
Quakertown	18951	215	8,982
Radnor Twp.	19087	610	28,705
Reading	*19612	610	78,380
Red Lion	17356	717	6,130
Richboro	18954	215	5,332
Ridley Park	19078	610	7,592
Ross Twp.	15237	412	33,482
Saint Marys	15857	814	5,511
Sayre	18840	717	5,791
Schuylkill Haven	17972	717	5,610
Scottdale	15683	412	5,184
Scott Twp.	15106	412	17,118
Scranton	*18505	717	81,805
Selinsgrove	17870	717	5,384
Shaler Twp.	15116	412	30,533
Shamokin	17872	717	9,184
Sharon	16146	412	17,533
Sharon Hill	19079	610	5,771
Shenandoah	17976	717	6,221
Shillington	19607	610	5,062
Shiloh	17404	717	8,245
Shippensburg	17257	717	5,331
Somerset	15501	814	6,454
Souderton	18964	215	5,957
South Williamsport	17701	717	6,496
Springfield	19064	717	24,160
State College	16804	814	38,981
Steelton	17113	717	5,152
Stowe Twp.	15136	412	7,681
Stroudsburg	18360	717	5,312
Sugar Creek	16323	717	5,532
Sunbury	17801	717	11,591
Swarthmore	19081	610	6,157
Swissvale	15218	412	10,637
Swoyersville	18704	717	5,630
Tamaqua	18252	717	7,943
Tarentum	15084	412	5,674
Taylor	18517	717	6,941

Place	ZIP Code	Area Code	1990 Population
Titusville	16354	814	6,434
Trooper	19401	610	5,137
Turtle Creek	15145	412	6,556
Tyrone	16686	814	5,743
Uniontown	15401	412	12,034
Upper Providence Twp.	19063	610	9,727
Upper Saint Clair	15241	412	19,692
Vandergrift	15690	412	5,904
Warren	16365	814	11,122
Washington	15301	412	15,864
Waynesboro	17268	717	9,578
West Chester	*19380	610	18,041
West Goshen	19380	610	8,948
West Mifflin	15122	412	23,644
Westmont	15905	814	5,789
West Norriton	19401	610	15,209
West Pittston	18643	717	5,590
West View	15229	412	7,734
Whitehall	18052	610	14,451
White Oak	15131	717	8,761
Wilkes-Barre	*18703	717	47,523
Wilkinsburg	15221	412	21,080
Wilkins Twp.	15145	412	7,487
Williamsport	17701	717	31,933
Willow Grove	19090	610	16,325
Wilson	15025	412	7,830
Woodlyn	19094	610	10,151
Wyndmoor	19118	215	5,682
Wyomissing	19610	610	7,332
Yeadon	19050	610	11,980
York	*17405	717	42,192
Rhode Island			
Barrington	02806	401	15,849
Bristol	02809	401	21,625
Burrillville	02830	401	16,230
Central Falls	02863	401	17,638
Charlestown	02813	401	6,478
Coventry	02816	401	31,083
Cranston	02910	401	76,060
Cumberland	02864	401	29,038
Cumberland Hill	02864	401	6,379
East Greenwich	02818	401	11,865
East Providence	02914	401	50,380
Exeter	02822	401	5,461
Glocester	02814	401	9,227

Place	ZIP Code	Area Code	1990 Population	Place	ZIP Code	Area Code	1990 Population
Rhode Island (continued)				Clinton	29325	864	9,603
Greenville	02828	401	8,303	Columbia	*29201	803	103,477
Hopkinton	02833	401	6,873	Conway	*29526	803	9,819
Johnston	02919	401	26,542	Darlington	29532	803	7,310
Kingston	02881	401	6,504	Dentsville	29204	803	11,839
Lincoln	02865	401	18,045	Dillon	29536	803	6,829
Middletown	02840	401	19,460	Easley	*29640	864	15,179
Narragansett	02882	401	15,004	Florence	*29501	803	29,913
Newport	02840	401	28,227	Forest Acres	29206	803	7,181
Newport East	02843	401	11,080	Gaffney	*29341	864	13,149
North Kingstown	02852	401	23,786	Gantt	29605	864	13,891
North Providence	02908	401	32,090	Georgetown	*29440	803	9,517
North Smithfield	02876	401	10,497	Goose Creek	29445	803	24,692
Pascoag	02859	401	5,011	Greenville	*29602	864	58,256
Pawtucket	*02860	401	72,644	Greenwood	*29646	864	20,807
Portsmouth	02871	401	16,857	Greer	*29650	864	10,322
Providence	*02904	401	160,728	Hanahan	29410	803	13,176
Scituate	02857	401	9,796	Hartsville	*29550	803	8,372
Smithfield	02917	401	19,163	Hilton Head Island	*29928	803	23,694
South Kingstown	02879	401	24,631	Homeland Park	29621	864	6,569
Tiverton	02878	401	14,312	Ladson	29456	803	13,540
Valley Falls	02864	401	11,175	Lake City	29560	803	7,153
Wakefield-Peacedale	*02880	401	7,134	Lancaster	*29720	803	8,914
				Laurens	29360	864	9,694
Warren	02885	401	11,385	Marion	29571	803	7,658
Warwick	*02887	401	85,427	Mauldin	29662	864	11,662
Westerly	02891	401	21,605	Moncks Corner	29461	803	5,599
Westerly Center	02891	401	16,477	Mount Pleasant	*29464	803	30,108
West Warwick	02893	401	29,268	Mullins	29574	803	5,910
Woonsocket	02895	401	43,877	Myrtle Beach	*29577	803	24,848
South Carolina				Newberry	29108	803	10,543
Abbeville	29620	864	5,778	North Augusta	29841	803	15,684
Aiken	*29801	803	20,386	North Charleston	29410	803	70,304
Anderson	*29621	864	26,385	North Myrtle Beach	*29582	803	8,731
Barnwell	29812	803	5,255	Oak Grove	29565	803	7,173
Beaufort	*29902	803	9,576	Orangeburg	29115	803	13,772
Belvedere	29841	803	6,133	Parris Island	29905	803	7,172
Bennettsville	29512	803	10,095	Rock Hill	*29730	803	41,610
Berea	29611	864	13,535	Saint Andrews	29417	803	25,692
Burton	29902	803	6,917	Sans Souci	29609	864	7,612
Camden	29020	803	6,696	Seneca	*29678	864	7,726
Cayce	29033	803	10,824	Simpsonville	29681	864	11,744
Charleston	*29402	803	79,925	Socastee	29577	803	10,426
Cheraw	29520	803	5,553	Spartanburg	*29306	864	43,479
Chester	29706	803	7,158	Summerville	*29483	803	22,519
Clemson	*29631	864	11,145	Sumter	*29150	803	40,977

Place	ZIP Code	Area Code	1990 Population	Place	ZIP Code	Area Code	1990 Population
South Carolina (continued)				Eagleton Village	37801	615	5,169
Taylors	29687	864	19,619	East Ridge	37412	423	21,101
Union	29379	864	9,840	Elizabethton	*37643	423	11,931
Walterboro	29488	803	5,595	Erwin	37650	423	5,017
Welcome	29611	864	6,560	Farragut	37922	423	12,802
West Columbia	*29169	803	10,944	Fayetteville	37334	615	7,158
Woodfield	29206	803	8,862	Franklin	*37064	615	20,098
York	29745	803	6,709	Gallatin	37066	615	18,794
South Dakota				Germantown	38138	901	33,016
Aberdeen	*57401	605	24,995	Goodlettsville	*37072	615	11,219
Brookings	57006	605	16,270	Greeneville	*37743	423	13,532
Ellsworth AFB	57706	605	7,017	Halls	38040	901	6,450
Huron	57350	605	12,448	Harriman	37748	423	7,119
Madison	57042	605	6,527	Harrison	37341	423	7,191
Mitchell	57301	605	13,798	Hendersonville	*37075	615	32,188
Pierre	57501	605	12,906	Humboldt	38343	901	9,651
Rapid City	*57701	605	54,523	Jackson	*38301	901	49,145
Sioux Falls	*57101	605	100,836	Jefferson City	37760	423	5,522
Spearfish Canyon	57783	605	6,966	Johnson City	*37601	423	49,479
Sturgis	57785	605	5,330	Kingsport	*37662	423	36,353
Vermillion	57069	605	10,034	Knoxville	*37950	423	165,039
Watertown	57201	605	17,632	La Follette	37766	423	7,201
Yankton	57078	605	12,703	LaVergne	37086	615	7,499
Tennessee				Lawrenceburg	38464	615	10,397
Alcoa	37701	423	6,400	Lebanon	*37087	615	15,208
Athens	*37303	423	12,054	Lenoir City	37771	423	6,147
Bartlett	38134	901	26,989	Lewisburg	37091	615	9,879
Bloomingdale	37660	423	10,953	Lexington	38351	901	5,810
Bolivar	38008	901	5,969	McKenzie	38201	901	5,168
Brentwood	*37027	615	16,392	McMinnville	37110	615	11,194
Bristol	*37621	423	23,421	Manchester	37355	615	7,709
Brownsville	38012	901	10,017	Martin	38237	901	8,588
Chattanooga	*37401	423	152,393	Maryville	*37804	423	19,208
Clarksville	*37040	615	75,542	Memphis	*38101	901	610,337
Cleveland	*37311	423	30,354	Middle Valley	37343	423	12,255
Clinton	*37716	423	8,960	Milan	38358	901	7,512
Collegedale	37315	423	5,048	Millington	*38053	901	17,866
Collierville	*38017	901	14,501	Morristown	*37813	423	21,316
Colonial Heights	37663	423	6,716	Murfreesboro	*37130	615	44,922
Columbia	*38401	615	28,583	Nashville	*37202	615	488,374
Cookeville	*38501	615	21,744	Newport	37821	423	7,123
Covington	38019	901	7,487	Oak Ridge	*37830	423	27,310
Crossville	*38555	423	6,930	Paris	38242	901	9,332
Dayton	37321	423	5,671	Portland	37148	615	5,165
Dickson	37055	615	8,783	Powell	37849	423	7,534
Dyersburg	*38024	901	16,321	Pulaski	38478	615	7,916

Place	ZIP Code	Area Code	1990 Population
Tennessee (continued)			
Red Bank	37415	423	12,320
Ripley	38063	901	6,188
Rockwood	37854	423	5,348
Savannah	38372	901	6,547
Sevierville	*37862	423	7,178
Shelbyville	37160	615	14,042
Signal Mountain	37377	423	7,034
Smyrna	37167	615	13,647
Soddy-Daisy	37379	423	8,240
Springfield	37172	615	11,227
Sweetwater	37874	423	5,066
Tullahoma	37388	615	16,761
Union City	38261	901	10,513
Winchester	37398	615	6,305
Texas			

New phone numbers in Houston are being assigned the 281 area code. New phone numbers in Dallas will receive the 972 area code.

Place	ZIP Code	Area Code	1990 Population
Abilene	*79604	915	106,707
Addison	75001	214	8,783
Alamo	78516	210	8,352
Alamo Heights	78209	210	6,502
Aldine	77039	713	11,133
Alice	*78332	512	19,788
Allen	75002	214	19,315
Alpine	*79830	915	5,622
Alvin	*77511	713	19,220
Amarillo	*79105	806	157,571
Andrews	79714	915	10,678
Angleton	*77515	409	17,140
Aransas Pass	*78336	512	7,180
Arlington	*76004	817	261,717
Athens	75751	903	10,982
Atlanta	75551	214	6,118
Austin	*78767	512	465,648
Azle	*76020	817	8,868
Balch Springs	75180	214	17,406
Bay City	*77414	409	18,170
Baytown	*77520	713	63,843
Beaumont	*77707	409	114,323
Bedford	*76021	817	43,762
Beeville	*78102	512	13,547
Bellaire	*77401	713	13,844
Bellmead	76704	817	8,336
Belton	76513	817	12,463
Benbrook	76126	817	19,564

Place	ZIP Code	Area Code	1990 Population
Big Spring	*79720	915	23,093
Bonham	75418	903	6,688
Borger	*79007	806	15,675
Brady	76825	915	5,946
Breckenridge	76424	817	5,665
Brenham	*77833	409	11,952
Bridge City	77611	409	8,010
Brownfield	79316	806	9,560
Brownsville	*78520	210	98,962
Brownwood	*76801	915	18,387
Bryan	*77801	409	55,002
Burkburnett	76354	817	10,145
Burleson	*76028	817	16,113
Cameron	76520	817	5,635
Canyon	79015	806	11,365
Canyon Lake	78132	210	9,975
Carrizo Springs	78834	210	5,745
Carrolton	*75006	214	82,169
Carthage	75633	903	6,496
Cedar Hill	75104	214	19,988
Cedar Park	*78613	512	5,121
Channelview	77530	713	25,564
Childress	79201	817	5,055
Cleburne	*76031	817	22,205
Cleveland	*77327	713	7,124
Cloverleaf	77015	713	18,230
Clute	77531	409	9,467
Coleman	76834	915	5,410
College Station	*77840	409	52,443
Colleyville	76034	817	12,724
Commerce	*75428	903	6,825
Conroe	77301	409	27,675
Converse	78109	512	8,887
Copperas Cove	76522	817	24,079
Corpus Christi	*78469	512	257,428
Corsicana	75110	903	22,911
Crockett	75835	409	7,024
Crowley	76036	817	6,974
Crystal City	78839	512	8,263
Cuero	77954	512	6,700
Dalhart	79022	806	6,246
Dallas	*75221	214/ 972	1,007,618
Dayton	77535	409	5,042
Deer Park	77536	713	27,424
Del Rio	*78840	210	30,705

Place	ZIP Code	Area Code	1990 Population	Place	ZIP Code	Area Code	1990 Population
Texas (continued)				Hereford	79045	806	14,745
Denison	*75020	903	21,505	Hewitt	76643	817	8,983
Denton	*76201	817	66,270	Highland Park	75205	214	8,739
De Soto	75115	214	30,544	Highlands	77562	713	6,632
Dickinson	77539	713	9,497	Hillsboro	76645	817	7,072
Donna	78537	210	12,652	Hitchcock	77563	409	5,868
Dumas	79029	806	12,871	Hondo	78861	210	6,018
Duncanville	75138	214	35,008	Houston	*77052	713/281	1,629,902
Eagle Pass	*78852	210	20,651				
Edinburg	78539	210	29,885	Humble	*77338	713	12,060
Edna	77957	512	5,343	Huntsville	*77340	409	27,925
El Campo	77437	409	10,511	Hurst	76053	817	33,574
El Paso	*79910	915	515,342	Ingleside	78362	512	5,696
Elsa	78543	210	5,242	Iowa Park	76367	817	6,072
Ennis	75119	214	13,869	Irving	*75015	214	155,037
Euless	76039	817	38,149	Jacinto City	77029	713	9,343
Everman	76140	817	5,672	Jacksonville	75766	214	12,765
Falfurrias	78355	512	5,788	Jasper	75951	409	7,160
Farmers Branch	75234	214	24,250	Jollyville	78729	512	15,206
Floresville	78114	210	5,247	Katy	77449	713	8,004
Forest Hill	76119	817	11,482	Kaufman	75142	214	5,251
Fort Bliss	79906	915	13,915	Keller	*76248	817	13,683
Fort Hood	76544	817	35,580	Kermit	79745	915	6,875
Fort Stockton	79735	915	8,524	Kerrville	*78028	210	17,384
Fort Worth	*76161	817	447,619	Kilgore	75662	903	11,066
Fredericksburg	78624	210	6,934	Killeen	*76540	817	63,535
Freeport	77541	409	11,389	Kingsville	*78363	512	25,276
Friendswood	77546	713	22,814	Kingwood	77325	713	37,397
Gainesville	76240	817	14,256	Kirby	78219	210	8,326
Galena Park	77547	713	10,033	Lackland AFB	78236	210	9,352
Galveston	77550	409	59,067	Lake Jackson	77566	409	22,771
Garland	*75040	214	180,635	La Marque	77568	409	14,120
Gatesville	76528	817	11,492	Lamesa	79331	806	10,809
Georgetown	*78626	512	14,840	Lampassas	76550	512	6,382
Gladewater	75647	903	6,027	Lancaster	75146	214	22,117
Gonzales	78629	210	6,527	La Porte	*77571	713	27,910
Graham	76450	817	8,986	Laredo	*78041	210	122,893
Grand Prairie	*75051	214	99,606	League City	*77573	713	30,159
Grapevine	*76051	817	29,198	Leon Valley	78268	210	9,581
Greenville	*75401	903	23,071	Levelland	*79336	806	13,986
Groves	77619	409	16,744	Lewisville	*75067	214	46,521
Haltom City	76117	817	32,856	Liberty	77575	713	7,690
Harker Heights	76543	817	12,932	Littlefield	79339	806	6,489
Harlingen	*78550	210	48,746	Live Oak	78233	210	10,023
Hearne	77859	409	5,132	Livingston	77351	409	5,019
Henderson	*75652	903	11,139	Lockhart	78644	512	9,205

Place	ZIP Code	Area Code	1990 Population	Place	ZIP Code	Area Code	1990 Population
Texas (continued)				Richmond	*77469	713	10,042
Longview	*75606	903	70,311	Rio Grande City	78582	210	9,891
Lubbock	*79408	806	186,206	River Oaks	77019	817	6,580
Lufkin	*75901	409	30,210	Robinson	76701	817	7,111
McAllen	*78501	210	84,021	Robstown	78380	512	12,849
McKinney	75070	214	21,283	Rockdale	76567	512	5,235
Mansfield	76063	817	15,615	Rockwall	75087	214	10,486
Marlin	76661	817	6,386	Roma	78584	210	8,059
Marshall	*75670	903	23,682	Rosenberg	77471	713	20,183
Mathis	78368	512	5,423	Round Rock	*78681	512	30,923
Mercedes	78570	210	12,694	Rowlett	*75088	214	23,260
Mesquite	*75149	214	101,484	Saginaw	76179	817	8,551
Mexia	76667	817	6,933	San Angelo	*76902	915	84,462
Midland	*79701	915	89,343	San Antonio	*78265	210	935,393
Midlothian	76065	214	5,040	San Benito	78586	210	20,125
Mineral Wells	*76067	817	14,935	San Juan	78589	210	10,815
Mission	*78572	210	28,653	San Marcos	*78666	512	28,738
Missouri City	77489	713	36,178	Santa Fe	*77510	713	8,429
Monahans	79756	915	8,101	Schertz	78154	210	10,597
Mount Pleasant	75455	903	12,291	Seabrook	77586	713	6,685
Nacogdoches	*75961	409	30,872	Seagoville	75159	214	8,969
Navasota	77868	409	6,296	Seguin	*78155	210	18,692
Nederland	77627	409	16,192	Seminole	79360	915	6,342
New Braunfels	*78130	210	27,334	Sherman	*75090	903	31,584
North Richland Hills	76118	817	45,895	Silsbee	77656	409	6,368
Odessa	*79761	915	89,699	Sinton	78387	512	5,549
Orange	*77630	409	19,370	Slaton	79364	806	6,078
Palestine	*75801	903	18,042	Snyder	*79549	915	12,195
Pampa	*79065	806	19,959	Socorro	79910	915	22,995
Paris	*75460	903	24,799	South Houston	77587	713	14,207
Pasadena	*77501	713	119,604	Spring	*77373	713	33,111
Pearland	*77581	713	18,927	Stafford	*77477	713	8,395
Pearsall	78061	210	6,924	Stephenville	76401	817	13,502
Pecos	79772	915	12,069	Sugar Land	*77478	713	24,549
Perryton	79070	806	7,619	Sulphur Springs	*75482	903	14,062
Pharr	78577	210	32,921	Sweetwater	79556	915	11,967
Plainview	*79072	806	21,698	Taylor	76574	512	11,472
Plano	*75074	214	127,885	Temple	*76501	817	46,150
Pleasanton	78064	210	7,678	Terrell	75160	214	12,490
Port Arthur	*77640	409	58,551	Texarkana	*75501	903	31,658
Portland	78374	512	12,224	Texas City	*77590	409	40,822
Port Lavaca	77979	512	10,886	The Colony	75056	214	22,113
Port Neches	77651	409	12,908	The Woodlands	77387	713	29,205
Raymondville	78580	210	8,880	Tomball	*77335	713	6,370
Richardson	*75080	214	74,840	Tyler	*75702	903	75,450
Richland Hills	76118	817	7,978	Universal City	78148	512	13,057

Place	ZIP Code	Area Code	1990 Population	Place	ZIP Code	Area Code	1990 Population
Texas (continued)				North Ogden	84404	801	11,593
University Park	76308	214	22,259	North Salt Lake	84054	801	6,464
Uvalde	*78801	210	14,729	Ogden	*84401	801	63,943
Vernon	*76384	817	12,001	Orem	*84057	801	67,561
Victoria	*77901	512	55,076	Payson	84651	801	9,510
Vidor	*77662	409	10,935	Pleasant Grove	84062	801	13,476
Waco	*76702	817	103,590	Price	84501	801	8,712
Watauga	76148	817	20,009	Provo	*84601	801	86,835
Waxahachie	75165	214	17,984	Richfield	84701	801	5,593
Weatherford	*76086	817	14,804	Riverdale	84403	801	6,419
Weslaco	*78596	210	21,877	Riverton	84065	801	11,261
West Odessa	79764	915	16,568	Roy	84067	801	24,595
West University Place	77005	713	12,920	Saint George	*84770	801	28,572
Wharton	77488	409	9,011	Salt Lake City	*84101	801	159,928
White Settlement	76108	817	15,472	Sandy	*84070	801	75,240
Wichita Falls	*76307	817	96,259	Smithfield	84335	801	5,566
Windcrest	78239	210	5,331	South Jordan	84095	801	12,215
Woodway	76712	817	8,695	South Ogden	84403	801	12,105
Wylie	75098	214	8,716	South Salt Lake	84115	801	10,129
Yoakum	77995	512	5,611	Spanish Fork	84660	801	11,272
Zapata	78076	512	7,119	Springville	84663	801	13,950
Utah				Sunset	84015	801	5,128
American Fork	84003	801	15,722	Taylorsville-Bennion	84107	801	52,351
Bountiful	*84010	801	36,147	Tooele	84074	801	13,887
Brigham City	84302	801	15,644	Union	84047	801	13,684
Canyon Rim	84109	801	10,527	Vernal	*84078	801	6,640
Cedar City	*84720	801	13,443	Washington Terrace	84403	801	8,189
Centerville	84014	801	11,500	West Jordan	84084	801	42,915
Clearfield	*84015	801	21,435	West Valley City	*84119	801	86,969
Clinton	84015	801	7,945	White City	84070	801	6,506
Cottonwood Heights	84121	801	28,766	Woods Cross	84087	801	5,384
Draper	84020	801	7,143	**Vermont**			
East Millcreek	84109	801	21,184	Barre	05641	802	9,482
Farmington	84025	801	9,049	Bennington	05201	802	16,451
Highland	84004	801	5,007	Brattleboro Center	05301	802	8,612
Holladay-Cottonwood	84117	801	14,095	Brattleboro	*05301	802	12,241
Kaysville	84037	801	13,961	Burlington	*05401	802	39,127
Kearns	84118	801	28,374	Colchester	*05446	802	14,731
Layton	*84041	801	41,784	Essex	05451	802	16,498
Lehi	84043	801	8,475	Essex Junction	*05452	802	8,396
Logan	*84321	801	32,771	Hartford	05047	802	9,404
Magna	84044	801	17,829	Lyndon	05849	802	5,371
Midvale	84047	801	11,886	Middlebury	05753	802	8,034
Millcreek	84109	801	32,230	Milton	05468	802	8,404
Mount Olympus	84117	801	7,413	Montpelier	*05446	802	8,247
Murray	84107	801	31,274	Northfield	05663	802	5,610

Place	ZIP Code	Area Code	1990 Population	Place	ZIP Code	Area Code	1990 Population
Vermont (continued)				Forest	24551	804	5,624
Rutland	*05701	802	18,230	Fort Belvoir	22060	703	8,590
Saint Albans	05478	802	7,339	Fort Hunt	22308	703	12,989
Saint Johnsbury	05819	802	7,608	Fort Lee	23801	804	6,895
Shelburne	05482	802	5,871	Franconia	22310	703	19,882
South Burlington	*05401	802	12,809	Franklin	23851	804	7,864
Springfield	05156	802	9,579	Fredericksburg	*22404	540	19,027
Swanton	05488	802	5,636	Front Royal	22630	540	11,880
Winooski	05404	802	6,649	Galax	24333	540	6,670
Virginia				Glen Allen	*23060	804	9,010
Abingdon	24210	540	7,003	Gloucester Point	23062	804	8,509
Alexandria	*22313	703	111,182	Great Falls	22066	703	6,945
Annandale	22003	703	50,975	Groveton	22306	540	19,997
Arlington	*22210	703	170,936	Hampton	*23670	804	133,811
Ashland	23005	804	5,864	Harrisonburg	22801	540	30,707
Bailey's Crossroads	22041	703	19,507	Herndon	*22070	703	16,139
Bedford	24523	540	6,073	Highland Springs	23075	804	13,823
Belle Haven	22306	804	6,427	Hollins	24019	540	13,305
Bellwood	23234	804	6,178	Hopewell	23860	804	23,101
Bensley	23234	804	5,093	Huntington	22303	703	7,489
Blacksburg	*24060	540	34,590	Hybla Valley	22306	703	15,491
Bluefield	24605	540	5,363	Idylwood	22043	703	14,710
Bon Air	23235	804	16,413	Jefferson	22042	804	25,782
Bristol	*24203	540	18,426	Lake Barcroft	22041	703	8,686
Buena Vista	24416	540	6,406	Lake Ridge	22191	703	23,862
Burke	*22015	703	57,734	Lakeside	23228	804	12,081
Cave Spring	24018	540	24,053	Laurel	23060	804	13,011
Centreville	22020	703	26,585	Leesburg	22075	540	16,202
Chantilly	*22021	703	29,337	Lexington	24450	540	6,959
Charlottesville	*22906	804	40,475	Lincolnia	22312	703	13,041
Chesapeake	*23320	804	151,982	Lorton	*22079	703	15,385
Chester	23831	804	14,986	Lynchburg	*24506	804	66,049
Christiansburg	*24073	540	15,004	McLean	*22101	703	38,168
Collinsville	24078	540	7,280	Madison Heights	24572	804	11,700
Colonial Heights	23834	804	16,064	Manassas	*22110	703	27,957
Covington	24426	540	6,991	Manassas Park	22110	703	6,734
Culpeper	22701	540	8,581	Mantua	22030	703	6,804
Dale City	22191	703	47,170	Marion	24354	540	6,630
Danville	*24541	804	53,056	Martinsville	*24112	540	16,162
Dumbarton	23228	804	8,526	Mechanicsville	23111	804	22,027
Dunn Loring	22027	703	6,509	Merrifield	22081	703	8,399
East Highland Park	23222	804	11,850	Montrose	23231	804	6,405
Emporia	23847	804	5,479	Mount Vernon	22121	703	27,485
Fairfax	*22030	703	19,629	Newington	22122	703	17,965
Falls Church	*22046	703	9,522	Newport News	*23607	804	171,439
Farmville	23901	804	6,046	Norfolk	*23503	804	261,250

Place	ZIP Code	Area Code	1990 Population
Virginia (continued)			
North Springfield	22151	703	8,996
Oakton	22124	703	24,610
Petersburg	*23804	804	37,027
Pimmit Hills	22043	703	6,019
Poquoson	23662	804	11,005
Portsmouth	*23705	804	103,910
Pulaski	24301	540	9,985
Quantico Station	22134	703	7,425
Radford	*24141	540	15,940
Reston	*22090	703	48,556
Richmond	*23232	804	202,798
Roanoke	*24022	540	96,509
Rose Hill	24281	540	12,675
Salem	24153	540	23,797
Seven Corners	22044	703	7,280
South Boston	24592	804	6,997
Springfield	*22150	703	23,706
Staunton	*24401	540	24,461
Sterling	20164	703	20,512
Suffolk	*23434	804	52,143
Sugarland Run	22170	703	9,357
Timberlake	24502	804	10,314
Tuckahoe	23229	804	42,629
Tysons Corner	22101	703	13,124
Vienna	*22180	703	14,852
Vinton	24179	540	7,643
Virginia Beach	*23458	804	393,089
Waynesboro	22980	540	18,549
West Gate	22110	703	6,565
West Springfield	22152	703	28,126
Williamsburg	*23185	804	11,409
Winchester	*22601	540	21,947
Wolf Trap	24592	804	13,133
Woodbridge	22191	703	26,401
Wytheville	24382	540	8,036
Washington			
Aberdeen	98520	360	16,565
Alderwood Manor-Bothell North	98036	206	22,945
Anacortes	98221	360	11,451
Auburn	*98002	206	33,650
Bellevue	*98009	206	86,872
Bellingham	*98225	360	52,179
Bonney Lake	98390	206	7,494
Bothell	*98011	206	12,345

Place	ZIP Code	Area Code	1990 Population
Bremerton	*98310	206	38,142
Bryn Mawr-Skyway	98178	206	12,514
Burien	98166	206	25,089
Camas	98607	360	6,442
Cascade-Fairwood	98055	206	30,107
Centralia	98531	360	12,101
Chehalis	98532	360	6,527
Cheney	99004	509	7,723
Clarkston	99403	509	6,753
College Place	99324	509	6,308
Des Moines	98198	206	17,283
Dishman	99213	509	9,671
East Renton Highlands	98056	206	13,218
East Wenatchee Bench	98801	509	12,539
Edmonds	*98020	206	30,743
Ellensburg	98926	509	12,360
Enumclaw	98022	360	7,227
Ephrata	98823	509	5,349
Esperance	99210	509	11,236
Everett	*98201	360	69,974
Fairwood	98055	206	5,807
Federal Way	*98063	206	67,554
Ferndale	98248	360	5,398
Fircrest	98466	206	5,258
Fort Lewis	98433	206	22,224
Grandview	98930	509	7,169
Hazel Dell North	98660	206	6,924
Hazel Dell South	98665	206	5,796
Hoquiam	98550	206	8,972
Inglewood-Finn Hill	98011	206	29,132
Issaquah	98027	206	7,786
Kelso	98626	360	11,767
Kenmore	98028	206	8,917
Kennewick	*99336	509	42,148
Kent	*98031	206	37,960
Kingsgate	98033	206	14,259
Kirkland	*98033	206	40,059
Lacey	98503	360	19,279
Lake Forest North	98155	206	8,002
Lakewood	98259	206	58,412
Longview	98632	360	31,499
Lynden	98264	360	5,709
Lynnwood	*98036	206	28,637
Marysville	98270	360	10,328
Mercer Island	98040	206	20,816

Place	ZIP Code	Area Code	1990 Population
Washington (continued)			
Moses Lake	98837	509	11,235
Mountlake Terrace	98043	206	19,320
Mount Vernon	98273	360	17,647
Mukilteo	98275	206	6,982
Newport Hills	98006	206	14,736
Normandy Park	98166	206	6,709
North City-Ridgecrest	98155	206	13,832
North Marysville	98270	206	18,711
Oak Harbor	98277	360	17,176
Olympia	*98501	360	33,729
Opportunity	99214	509	22,326
Orchards North	98662	360	6,479
Orchards South	98662	360	12,956
Otis Orchards-East Farms	99027	360	5,811
Parkland	98444	206	20,882
Pasco	99302	509	20,337
Port Angeles	98362	360	17,710
Port Townsend	98368	360	7,001
Pullman	99163	509	23,478
Puyallup	*98371	206	23,878
Redmond	*98052	206	35,800
Renton	*98058	206	41,688
Richland	99352	509	32,315
Richmond Beach-Innis Arden	98160	206	7,242
Richmond Highlands	98113	206	26,037
Riverton-Boulevard Park	98188	206	15,337
Seatac	*98148	206	22,694
Seattle	*98101	206	516,259
Sedro Woolley	98284	360	6,333
Selah	98942	509	5,113
Shelton	98584	360	7,241
Sheridan Beach	98155	206	6,518
Silverdale	98383	360	7,660
Silver Lake-Fircrest	98201	360	24,474
Snohomish	*98290	360	6,499
Spanaway	98387	206	15,001
Spokane	*99210	509	177,165
Steilacoom	98303	206	5,728
Sumner	98390	206	6,459
Sunnyside	98944	509	11,238
Tacoma	*98402	206	176,664
Toppenish	98948	509	7,419

Place	ZIP Code	Area Code	1990 Population
Tukwila	98138	206	11,874
Tumwater	98502	360	9,976
University Place	98464	206	27,701
Vancouver	*98661	360	46,380
Vancouver Mall	98662	360	6,938
Veradale	99037	509	7,836
Walla Walla	99362	509	26,482
Wenatchee	*98801	509	21,746
West Lake Stevens	98258	206	12,453
West Pasco	99301	509	7,312
White Center-Shorewood	98166	206	20,531
Yakima	*98903	509	54,843
West Virginia			
Beckley	*25801	304	18,274
Bluefield	24701	304	12,756
Bridgeport	26330	304	6,695
Buckhannon	26201	304	5,909
Charleston	*25301	304	57,287
Clarksburg	*26301	304	17,970
Cross Lanes	25301	304	10,878
Dunbar	25064	304	8,697
Elkins	26241	304	7,494
Fairmont	*26554	304	20,210
Grafton	26354	304	5,524
Huntington	*25704	304	54,844
Keyser	26726	304	5,870
Martinsburg	25401	304	14,073
Morgantown	*26505	304	25,879
Moundsville	26041	304	10,753
New Martinsville	26155	304	6,705
Nitro	25143	304	6,851
Oak Hill	25901	304	6,812
Parkersburg	26101	304	33,862
Princeton	*24740	304	7,043
St. Albans	25177	304	11,257
South Charleston	25303	304	13,645
Teays Valley	25560	304	8,436
Vienna	26105	304	10,862
Weirton	26062	304	22,124
Wheeling	26003	304	34,882
Wisconsin			
Allouez	54301	414	14,431
Antigo	54409	715	8,284
Appleton	59411	414	65,695
Ashland	54806	715	8,695

Place	ZIP Code	Area Code	1990 Population	Place	ZIP Code	Area Code	1990 Population
Wisconsin (continued)				Menasha	54952	414	14,711
Ashwaubenon	54304	414	16,376	Menomonee Falls	*53051	414	26,840
Baraboo	53913	608	9,203	Menomonie	54751	715	13,547
Beaver Dam	53916	414	14,196	Mequon	53092	414	18,885
Beloit	*53511	608	35,571	Merrill	54452	715	9,860
Berlin	54923	414	5,371	Middleton	53562	608	13,785
Brookfield	53045	414	35,184	Milwaukee	*53201	414	628,088
Brown Deer	53209	414	12,236	Monona	53716	608	8,637
Burlington	53105	414	8,855	Monroe	53566	608	10,241
Cedarburg	53012	414	10,086	Muskego	53150	414	16,813
Chippewa Falls	54729	715	12,727	Neenah	*54956	414	23,219
Cudahy	53110	414	18,659	New Berlin	53151	414	33,592
Delavan	53115	414	6,073	New London	54961	414	6,658
De Pere	54115	414	16,569	New Richmond	54017	715	5,106
Eau Claire	54703	715	56,806	Oak Creek	53154	414	19,513
Elkhorn	53121	414	5,337	Oconomowoc	53066	414	10,993
Elm Grove	53122	414	6,261	Onalaska	54650	608	11,414
Fitchburg	53714	608	15,648	Oshkosh	*54901	414	55,006
Fond du Lac	*54935	414	37,757	Platteville	53818	608	9,862
Fort Atkinson	53538	414	10,213	Pleasant Prairie	53158	414	12,037
Fox Point	53217	414	7,238	Plover	54467	715	8,176
Franklin	53132	414	21,855	Plymouth	53073	414	6,769
Germantown	53022	414	13,658	Portage	53901	608	8,640
Glendale	53209	414	14,088	Port Washington	53074	414	9,338
Grafton	53024	414	9,340	Prairie du Chien	53821	608	5,657
Green Bay	*54303	414	96,466	Racine	*53401	414	84,298
Greendale	53129	414	15,128	Reedsburg	53959	608	5,834
Greenfield	53220	414	33,403	Rhinelander	54501	715	7,382
Hales Corners	53130	414	7,623	Rice Lake	54868	715	7,998
Hartford	53027	414	8,188	Richland Center	53581	608	5,018
Hartland	53029	414	6,906	Ripon	54971	414	7,241
Howard	54303	414	9,874	River Falls	54022	715	10,610
Hudson	54016	715	6,378	Saint Francis	53207	414	9,245
Janesville	*53545	608	52,210	Shawano	54166	715	7,598
Jefferson	53549	414	6,078	Sheboygan	*53081	414	49,587
Kaukauna	54130	414	11,982	Sheboygan Falls	53085	414	5,823
Kenosha	*53140	414	80,426	Shorewood	53211	414	14,116
Kimberly	54136	414	5,406	South Milwaukee	53172	414	20,958
La Crosse	*54601	608	51,120	Sparta	54656	608	7,788
Lake Geneva	53147	414	5,979	Stevens Point	54481	715	23,006
Little Chute	54140	414	9,207	Stoughton	53589	608	8,786
McFarland	53558	608	5,232	Sturgeon Bay	54235	414	9,176
Madison	*53714	608	190,766	Sun Prairie	53590	608	15,333
Manitowoc	*54220	414	32,521	Superior	54880	715	27,134
Marinette	54143	715	11,843	Tomah	54660	608	7,570
Marshfield	54449	715	19,291	Two Rivers	54241	414	13,030

Place	ZIP Code	Area Code	1990 Population
Wisconsin (continued)			
Watertown	*53094	414	19,142
Waukesha	*53186	414	56,958
Waupun	53963	414	8,844
Wausau	*54403	715	37,060
Wauwatosa	53213	414	49,366
West Allis	53214	414	63,221
West Bend	53095	414	24,470
Weston	54476	715	9,714
Whitefish Bay	53217	414	14,272
Whitewater	53190	414	12,636
Wisconsin Rapids	*54494	715	18,245
Wyoming			
Casper	*82601	307	46,765
Cheyenne	*82001	307	50,008

Place	ZIP Code	Area Code	1990 Population
Cody	82414	307	7,897
Douglas	82633	307	5,076
Evanston	*82930	307	10,904
Gillette	*82716	307	17,545
Green River	82935	307	12,711
Lander	82520	307	7,023
Laramie	*82071	307	26,687
Powell	82435	307	5,292
Rawlins	82301	307	9,380
Riverton	82501	307	9,202
Rock Springs	*82901	307	19,050
Sheridan	82801	307	13,904
Torrington	82240	307	5,651
Worland	82401	307	5,742

6

Transportation

Cars and Trucks

Buying a Car or Truck

The most important number to know before buying a new car is the dealer's invoice cost. These services will provide that number, for a small fee, usually under $15; have your credit card and the make and model numbers ready when you call:

Automobile Consumer Services
(800) 223-4882

Consumer Reports New Car Price Service
(800) 933-5555

Nationwide Auto Brokers
(800) 521-7257

Buying services may also help you find the best price on a car and free you from the hassle of negotiating with car dealers. Here are a few; they all charge a fee:

AutoAdvisor
(800) 326-1976

Automobile Consumer Services
(800) 223-4882

Car Bargains
(800) 475-7283

Nationwide Auto Brokers
(800) 521-7257

USAA Auto Acquisition Services (available to military officers, retirees, and eligible government employees only)
(800) 531-8905

To check prices on used cars, refer to one of these guides, which can be found in libraries and bookstores:

- The Kelley *Blue Book*
- The NADA *Retail Consumer Guide,* available by subscription; call (800) 248-6232

Two telephone services provide advice on buying used cars:

Consumers Digest Used Car Price Service
(900) 884-2277 ($1.50 per minute)

Consumers Union Used Car Buying Service
(900) 446-0500 ($1.75 per minute)

Rule of Thumb: New Car Markups The list, or sticker, price on most new cars includes the dealer's markup. That markup is usually 10 to 15 percent on economy, subcompact, compact, and sporty cars; 18 to 20 percent on intermediate and small luxury cars; 22 percent on full-size cars; and 25 percent on luxury cars, vans, and pickup trucks. Also assume a 30-percent markup on extras like vinyl roofs, air conditioning, stereo systems, and power windows.

Buying vs. Leasing

In early 1995, 31 percent of new cars and 22 percent of new trucks were being leased rather than purchased in the United States. The following worksheet will help you decide whether it is better to buy or lease a new car. First, however, obtain these numbers:

1. Negotiate the *best price* for a car before you decide whether to buy or lease, deducting any rebates.

2. Ask the dealer to tell you the *rate or money factor* of the lease. This number is similar to the interest rate you would pay if financing the car. It may be a decimal (for example, 0.0041). Multiply that number by 2,400 to get the percentage rate; then compare that rate to the current financing rate.

3. Find out the *residual value*—the amount the car will be worth at the end of the lease period. The higher the number, the lower your monthly payments will be because the car will be worth more at the end of the lease.

4. Find out if there are any *fees, end-of-lease charges, or security deposits*.

5. Decide whether the *yearly mileage limit* is right for you; if it isn't, find out the *charges for additional miles*.

Worksheet

Costs	Buying	Leasing	
Initial costs			
Down payment/security deposit	_____	_____	
Fees, license	_____	_____	
Sales tax	_____	_____	
Subtotal	_____	_____	(A)
Monthly payment × number of months of loan/lease	_____	_____	(B)
Other costs			
Interest lost on down payment if invested elsewhere (find out current bank savings rate and refer to interest tables in Chapter 1)	_____	_____	
Lease termination fee	_____	_____	
Mileage charges	_____	_____	
Subtotal	_____	_____	(C)
Add A, B, C for cost after finance/lease period	_____	_____	(D)
Subtract estimated value of owned car at end of loan *or* refundable security deposit at end of lease	_____	_____	(F)
TOTAL COST (subtract E from D)	_____	_____	

Car Color Preferences   Nearly 25 percent of all cars are painted some shade of red, with one exception: almost 30 percent of luxury models are white or silver. In second place for all types of cars is white, at 20 percent.

Car and Truck Hotlines

To report—	Call—
Problems with your car or truck operating correctly, or to find out about safety recalls	Auto and Truck Safety Hotline National Highway Traffic Safety Administration (800) 424-9393
Problems with dealers or car warranties	Federal Trade Commission (202) 326-2222

Numbers to Know: Oil and Gasoline

Oil is measured in terms of its *viscosity*, which is a measure of a liquid's ability to flow. There are 10 grades, from 0W to 25W for oils tested to work at 0°F and meant for winter weather use (the W stands for winter), and from 20 to 60 for oils rated to work at 212°F. The lower the number, the thinner the oil. Multigrade oils, like 10W-30, were developed to stay thin at low temperatures and still work well at high temperatures.

Most experts recommend 5W-30 for very cold weather and 10W-30 for warmer weather. They don't recommend 10W-40 oil because it can leave deposits in newer cars. Check your owner's manual for specific recommendations for your car.

Gasoline is measured in terms of *octane*, which is not a measure of how much octane it contains (it may contain none), but rather a way to compare the performance of that blend of gasoline with pure iso-octane, a very expensive high-performance fuel. Experts advise that you use the lowest rating of octane your car will take without knocking. Note that words like "premium" and "regular" don't necessarily mean a fixed octane rating; some states set minimum standards for these ratings, and others do not regulate octane ratings at all.

Car Maintenance Schedule

These are general recommendations only; in some categories, such as spark plugs, newer cars may need less frequent upkeep. Check your owner's manual.

Item	How Often	What to Do
Air filter	12,000–24,000 miles	Check; replace as necessary
Antifreeze/coolant	24,000 miles	Replace
Battery water	2,000 miles	Check level; fill
Brake fluid	1,000 miles	Check level; fill
Brake pads	12,000 miles	Check for damage and wear
Coolant hoses, drive belts	6,000 miles	Check; replace as necessary or every 4 years
Distributor cap/rotor	30,000 miles	Replace
Fuel filter	30,000 miles	Replace
Oil	3,000 miles	Replace
Oil filter	7,500 miles	Replace
Power steering fluid	3,000 miles	Check level; fill; replace every 2 years
Shock absorbers	15,000 miles	Check; replace if leaking or ineffective
Spark plugs	30,000 miles	Replace
Tires	6,000–12,000 miles	Rotate
Transmission filter	30,000 miles	Replace
Transmission fluid	1,000 miles	Check level; replace at filter change
Wheels	15,000 miles	Check alignment

Average Life Spans of Car Parts

Alternator	2–6 years
Automatic transmission	50,000–70,000 miles
Battery	30,000–50,000 miles
Brakes	
Disc caliper	60,000–80,000 miles
Disc pad	30,000–45,000 miles
Drum	100,000 miles
Carburetor	30,000–75,000 miles
Clutch	50,000–60,000 miles
Distributor	3–5 years
Exhaust pipe	3–4 years
Manual transmission	60,000–100,000 miles
Muffler	2–3 years
Oil pump	60,000–70,000 miles
Shock absorbers	20,000 miles
Valves	100,000 miles
Water pump	2–5 years

Numbers to Know: Tires

The numbers used to identify tire types and sizes—for example, P205/70R14—can be translated as follows:

P = Passenger vehicle; LT means light truck.

205 = Width in millimeters of the tire, sidewall to sidewall

70 = Aspect ratio, or the relationship between the height of the tire from the road to the rim in relation to the tire's width; in this example, the height is 70 percent of the width.

R = Type of tire, in this case radial. B means belted bias and D stands for bias. The letters S, T, U, H, V, or Z may appear before the R and represent speed ratings for high-performance tires, with S being the lowest (up to 112 mph) and Z the highest speed (over 149 mph) at which the tires can be driven safely.

14 = Rim diameter in inches.

Tires for light trucks may be numbered like this: LT 31X10.5R15. Here, the 31 is the tire's diameter in inches (not in millimeters, as for cars), and the 10.5 is the sidewall to sidewall width in inches.

The federal government requires that manufacturers label each tire with information on treadwear, traction, and heat resistance. The *treadwear* grade indicates how much mileage you can expect from the tire. Multiply the treadwear grade by 200 to get an estimated number of miles the tire should last; for example, a grade of 150 (multiply 150 × 200) means that the tire should last for about 30,000 miles.

Traction is rated as A, B, or C, with A meaning that the tires will stop sooner on a wet road in a shorter distance than tires rated B or C.

Heat resistance is also rated as A, B, or C, with A indicating a tire that will run cooler and be safer for driving long distances at highway speeds than B- or C-rated tires. Call the National Highway Traffic Safety Administration at (800) 424-9393 or (202) 366-0123 for a complete list of tire gradings.

The *load range* is printed on the side of the tire and indicates the maximum load the tire is meant to carry. To see if you are exceeding this amount, add up the weight of your car and its baggage and passengers, and divide by 4. The result should be less than the load number on the tire.

Finally, the last three digits of the Department of Transportation's (DOT's) serial number, printed on the side of every tire, will tell you the *tire manufacture date*. The first two numbers tell you the week of the year, and the last number tells you the year; for example, 046 means the tire was made in the fourth week of 1996. Try to select a tire made within the past year.

Tire pressure can increase as much as 6 pounds per square inch when the tire is hot; always check tire pressure when tires are cold.

Experts recommend rotating tires every 6,000 miles to maintain even treadwear and help tires last longer.

For information on tire recycling, call the Scrap Tire Management Council at (202) 408-7781.

Numbers to Know: Fuel Efficiency

Here are some miscellaneous facts and figures on fuel efficiency:

- Keep car windows closed; wind resistance can consume up to 5 miles a gallon in fuel efficiency.
- The average car loses 2 miles per gallon in fuel efficiency for every 5 miles per hour over 50.
- Don't let a car idle for more than 1 minute—after that time, it's more fuel efficient to shut off the engine and start it up again.
- A dirty air filter can increase fuel consumption by as much as 10 percent.
- New spark plugs can increase fuel economy by as much as 13 percent.
- Underinflated tires can decrease fuel economy by as much as 5 percent.

Determining the Cost of Owning Your Car

Item	Yearly Costs
Gasoline	
Divide cost per gallon by miles per gallon to get cost per mile driven; then multiply that figure times number of miles driven a year (e.g., $1.20/gallon ÷ 30 miles per gallon = $0.04/gallon × 10,000 miles/year = $400)	$_____
Oil (adding, changing)	_____
Other maintenance, repairs	_____
Tires	_____
Insurance	_____
Taxes	_____
Depreciation (initial cost of car divided by number of years you expect to own it)	_____
License/registration fees	_____
Finance charges	_____
TOTAL COST PER YEAR	$_____
TOTAL COST PER MILE (divide total by miles driven per year)	$_____

Car Costs On average, it costs the average American about $12.78 a day, or 46.65 cents per mile, to own and operate a 1994 model intermediate-size car, according to the American Automobile Association. This estimate assumes the car is owned for 6 years or 60,000 miles and includes costs of insurance, taxes and fees, finance charges, and depreciation.

Braking Distances

This table shows how far your car will continue to travel, from the time you start to apply the brakes to the time the car comes to a stop.

Speed (mph)	Total Braking Distance (feet)
10	20
20	45
30	78
40	125
50	188
60	272
70	381

Car and Truck Weights

Vehicle Type	Weight (lbs.)
Subcompact	Less than 2,500
Compact	2,501 to 3,000
Intermediate	3,001 to 3,500
Large	3,501 and over
Passenger van	Less than 5,000
Medium truck	14,001 to 33,000
Heavy truck	33,001 and over

Mean Commuting Time to Work, 1960 through 1990

The Texas Transportation Institute at Texas A&M University estimates that traffic congestion costs the United States $40 billion in lost productivity a year. The institute estimates that during rush hours, two-thirds of cars on interstate highways are traveling at speeds of less than 35 miles per hour.

Commuting Fact	1960	1970	1980	1990
Mean travel time to work (minutes)	—	—	21.7	22.4
Percentage using private vehicle	69.48	80.63	85.92	88.02
Percentage using public transportation	12.62	8.48	6.22	5.12
Percentage who walked to work	10.37	7.40	5.60	3.90
Percentage who worked at home	7.54	3.49	2.26	2.96

Source: Federal Highway Administration, U.S. Department of Transportation.

Percentage of Households with Cars, 1960 through 1990

Number of Vehicles per Household	1960	1970	1980	1990
None	21.53	17.47	12.92	11.53
One	56.94	47.71	35.53	33.74
Two	19.00	29.32	34.02	37.35
Three or more	2.53	5.51	17.52	17.33

Source: Federal Highway Administration, U.S. Department of Transportation.

Horsepower Defined

One horsepower is equal to 33,000 foot-pounds of work per minute, or 550 foot-pounds per second. The term came about when James Watt, who had invented a steam engine, wanted to show the machine's power. He determined that a strong horse could lift 150 pounds 220 feet in 1 minute. One horsepower then became 150 x 220/1, or 33,000 foot-pounds per minute, or 550 foot-pounds per second, which the *Oxford English Dictionary* states is about 1$\frac{1}{3}$ times the actual power of a horse. It quotes a writer in the Glasgow *Herald* in 1897: "The term horsepower has probably seen its best days.... As a scientific term it has been much abused, and as a commercial term it conveys no meaning."

Motor Vehicle Traffic Fatalities

Year	Number of Registered Vehicles				Number of Traffic Fatalities	Traffic Fatality Rate*
	Automobiles	**Trucks**	**Buses**	**Motorcycles**		
1960	61,684,000	11,914,000	272,000	574,000	36,399	5.06
1965	75,261,000	14,786,000	314,000	1,312,000	47,089	5.30
1970	89,244,000	18,797,000	378,000	2,824,000	54,180	4.92
1975	106,705,000	25,781,000	462,000	4,967,000	44,525	3.36
1980	121,601,000	33,667,000	529,000	5,694,000	51,091	3.35
1985	131,864,000	39,196,000	594,000	5,444,000	43,825	2.47
1990	143,453,000	44,718,000	627,000	4,259,000	44,599	2.08
1991	142,956,000	44,785,000	631,000	4,177,000	41,508	1.91
1992	144,213,000	45,504,000	645,000	4,065,000	39,250	1.75
1993**	145,740,000	47,125,000	(a)	4,001,000	40,115	1.75

Source: Federal Highway Administration and National Highway Traffic Safety Administration, U.S. Department of Transportation.

*Per 100 million vehicle-miles.

**Estimate.

a. Included in truck figure.

Ships

The Size of a Ship: Tonnage

Rating Method	What It Means
Displacement tonnage	The weight of the water displaced by the ship, arrived at by dividing the cubic footage of the submerged area by 35 (the number of cubic feet in a ton of sea water), then converting the result into long tons (2,240 pounds). Used for warships and U.S. merchant ships.
Gross registered tonnage	The enclosed capacity of the ship, arrived at by dividing the vessel's total cubic footage by 100 (100 cubic feet in this case is 1 ton). Used for passenger and merchant ships.
Deadweight tonnage	The total weight in long tons of a fully loaded ship. Used for freighters and tankers.
Net registered tonnage	The result of subtracting the space that can't be used for passengers or cargo from the gross registered tonnage.

Numbers to Know: Ships

The largest cruise ship in the world is the *Norway*, with a gross registered tonnage of 76,049 and a length of 1,035.5 feet.

The largest tanker is the *Jahre Viking*, at 622,420 tons deadweight and 1,471 feet long. The largest dry cargo ship is the Norwegian carrier *Berge Stahl*, which is 402,082 tons deadweight and 1,125 feet long.

For information on numbers to consider when choosing a cruise ship, as well as cruise line phone numbers, see Chapter 4, Travel.

Airplanes

Passenger Airline Facts: Seats and Sizes

Aircraft	Length	Cruising Speed (mph)	Number of Seats*
Airbus 300	177'5"	543	267
Airbus 320	123'3"	515	179
Airbus 340	194'10"	605	262
Aerospatiale/Concorde	203'9"	1,336	128
Boeing 727	153'2"	570	189
Boeing 737-300	109'7"	564	128–149
Boeing 747-400	231'10"	604	421–568
Boeing 757	155'3"	494	178–208
Boeing 767-200	159'2"	494	240–300
Boeing 777	209'1"	494	313–375
British Aerospace BAe 146-100	93'8"	440	82–100
Fokker F50	82'10"	325	46–58
Fokker F100	116'6"	508	107–122
Lockheed L1011	177'8"	558	345
McDonnell Douglas DC-9	147'10"	565	158
McDonnell Douglas DC-10			
MD-10	182'1"	540	255–380
MD-11	200'10"	578	277–323
McDonnell Douglas MD-80	147'10"	528	172

*Seating capacity can vary depending on arrangement of seats.

**Boeing 777
Statistics**
The Boeing 777 was first flown commercially in June 1995; its length of 209 feet and wingspan of 199 feet make it the largest 2-engine plane in the world. It contains 132,500 engineered parts; adding fasteners such as rivets and bolts brings the total number of parts in the plane to 3 million. Boeing spent $4.3 billion on its research and development and another $4 billion on plant and equipment costs.

Leading U.S. Airports, 1994

Airport	Passengers (arriving & departing)
Chicago O'Hare	66,435,252
Atlanta	54,090,579
Dallas/Ft. Worth	52,601,125
Los Angeles	51,050,275
San Francisco	34,550,652
Denver	33,129,126
Miami	30,203,269
New York Kennedy	28,799,275
Newark	27,995,721
Las Vegas	26,810,020
Detroit	26,797,876
Phoenix	25,619,704
Boston	25,360,104
Minneapolis/St. Paul	24,513,419
St. Louis	23,362,671
Houston Intercontinental	22,521,655
Honolulu	22,473,611
Orlando	22,347,412
Seattle	20,972,819
New York LaGuardia	20,808,242

Source: Air Transport Association, *1995 Annual Report.*

Commercial Aircraft Accident Rate, by Aircraft Type

Aircraft	Accident Rate per Million Departures*
Fokker F28	3.82
McDonnell Douglas DC-10	2.67
Airbus A320/A321	2.50
Boeing 747-400	1.86
Boeing 747-100/200/300	1.71
McDonnell Douglas DC9	1.18
Boeing 737-100/200	1.17
Airbus A300	0.98
Lockheed L1011	0.94
Boeing 727	0.87
Boeing 737-300/400/500	0.61
McDonnell Douglas MD-80	0.60
Boeing 767	0.29
Boeing 757	0

Source: Boeing Corporation.
*As of September 1994.

Accidents and Fatalities for U.S. Air Carriers

Year	Departures (millions)	Fatal Accidents	Fatalities
1984	5.4	1	4
1985	5.8	4	197
1986	6.4	2	5
1987	6.6	4	231
1988	6.7	3	285
1989	6.6	8	131
1990	6.9	6	39
1991	6.8	4	62
1992	7.1	4	33
1993	7.2	1	1
1994	7.5	4	237

Source: National Transportation Safety Board, U.S. Department of Transportation.
Note: For scheduled service aircraft with 30 seats or more.

**Mach 1
Defined**
Ernst Mach, an Austrian scientist in the 1800s, developed a scale that accounts for the fact that the speed of sound varies depending on the condition of the air it travels through. Sound travels faster through warm air than cold, and through humid air faster than through dry air. Mach 1, then, is the speed of an object traveling at the local speed of sound; that is, an object's speed in relation to that of sound under the same atmospheric conditions. Mach 2 measures the speed of an object that is traveling twice the speed at which sound travels under the same conditions.

Top 30 Domestic Airline Markets
(for 12 months ended June 1994)

Rank	Between—	And—	Passengers (outbound plus inbound)
1	New York	Los Angeles	2,821,910
2	New York	Chicago	2,664,240
3	New York	Boston	2,511,250
4	New York	Miami	2,474,220
5	Honolulu	Kahului, HI	2,316,680
6	New York	Washington, DC	2,247,840
7	Dallas/Ft. Worth	Houston	2,224,240
8	Los Angeles	San Francisco	2,164,500
9	New York	San Francisco	1,926,970
10	New York	Orlando	1,883,570
11	New York	Ft. Lauderdale	1,753,460
12	New York	Atlanta	1,716,300
13	New York	San Juan	1,657,460
14	Los Angeles	Las Vegas	1,601,880
15	Chicago	Detroit	1,365,900
16	Honolulu	Lihue, HI	1,323,120
17	Los Angeles	Phoenix	1,322,930
18	New York	West Palm Beach	1,289,270
19	Los Angeles	Honolulu	1,255,820
20	Los Angeles	Oakland	1,216,900
21	Honolulu	Kona, HI	1,158,500

Top 30 Domestic Airline Markets
(for 12 months ended June 1994) (continued)

Rank	Between—	And—	Passengers (outbound plus inbound)
22	Chicago	Los Angeles	1,132,240
23	Boston	Washington, DC	1,100,120
24	Honolulu	Hilo, HI	1,082,210
25	San Francisco	San Diego	1,049,700
26	Chicago	St. Louis	1,023,190
27	New York	Tampa	980,720
28	New York	Dallas/Ft. Worth	975,910
29	Chicago	Minneapolis	938,070
30	Chicago	Atlanta	922,680

Source: Air Transport Association, *1995 Annual Report*.

Note: Includes all commercial airports in a metropolitan area. Does not include connecting passengers.

Overbooking: Number of Passengers Denied Boarding
by U.S. Airlines, 1986 through 1993

Year	Voluntary	Involuntary	Total Passengers Denied Boarding
1986	557,000	167,000	724,000
1987	705,000	169,000	874,000
1988	648,000	128,000	776,000
1989	616,000	107,000	723,000
1990	561,000	67,000	628,000
1991	599,000	47,000	646,000
1992	718,000	46,000	764,000
1993	632,000	51,000	683,000

Source: U.S. Department of Transportation.

Number of Weapons Found During Airline Passenger Screenings, 1975 through 1993

Year	Persons Screened	Handguns	Long Guns	Explosive/ Incendiary Devices	Other*
1975	202,000,000	1,993	—	—	—
1980	585,000,000	1,878	36	8	108
1985	993,000,000	2,823	90	12	74
1990	1,145,000,000	2,490	59	15	304
1991	1,015,000,000	1,597	47	94	275
1992	1,111,000,000	2,503	105	167	2,341
1993	1,150,000,000	2,707	91	251	3,867

Source: U.S. Department of Transportation.

*Beginning in 1992 other dangerous articles included stunning devices, chemical agents, martial arts equipment, knives, and bludgeons.

7

Time

Clocks

Types of Time

Type	What It Is
Apparent solar	The sun at its highest point in the sky; sundial time; can vary due to variations in orbit and rotation. Mean solar time is an average of all apparent solar days in an orbital year. A mean solar day is 24 hours, 3 minutes, and 56 seconds.
Apparent sidereal	Measures the rotation of the earth in relation to stars; used by astronomers. A mean sidereal day is 23 hours, 56 minutes, and 4 seconds. A sidereal month is 27.3217 days. A sidereal year is 365.2564 days.
Ephemeris	Based on idealized motions of the sun and moon; also used by astronomers; uses 1 orbit of the earth around the sun, a tropical year (365.2422 days). One second of ephemeris time is a fraction of the tropical year 1900 (which equals 31,556,925.9747 seconds).
Atomic	Uses atomic clocks, which measure the vibrations of energized cesium atoms; kept by the National Institute for Standards and Technology, which must add leap seconds every few years. The International System (SI) second is the time it takes cesium atoms to oscillate 9,192,631,770 times.
Universal time (UT)	Also known as Greenwich mean time (GMT), based on earth's rotation and uses atomic time to standardize time worldwide.
Universal time coordinated (UTC)	Coordinates atomic clock time with the U.S. Naval Observatory's sidereal time.
U.S. standard time	Based on UTC, translated into U.S. standard time zone time.
Daylight saving time	U.S. standard time plus 1 hour; begins first Sunday in April and ends on last Sunday in October.

The Almost Universal Week

The 7-day week has been used by almost every world society, yet the choice of 7 days is the most arbitrary of all numbers used to express time. The number has held great significance for many societies, perhaps because it was the number of close celestial bodies—the 5 known planets of early times plus the sun and the moon. In the Babylonian calendar, every 7th day was set aside for worship to God.

World Time Zones

The idea of time zones was introduced by U.S. railroad companies in 1883. Four zones were set across the United States, each of which was 15° longitude wide, or the distance it takes the sun to travel 1 hour. International time zones were set in the following year; the earth was divided into 24 zones using Greenwich, England, as the baseline and the 180° longitude line as the International Date Line. If you travel west, toward the Far East, across this line you will lose a day; if you travel east across it you will gain a day.

The lines are numbered +1 to +11 to the east of the Greenwich line, and numbered −1 to −11 to the west of that line. The International Date Line is +12. The lines themselves may bend and curve to avoid splitting cities and regions.

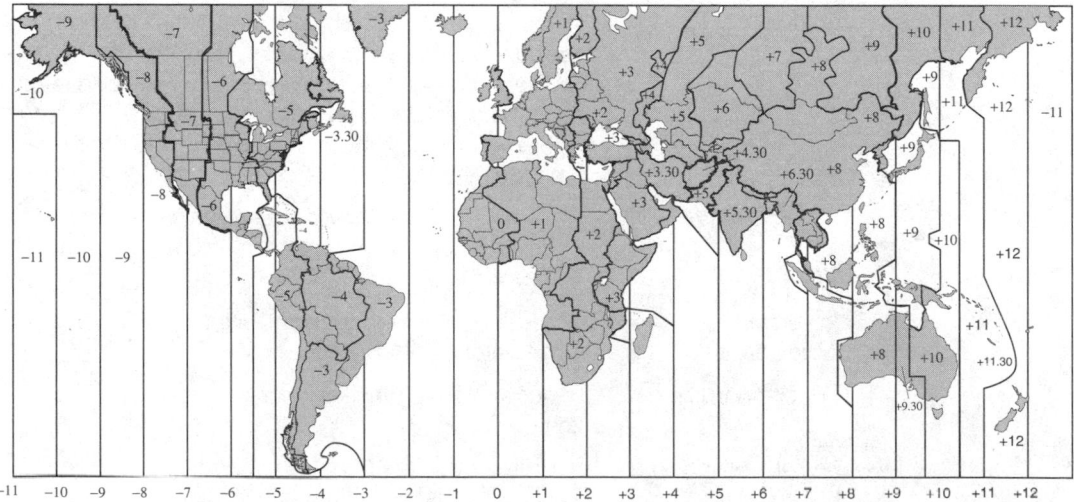

The 24-Hour Clock

Minutes and hours are expressed as follows: 10:30 a.m. is 1030 hours; 9:47 p.m. is 2147 hours. See Chapter 4, Travel, for an illustration of a 24-hour clock superimposed on a 12-hour clock.

Conventional Time	24-Hour Time	Conventional Time	24-Hour Time
1 a.m.	0100	1 p.m.	1300
2 a.m.	0200	2 p.m.	1400
3 a.m.	0300	3 p.m.	1500
4 a.m.	0400	4 p.m.	1600
5 a.m.	0500	5 p.m.	1700
6 a.m.	0600	6 p.m.	1800
7 a.m.	0700	7 p.m.	1900
8 a.m.	0800	8 p.m.	2000
9 a.m.	0900	9 p.m.	2100
10 a.m.	1000	10 p.m.	2200
11 a.m.	1100	11 p.m.	2300
12 noon	1200	12 midnight	2400

Up-to-the-Second Time

To get the current time in milliseconds, call the U.S. Naval Observatory at (202) 653-1800. Here is how a millisecond and some other very small time intervals are defined:

millisecond (ms)	=	0.001 second (10^{-3})
microsecond (µs)	=	0.000001 second (10^{-6})
nanosecond (ns)	=	0.000000001 second (10^{-9})
picosecond (ps)	=	0.000000000001 second (10^{-12})

If you have an all-band radio, tune to WWWV and WWVH at 2.5, 5.0, 10, 15, or 20 MHz to get the exact time. This station, operated by the National Institute of Standards and Technology, broadcasts the seconds as they tick off—86,400 a day.

Universal Time Converted to U.S. Standard Time, by Time Zone

Universal time is based on international atomic time and is the same as Greenwich mean time. It is expressed using the 24-hour clock.

Universal Time	Eastern Daylight Time	Eastern Standard Time	Central Standard Time	Mountain Standard Time	Pacific Standard Time
0100	9 p.m.*	8 p.m.*	7 p.m.*	6 p.m.*	5 p.m.*
0200	10 p.m.*	9 p.m.*	8 p.m.*	7 p.m.*	6 p.m.*
0300	11 p.m.*	10 p.m.*	9 p.m.*	8 p.m.*	7 p.m.*
0400	midnight	11 p.m.*	10 p.m.*	9 p.m.*	8 p.m.*
0500	1 a.m.	midnight	11 p.m.*	10 p.m.*	9 p.m.*
0600	2 a.m.	1 a.m.	midnight	11 p.m.*	10 p.m.*
0700	3 a.m.	2 a.m.	1 a.m.	midnight	11 p.m.*
0800	4 a.m.	3 a.m.	2 a.m.	1 a.m.	midnight
0900	5 a.m.	4 a.m.	3 a.m.	2 a.m.	1 a.m.
1000	6 a.m.	5 a.m.	4 a.m.	3 a.m.	2 a.m.
1100	7 a.m.	6 a.m.	5 a.m.	4 a.m.	3 a.m.
1200	8 a.m.	7 a.m.	6 a.m.	5 a.m.	4 a.m.
1300	9 a.m.	8 a.m.	7 a.m.	6 a.m.	5 a.m.
1400	10 a.m.	9 a.m.	8 a.m.	7 a.m.	6 a.m.
1500	11 a.m.	10 a.m.	9 a.m.	8 a.m.	7 a.m.
1600	noon	11 a.m.	10 a.m.	9 a.m.	8 a.m.
1700	1 p.m.	noon	11 a.m.	10 a.m.	9 a.m.
1800	2 p.m.	1 p.m.	noon	11 a.m.	10 a.m.
1900	3 p.m.	2 p.m.	1 p.m.	noon	11 a.m.
2000	4 p.m.	3 p.m.	2 p.m.	1 p.m.	noon
2100	5 p.m.	4 p.m.	3 p.m.	2 p.m.	1 p.m.
2200	6 p.m.	5 p.m.	4 p.m.	3 p.m.	2 p.m.
2300	7 p.m.	6 p.m.	5 p.m.	4 p.m.	3 p.m.
2400	8 p.m.*	7 p.m.*	6 p.m.*	5 p.m.*	4 p.m.*

* Denotes previous day.

The Doomsday Clock	Since 1947, the *Bulletin of the Atomic Scientists* has used a clock setting to graphically display the danger of worldwide nuclear disaster. On this clock, midnight represents the time of destruction. Its first setting in 1947 was at 11:53 p.m.; the closest it has been set to midnight was at 11:58 p.m., in 1953. In December 1995 the scientists set the clock forward to 11:46 p.m. from 11:43 p.m., where it had been set since 1991.

When It's Noon in New York: A Sampling of World Times

The following times are based on a starting point of 1200 (12:00 noon) eastern standard time in the United States. These times do not reflect any daylight saving time programs that may be in effect.

Location	Time	Location	Time
Addis Ababa, Ethiopia	2000	Cape Town, S. Africa	1900
Alexandria, Egypt	1900	Caracas, Venezuela	1300
Algiers, Algeria	1800	Casablanca, Morocco	1700
Amsterdam, Netherlands	1800	Chicago, IL	1100
Athens, Greece	1900	Copenhagen, Denmark	1800
Auckland, New Zealand	0500*	Dakar, Senegal	1700
Azores	1600	Delhi, India	2230
Baghdad, Iraq	2000	Denver, CO	1000
Bangkok, Thailand	0000	Dublin, Ireland	1700
Beijing, China	0100*	Edmonton, Canada	1000
Belfast, N. Ireland	1700	Gdansk, Poland	1800
Belgrade, Yugoslavia	1800	Geneva, Switzerland	1800
Berlin, Germany	1800	Guam	0300*
Bogotá, Colombia	1200	Havana, Cuba	1200
Bombay, India	2230	Helsinki, Finland	1900
Brisbane, Australia	0300*	Hong Kong	0100*
Brussels, Belgium	1800	Honolulu, HI	0700
Bucharest, Romania	1900	Istanbul, Turkey	2000
Budapest, Hungary	1800	Jakarta, Indonesia	0000
Buenos Aires, Argentina	1400	Jerusalem, Israel	1900
Cairo, Egypt	1900	Jidda, Saudi Arabia	2000
Calcutta, India	2230	Johannesburg, S. Africa	1900
Calgary, Canada	1000	Juneau, AK	0800

When It's Noon in New York: A Sampling of World Times (continued)

Location	Time	Location	Time
Karachi, Pakistan	2200	Reykjavik, Iceland	1700
La Paz, Bolivia	1300	Rio de Janeiro, Brazil	1400
Lima, Peru	1200	Rome, Italy	1800
Lisbon, Portugal	1700	St. John's, Newfoundland	1330
London, England	1700	Santiago, Chile	1300
Los Angeles, CA	0900	Seoul, S. Korea	0200*
Madrid, Spain	1800	Shanghai, China	0100*
Manila, Philippines	0100*	Singapore	0100*
Melbourne, Australia	0300*	Stockholm, Sweden	1800
Mexico City, Mexico	1100	Sydney, Australia	0300*
Montevideo, Uruguay	1500	Tashkent, Russia	2300
Montreal, Canada	1200	Teheran, Iran	2030
Moscow, Russia	2000	Tel Aviv, Israel	1900
Nagasaki, Japan	0200*	Tokyo, Japan	0200*
Nairobi, Kenya	2000	Toronto, Canada	1200
New York, NY	1200	Valparaiso, Chile	1300
Oslo, Norway	1800	Vancouver, Canada	0900
Panama City, Panama	1200	Vienna, Austria	1800
Paris, France	1800	Vladivostok, Russia	0300*
Perth, Australia	0100*	Warsaw, Poland	1800
Port Moresby, Papua New Guinea	0300*	Wellington, New Zealand	0500*
Prague, Czech Republic	1800	Yangon, Myanmar (Burma)	2330
Quito, Ecuador	1200	Yokohama, Japan	0200*
Regina, Canada	1100	Zurich, Switzerland	1800

* Next day.

When Time Stands Still

Why is it that when pictures of watches and clocks appear in advertisements, the time is often 10:10? One theory is that the shape created by the hands suggests an upbeat, "happy face" kind of feeling. Timex claims to be the first to use the 10:10 setting, back in the 1970s, perhaps to counter the "frown" that resulted from the previous standard setting of 8:20.

Ship's Time

The tradition of ship's bells and watches dates back to the 1700s, when a 30-minute hourglass was used to keep time on board ships. In each 4-hour watch, the hourglass was turned 8 times, and a bell was struck at each turn, once for each half-hour that had passed since the start of the watch.

Watch Hours	Watch Name
midnight–4 a.m.	middle
4 a.m.–8 a.m.	morning
8 a.m.–noon	forenoon
noon–4 p.m.	afternoon
4 p.m.–6 p.m.	first dog
6 p.m.–8 p.m.	second dog
8 p.m.–midnight	first

Calendars

Perpetual Calendar, 1775 Through 2076

To determine on which day of the week a particular date has fallen, or will fall, find the year and calendar number in the list below and refer to the corresponding numbered calendar on the following pages.

Year	Calendar No.	Year	Calendar No.	Year	Calendar No.	Year	Calendar No.
1775	1	1800	4	1825	7	1850	3
1776	9	1801	5	1826	1	1851	4
1777	4	1802	6	1827	2	1852	12
1778	5	1803	7	1828	10	1853	7
1779	6	1804	8	1829	5	1854	1
1780	14	1805	3	1830	6	1855	2
1781	2	1806	4	1831	7	1856	10
1782	3	1807	5	1832	8	1857	5
1783	4	1808	13	1833	3	1858	6
1784	12	1809	1	1834	4	1859	7
1785	7	1810	2	1835	5	1860	8
1786	1	1811	3	1836	13	1861	3
1787	2	1812	11	1837	1	1862	4
1788	10	1813	6	1838	2	1863	5
1789	5	1814	7	1839	3	1864	13
1790	6	1815	1	1840	11	1865	1
1791	7	1816	9	1841	6	1866	2
1792	8	1817	4	1842	7	1867	3
1793	3	1818	5	1843	1	1868	11
1794	4	1819	6	1844	9	1869	6
1795	5	1820	14	1845	4	1870	7
1796	13	1821	2	1846	5	1871	1
1797	1	1822	3	1847	6	1872	9
1798	2	1823	4	1848	14	1873	4
1799	3	1824	12	1849	2	1874	5

Perpetual Calendar, 1775 Through 2076 (continued)

Year	Calendar No.	Year	Calendar No.	Year	Calendar No.	Year	Calendar No.
1875	6	1912	9	1949	7	1986	4
1876	14	1913	4	1950	1	1987	5
1877	2	1914	5	1951	2	1988	13
1878	3	1915	6	1952	10	1989	1
1879	4	1916	14	1953	5	1990	2
1880	12	1917	2	1954	6	1991	3
1881	7	1918	3	1955	7	1992	11
1882	1	1919	4	1956	8	1993	6
1883	2	1920	12	1957	3	1994	7
1884	10	1921	7	1958	4	1995	1
1885	5	1922	1	1959	5	1996	9
1886	6	1923	2	1960	13	1997	4
1887	7	1924	10	1961	1	1998	5
1888	8	1925	5	1962	2	1999	6
1889	3	1926	6	1963	3	2000	14
1890	4	1927	7	1964	11	2001	2
1891	5	1928	8	1965	6	2002	3
1892	13	1929	3	1966	7	2003	4
1893	1	1930	4	1967	1	2004	12
1894	2	1931	5	1968	9	2005	7
1895	3	1932	13	1969	4	2006	1
1896	11	1933	1	1970	5	2007	2
1897	6	1934	2	1971	6	2008	10
1898	7	1935	3	1972	14	2009	5
1899	1	1936	11	1973	2	2010	6
1900	2	1937	6	1974	3	2011	7
1901	3	1938	7	1975	4	2012	8
1902	4	1939	1	1976	12	2013	3
1903	5	1940	9	1977	7	2014	4
1904	13	1941	4	1978	1	2015	5
1905	1	1942	5	1979	2	2016	13
1906	2	1943	6	1980	10	2017	1
1907	3	1944	14	1981	5	2018	2
1908	11	1945	2	1982	6	2019	3
1909	6	1946	3	1983	7	2020	11
1910	7	1947	4	1984	8	2021	6
1911	1	1948	12	1985	3	2022	7

Perpetual Calendar, 1775 Through 2076 (continued)

Year	Calendar No.	Year	Calendar No.	Year	Calendar No.	Year	Calendar No.
2023	1	2037	5	2051	1	2065	5
2024	9	2038	6	2052	9	2066	6
2025	4	2039	7	2053	4	2067	7
2026	5	2040	8	2054	5	2068	8
2027	6	2041	3	2055	6	2069	3
2028	14	2042	4	2056	14	2070	4
2029	2	2043	5	2057	2	2071	5
2030	3	2044	13	2058	3	2072	13
2031	4	2045	1	2059	4	2073	1
2032	12	2046	2	2060	12	2074	2
2033	7	2047	3	2061	7	2075	3
2034	1	2048	11	2062	1	2076	11
2035	2	2049	6	2063	2		
2036	10	2050	7	2064	10		

Numbers

1

Month	S	M	T	W	T	F	S
January	1	2	3	4	5	6	7
	8	9	10	11	12	13	14
	15	16	17	18	19	20	21
	22	23	24	25	26	27	28
	29	30	31				
February				1	2	3	4
	5	6	7	8	9	10	11
	12	13	14	15	16	17	18
	19	20	21	22	23	24	25
	26	27	28				
March				1	2	3	4
	5	6	7	8	9	10	11
	12	13	14	15	16	17	18
	19	20	21	22	23	24	25
	26	27	28	29	30	31	
April							1
	2	3	4	5	6	7	8
	9	10	11	12	13	14	15
	16	17	18	19	20	21	22
	23	24	25	26	27	28	29
	30						
May		1	2	3	4	5	6
	7	8	9	10	11	12	13
	14	15	16	17	18	19	20
	21	22	23	24	25	26	27
	28	29	30	31			
June					1	2	3
	4	5	6	7	8	9	10
	11	12	13	14	15	16	17
	18	19	20	21	22	23	24
	25	26	27	28	29	30	
July							1
	2	3	4	5	6	7	8
	9	10	11	12	13	14	15
	16	17	18	19	20	21	22
	23	24	25	26	27	28	29
	30	31					
August			1	2	3	4	5
	6	7	8	9	10	11	12
	13	14	15	16	17	18	19
	20	21	22	23	24	25	26
	27	28	29	30	31		
September						1	2
	3	4	5	6	7	8	9
	10	11	12	13	14	15	16
	17	18	19	20	21	22	23
	24	25	26	27	28	29	30
October	1	2	3	4	5	6	7
	8	9	10	11	12	13	14
	15	16	17	18	19	20	21
	22	23	24	25	26	27	28
	29	30	31				
November				1	2	3	4
	5	6	7	8	9	10	11
	12	13	14	15	16	17	18
	19	20	21	22	23	24	25
	26	27	28	29	30		
December						1	2
	3	4	5	6	7	8	9
	10	11	12	13	14	15	16
	17	18	19	20	21	22	23
	24	25	26	27	28	29	30
	31						

2

Month	S	M	T	W	T	F	S
January		1	2	3	4	5	6
	7	8	9	10	11	12	13
	14	15	16	17	18	19	20
	21	22	23	24	25	26	27
	28	29	30	31			
February					1	2	3
	4	5	6	7	8	9	10
	11	12	13	14	15	16	17
	18	19	20	21	22	23	24
	25	26	27	28			
March					1	2	3
	4	5	6	7	8	9	10
	11	12	13	14	15	16	17
	18	19	20	21	22	23	24
	25	26	27	28	29	30	31
April	1	2	3	4	5	6	7
	8	9	10	11	12	13	14
	15	16	17	18	19	20	21
	22	23	24	25	26	27	28
	29	30					
May			1	2	3	4	5
	6	7	8	9	10	11	12
	13	14	15	16	17	18	19
	20	21	22	23	24	25	26
	27	28	29	30	31		
June						1	2
	3	4	5	6	7	8	9
	10	11	12	13	14	15	16
	17	18	19	20	21	22	23
	24	25	26	27	28	29	30
July	1	2	3	4	5	6	7
	8	9	10	11	12	13	14
	15	16	17	18	19	20	21
	22	23	24	25	26	27	28
	29	30	31				
August				1	2	3	4
	5	6	7	8	9	10	11
	12	13	14	15	16	17	18
	19	20	21	22	23	24	25
	26	27	28	29	30	31	
September							1
	2	3	4	5	6	7	8
	9	10	11	12	13	14	15
	16	17	18	19	20	21	22
	23	24	25	26	27	28	29
	30						
October		1	2	3	4	5	6
	7	8	9	10	11	12	13
	14	15	16	17	18	19	20
	21	22	23	24	25	26	27
	28	29	30	31			
November					1	2	3
	4	5	6	7	8	9	10
	11	12	13	14	15	16	17
	18	19	20	21	22	23	24
	25	26	27	28	29	30	
December							1
	2	3	4	5	6	7	8
	9	10	11	12	13	14	15
	16	17	18	19	20	21	22
	23	24	25	26	27	28	29
	30	31					

3

Month	S	M	T	W	T	F	S
January			1	2	3	4	5
	6	7	8	9	10	11	12
	13	14	15	16	17	18	19
	20	21	22	23	24	25	26
	27	28	29	30	31		
February						1	2
	3	4	5	6	7	8	9
	10	11	12	13	14	15	16
	17	18	19	20	21	22	23
	24	25	26	27	28		
March						1	2
	3	4	5	6	7	8	9
	10	11	12	13	14	15	16
	17	18	19	20	21	22	23
	24	25	26	27	28	29	30
	31						
April		1	2	3	4	5	6
	7	8	9	10	11	12	13
	14	15	16	17	18	19	20
	21	22	23	24	25	26	27
	28	29	30				
May				1	2	3	4
	5	6	7	8	9	10	11
	12	13	14	15	16	17	18
	19	20	21	22	23	24	25
	26	27	28	29	30	31	
June							1
	2	3	4	5	6	7	8
	9	10	11	12	13	14	15
	16	17	18	19	20	21	22
	23	24	25	26	27	28	29
	30						
July		1	2	3	4	5	6
	7	8	9	10	11	12	13
	14	15	16	17	18	19	20
	21	22	23	24	25	26	27
	28	29	30	31			
August					1	2	3
	4	5	6	7	8	9	10
	11	12	13	14	15	16	17
	18	19	20	21	22	23	24
	25	26	27	28	29	30	31
September	1	2	3	4	5	6	7
	8	9	10	11	12	13	14
	15	16	17	18	19	20	21
	22	23	24	25	26	27	28
	29	30					
October			1	2	3	4	5
	6	7	8	9	10	11	12
	13	14	15	16	17	18	19
	20	21	22	23	24	25	26
	27	28	29	30	31		
November						1	2
	3	4	5	6	7	8	9
	10	11	12	13	14	15	16
	17	18	19	20	21	22	23
	24	25	26	27	28	29	30
December	1	2	3	4	5	6	7
	8	9	10	11	12	13	14
	15	16	17	18	19	20	21
	22	23	24	25	26	27	28
	29	30	31				

4

Month	S	M	T	W	T	F	S
January				1	2	3	4
	5	6	7	8	9	10	11
	12	13	14	15	16	17	18
	19	20	21	22	23	24	25
	26	27	28	29	30	31	
February							1
	2	3	4	5	6	7	8
	9	10	11	12	13	14	15
	16	17	18	19	20	21	22
	23	24	25	26	27	28	
March							1
	2	3	4	5	6	7	8
	9	10	11	12	13	14	15
	16	17	18	19	20	21	22
	23	24	25	26	27	28	29
	30	31					
April			1	2	3	4	5
	6	7	8	9	10	11	12
	13	14	15	16	17	18	19
	20	21	22	23	24	25	26
	27	28	29	30			
May					1	2	3
	4	5	6	7	8	9	10
	11	12	13	14	15	16	17
	18	19	20	21	22	23	24
	25	26	27	28	29	30	31
June	1	2	3	4	5	6	7
	8	9	10	11	12	13	14
	15	16	17	18	19	20	21
	22	23	24	25	26	27	28
	29	30					
July			1	2	3	4	5
	6	7	8	9	10	11	12
	13	14	15	16	17	18	19
	20	21	22	23	24	25	26
	27	28	29	30	31		
August						1	2
	3	4	5	6	7	8	9
	10	11	12	13	14	15	16
	17	18	19	20	21	22	23
	24	25	26	27	28	29	30
	31						
September		1	2	3	4	5	6
	7	8	9	10	11	12	13
	14	15	16	17	18	19	20
	21	22	23	24	25	26	27
	28	29	30				
October				1	2	3	4
	5	6	7	8	9	10	11
	12	13	14	15	16	17	18
	19	20	21	22	23	24	25
	26	27	28	29	30	31	
November							1
	2	3	4	5	6	7	8
	9	10	11	12	13	14	15
	16	17	18	19	20	21	22
	23	24	25	26	27	28	29
	30						
December		1	2	3	4	5	6
	7	8	9	10	11	12	13
	14	15	16	17	18	19	20
	21	22	23	24	25	26	27
	28	29	30	31			

5

Month	S	M	T	W	T	F	S
January					1	2	3
	4	5	6	7	8	9	10
	11	12	13	14	15	16	17
	18	19	20	21	22	23	24
	25	26	27	28	29	30	31
February	1	2	3	4	5	6	7
	8	9	10	11	12	13	14
	15	16	17	18	19	20	21
	22	23	24	25	26	27	28
March	1	2	3	4	5	6	7
	8	9	10	11	12	13	14
	15	16	17	18	19	20	21
	22	23	24	25	26	27	28
	29	30	31				
April				1	2	3	4
	5	6	7	8	9	10	11
	12	13	14	15	16	17	18
	19	20	21	22	23	24	25
	26	27	28	29	30		
May						1	2
	3	4	5	6	7	8	9
	10	11	12	13	14	15	16
	17	18	19	20	21	22	23
	24	25	26	27	28	29	30
	31						
June		1	2	3	4	5	6
	7	8	9	10	11	12	13
	14	15	16	17	18	19	20
	21	22	23	24	25	26	27
	28	29	30				
July				1	2	3	4
	5	6	7	8	9	10	11
	12	13	14	15	16	17	18
	19	20	21	22	23	24	25
	26	27	28	29	30	31	
August							1
	2	3	4	5	6	7	8
	9	10	11	12	13	14	15
	16	17	18	19	20	21	22
	23	24	25	26	27	28	29
	30	31					
September			1	2	3	4	5
	6	7	8	9	10	11	12
	13	14	15	16	17	18	19
	20	21	22	23	24	25	26
	27	28	29	30			
October					1	2	3
	4	5	6	7	8	9	10
	11	12	13	14	15	16	17
	18	19	20	21	22	23	24
	25	26	27	28	29	30	31
November	1	2	3	4	5	6	7
	8	9	10	11	12	13	14
	15	16	17	18	19	20	21
	22	23	24	25	26	27	28
	29	30					
December			1	2	3	4	5
	6	7	8	9	10	11	12
	13	14	15	16	17	18	19
	20	21	22	23	24	25	26
	27	28	29	30	31		

6

Month	S	M	T	W	T	F	S
January						1	2
	3	4	5	6	7	8	9
	10	11	12	13	14	15	16
	17	18	19	20	21	22	23
	24	25	26	27	28	29	30
	31						
February		1	2	3	4	5	6
	7	8	9	10	11	12	13
	14	15	16	17	18	19	20
	21	22	23	24	25	26	27
	28						
March		1	2	3	4	5	6
	7	8	9	10	11	12	13
	14	15	16	17	18	19	20
	21	22	23	24	25	26	27
	28	29	30	31			
April					1	2	3
	4	5	6	7	8	9	10
	11	12	13	14	15	16	17
	18	19	20	21	22	23	24
	25	26	27	28	29	30	
May							1
	2	3	4	5	6	7	8
	9	10	11	12	13	14	15
	16	17	18	19	20	21	22
	23	24	25	26	27	28	29
	30	31					
June			1	2	3	4	5
	6	7	8	9	10	11	12
	13	14	15	16	17	18	19
	20	21	22	23	24	25	26
	27	28	29	30			
July					1	2	3
	4	5	6	7	8	9	10
	11	12	13	14	15	16	17
	18	19	20	21	22	23	24
	25	26	27	28	29	30	31
August	1	2	3	4	5	6	7
	8	9	10	11	12	13	14
	15	16	17	18	19	20	21
	22	23	24	25	26	27	28
	29	30	31				
September				1	2	3	4
	5	6	7	8	9	10	11
	12	13	14	15	16	17	18
	19	20	21	22	23	24	25
	26	27	28	29	30		
October						1	2
	3	4	5	6	7	8	9
	10	11	12	13	14	15	16
	17	18	19	20	21	22	23
	24	25	26	27	28	29	30
	31						
November		1	2	3	4	5	6
	7	8	9	10	11	12	13
	14	15	16	17	18	19	20
	21	22	23	24	25	26	27
	28	29	30				
December				1	2	3	4
	5	6	7	8	9	10	11
	12	13	14	15	16	17	18
	19	20	21	22	23	24	25
	26	27	28	29	30	31	

7

Month	S	M	T	W	T	F	S
January							1
	2	3	4	5	6	7	8
	9	10	11	12	13	14	15
	16	17	18	19	20	21	22
	23	24	25	26	27	28	29
	30	31					
February			1	2	3	4	5
	6	7	8	9	10	11	12
	13	14	15	16	17	18	19
	20	21	22	23	24	25	26
	27	28					
March			1	2	3	4	5
	6	7	8	9	10	11	12
	13	14	15	16	17	18	19
	20	21	22	23	24	25	26
	27	28	29	30	31		
April						1	2
	3	4	5	6	7	8	9
	10	11	12	13	14	15	16
	17	18	19	20	21	22	23
	24	25	26	27	28	29	30
May	1	2	3	4	5	6	7
	8	9	10	11	12	13	14
	15	16	17	18	19	20	21
	22	23	24	25	26	27	28
	29	30	31				
June				1	2	3	4
	5	6	7	8	9	10	11
	12	13	14	15	16	17	18
	19	20	21	22	23	24	25
	26	27	28	29	30		
July						1	2
	3	4	5	6	7	8	9
	10	11	12	13	14	15	16
	17	18	19	20	21	22	23
	24	25	26	27	28	29	30
	31						
August		1	2	3	4	5	6
	7	8	9	10	11	12	13
	14	15	16	17	18	19	20
	21	22	23	24	25	26	27
	28	29	30	31			
September					1	2	3
	4	5	6	7	8	9	10
	11	12	13	14	15	16	17
	18	19	20	21	22	23	24
	25	26	27	28	29	30	
October							1
	2	3	4	5	6	7	8
	9	10	11	12	13	14	15
	16	17	18	19	20	21	22
	23	24	25	26	27	28	29
	30	31					
November			1	2	3	4	5
	6	7	8	9	10	11	12
	13	14	15	16	17	18	19
	20	21	22	23	24	25	26
	27	28	29	30			
December					1	2	3
	4	5	6	7	8	9	10
	11	12	13	14	15	16	17
	18	19	20	21	22	23	24
	25	26	27	28	29	30	31

8

Month	S	M	T	W	T	F	S
January	1	2	3	4	5	6	7
	8	9	10	11	12	13	14
	15	16	17	18	19	20	21
	22	23	24	25	26	27	28
	29	30	31				
February				1	2	3	4
	5	6	7	8	9	10	11
	12	13	14	15	16	17	18
	19	20	21	22	23	24	25
	26	27	28	29			
March					1	2	3
	4	5	6	7	8	9	10
	11	12	13	14	15	16	17
	18	19	20	21	22	23	24
	25	26	27	28	29	30	31
April	1	2	3	4	5	6	7
	8	9	10	11	12	13	14
	15	16	17	18	19	20	21
	22	23	24	25	26	27	28
	29	30					
May			1	2	3	4	5
	6	7	8	9	10	11	12
	13	14	15	16	17	18	19
	20	21	22	23	24	25	26
	27	28	29	30	31		
June						1	2
	3	4	5	6	7	8	9
	10	11	12	13	14	15	16
	17	18	19	20	21	22	23
	24	25	26	27	28	29	30
July	1	2	3	4	5	6	7
	8	9	10	11	12	13	14
	15	16	17	18	19	20	21
	22	23	24	25	26	27	28
	29	30	31				
August				1	2	3	4
	5	6	7	8	9	10	11
	12	13	14	15	16	17	18
	19	20	21	22	23	24	25
	26	27	28	29	30	31	
September							1
	2	3	4	5	6	7	8
	9	10	11	12	13	14	15
	16	17	18	19	20	21	22
	23	24	25	26	27	28	29
	30						
October		1	2	3	4	5	6
	7	8	9	10	11	12	13
	14	15	16	17	18	19	20
	21	22	23	24	25	26	27
	28	29	30	31			
November					1	2	3
	4	5	6	7	8	9	10
	11	12	13	14	15	16	17
	18	19	20	21	22	23	24
	25	26	27	28	29	30	
December							1
	2	3	4	5	6	7	8
	9	10	11	12	13	14	15
	16	17	18	19	20	21	22
	23	24	25	26	27	28	29
	30	31					

9

Month	S	M	T	W	T	F	S
January		1	2	3	4	5	6
	7	8	9	10	11	12	13
	14	15	16	17	18	19	20
	21	22	23	24	25	26	27
	28	29	30	31			
February					1	2	3
	4	5	6	7	8	9	10
	11	12	13	14	15	16	17
	18	19	20	21	22	23	24
	25	26	27	28	29		
March						1	2
	3	4	5	6	7	8	9
	10	11	12	13	14	15	16
	17	18	19	20	21	22	23
	24	25	26	27	28	29	30
	31						
April		1	2	3	4	5	6
	7	8	9	10	11	12	13
	14	15	16	17	18	19	20
	21	22	23	24	25	26	27
	28	29	30				
May				1	2	3	4
	5	6	7	8	9	10	11
	12	13	14	15	16	17	18
	19	20	21	22	23	24	25
	26	27	28	29	30	31	
June							1
	2	3	4	5	6	7	8
	9	10	11	12	13	14	15
	16	17	18	19	20	21	22
	23	24	25	26	27	28	29
	30						
July		1	2	3	4	5	6
	7	8	9	10	11	12	13
	14	15	16	17	18	19	20
	21	22	23	24	25	26	27
	28	29	30	31			
August					1	2	3
	4	5	6	7	8	9	10
	11	12	13	14	15	16	17
	18	19	20	21	22	23	24
	25	26	27	28	29	30	31
September	1	2	3	4	5	6	7
	8	9	10	11	12	13	14
	15	16	17	18	19	20	21
	22	23	24	25	26	27	28
	29	30					
October			1	2	3	4	5
	6	7	8	9	10	11	12
	13	14	15	16	17	18	19
	20	21	22	23	24	25	26
	27	28	29	30	31		
November						1	2
	3	4	5	6	7	8	9
	10	11	12	13	14	15	16
	17	18	19	20	21	22	23
	24	25	26	27	28	29	30
December	1	2	3	4	5	6	7
	8	9	10	11	12	13	14
	15	16	17	18	19	20	21
	22	23	24	25	26	27	28
	29	30	31				

10 (leap year, January 1 = Tuesday)

```
       January                May                 September
 S  M  T  W  T  F  S    S  M  T  W  T  F  S    S  M  T  W  T  F  S
          1  2  3  4             1  2  3          1  2  3  4  5  6
 6  7  8  9 10 11 12    4  5  6  7  8  9 10    7  8  9 10 11 12 13
13 14 15 16 17 18 19   11 12 13 14 15 16 17   14 15 16 17 18 19 20
20 21 22 23 24 25 26   18 19 20 21 22 23 24   21 22 23 24 25 26 27
27 28 29 30 31         25 26 27 28 29 30 31   28 29 30

       February               June                October
 S  M  T  W  T  F  S    S  M  T  W  T  F  S    S  M  T  W  T  F  S
                1  2    1  2  3  4  5  6  7             1  2  3  4
 3  4  5  6  7  8  9    8  9 10 11 12 13 14    5  6  7  8  9 10 11
10 11 12 13 14 15 16   15 16 17 18 19 20 21   12 13 14 15 16 17 18
17 18 19 20 21 22 23   22 23 24 25 26 27 28   19 20 21 22 23 24 25
24 25 26 27 28 29      29 30                  26 27 28 29 30 31

       March                  July                November
 S  M  T  W  T  F  S    S  M  T  W  T  F  S    S  M  T  W  T  F  S
                   1             1  2  3  4  5                    1
 2  3  4  5  6  7  8    6  7  8  9 10 11 12    2  3  4  5  6  7  8
 9 10 11 12 13 14 15   13 14 15 16 17 18 19    9 10 11 12 13 14 15
16 17 18 19 20 21 22   20 21 22 23 24 25 26   16 17 18 19 20 21 22
23 24 25 26 27 28 29   27 28 29 30 31         23 24 25 26 27 28 29
30 31                                         30

       April                  August              December
 S  M  T  W  T  F  S    S  M  T  W  T  F  S    S  M  T  W  T  F  S
          1  2  3  4                   1  2             1  2  3  4  5  6
 6  7  8  9 10 11 12    3  4  5  6  7  8  9    7  8  9 10 11 12 13
13 14 15 16 17 18 19   10 11 12 13 14 15 16   14 15 16 17 18 19 20
20 21 22 23 24 25 26   17 18 19 20 21 22 23   21 22 23 24 25 26 27
27 28 29 30            24 25 26 27 28 29 30   28 29 30 31
                       31
```

11 (leap year, January 1 = Wednesday)

```
       January                May                 September
 S  M  T  W  T  F  S    S  M  T  W  T  F  S    S  M  T  W  T  F  S
          1  2  3  4                   1  2             1  2  3  4  5
 5  6  7  8  9 10 11    3  4  5  6  7  8  9    6  7  8  9 10 11 12
12 13 14 15 16 17 18   10 11 12 13 14 15 16   13 14 15 16 17 18 19
19 20 21 22 23 24 25   17 18 19 20 21 22 23   20 21 22 23 24 25 26
26 27 28 29 30 31      24 25 26 27 28 29 30   27 28 29 30
                       31

       February               June                October
 S  M  T  W  T  F  S    S  M  T  W  T  F  S    S  M  T  W  T  F  S
                   1       1  2  3  4  5  6                1  2  3
 2  3  4  5  6  7  8    7  8  9 10 11 12 13    4  5  6  7  8  9 10
 9 10 11 12 13 14 15   14 15 16 17 18 19 20   11 12 13 14 15 16 17
16 17 18 19 20 21 22   21 22 23 24 25 26 27   18 19 20 21 22 23 24
23 24 25 26 27 28 29   28 29 30               25 26 27 28 29 30 31

       March                  July                November
 S  M  T  W  T  F  S    S  M  T  W  T  F  S    S  M  T  W  T  F  S
 1  2  3  4  5  6  7             1  2  3  4    1  2  3  4  5  6  7
 8  9 10 11 12 13 14    5  6  7  8  9 10 11    8  9 10 11 12 13 14
15 16 17 18 19 20 21   12 13 14 15 16 17 18   15 16 17 18 19 20 21
22 23 24 25 26 27 28   19 20 21 22 23 24 25   22 23 24 25 26 27 28
29 30 31               26 27 28 29 30 31      29 30

       April                  August              December
 S  M  T  W  T  F  S    S  M  T  W  T  F  S    S  M  T  W  T  F  S
          1  2  3  4                      1             1  2  3  4  5
 5  6  7  8  9 10 11    2  3  4  5  6  7  8    6  7  8  9 10 11 12
12 13 14 15 16 17 18    9 10 11 12 13 14 15   13 14 15 16 17 18 19
19 20 21 22 23 24 25   16 17 18 19 20 21 22   20 21 22 23 24 25 26
26 27 28 29 30         23 24 25 26 27 28 29   27 28 29 30 31
                       30 31
```

12 (leap year, January 1 = Thursday)

```
       January                May                 September
 S  M  T  W  T  F  S    S  M  T  W  T  F  S    S  M  T  W  T  F  S
             1  2  3                      1             1  2  3  4
 4  5  6  7  8  9 10    2  3  4  5  6  7  8    5  6  7  8  9 10 11
11 12 13 14 15 16 17    9 10 11 12 13 14 15   12 13 14 15 16 17 18
18 19 20 21 22 23 24   16 17 18 19 20 21 22   19 20 21 22 23 24 25
25 26 27 28 29 30 31   23 24 25 26 27 28 29   26 27 28 29 30
                       30 31

       February               June                October
 S  M  T  W  T  F  S    S  M  T  W  T  F  S    S  M  T  W  T  F  S
 1  2  3  4  5  6  7          1  2  3  4  5                   1  2
 8  9 10 11 12 13 14    6  7  8  9 10 11 12    3  4  5  6  7  8  9
15 16 17 18 19 20 21   13 14 15 16 17 18 19   10 11 12 13 14 15 16
22 23 24 25 26 27 28   20 21 22 23 24 25 26   17 18 19 20 21 22 23
29                     27 28 29 30            24 25 26 27 28 29 30
                                              31

       March                  July                November
 S  M  T  W  T  F  S    S  M  T  W  T  F  S    S  M  T  W  T  F  S
    1  2  3  4  5  6                1  2  3       1  2  3  4  5  6
 7  8  9 10 11 12 13    4  5  6  7  8  9 10    7  8  9 10 11 12 13
14 15 16 17 18 19 20   11 12 13 14 15 16 17   14 15 16 17 18 19 20
21 22 23 24 25 26 27   18 19 20 21 22 23 24   21 22 23 24 25 26 27
28 29 30 31            25 26 27 28 29 30 31   28 29 30

       April                  August              December
 S  M  T  W  T  F  S    S  M  T  W  T  F  S    S  M  T  W  T  F  S
             1  2  3    1  2  3  4  5  6  7          1  2  3  4
 4  5  6  7  8  9 10    8  9 10 11 12 13 14    5  6  7  8  9 10 11
11 12 13 14 15 16 17   15 16 17 18 19 20 21   12 13 14 15 16 17 18
18 19 20 21 22 23 24   22 23 24 25 26 27 28   19 20 21 22 23 24 25
25 26 27 28 29 30      29 30 31               26 27 28 29 30 31
```

13 (leap year, January 1 = Friday)

```
       January                May                 September
 S  M  T  W  T  F  S    S  M  T  W  T  F  S    S  M  T  W  T  F  S
                1  2    1  2  3  4  5  6  7             1  2  3
 3  4  5  6  7  8  9    8  9 10 11 12 13 14    4  5  6  7  8  9 10
10 11 12 13 14 15 16   15 16 17 18 19 20 21   11 12 13 14 15 16 17
17 18 19 20 21 22 23   22 23 24 25 26 27 28   18 19 20 21 22 23 24
24 25 26 27 28 29 30   29 30 31               25 26 27 28 29 30
31

       February               June                October
 S  M  T  W  T  F  S    S  M  T  W  T  F  S    S  M  T  W  T  F  S
    1  2  3  4  5  6             1  2  3  4                      1
 7  8  9 10 11 12 13    5  6  7  8  9 10 11    2  3  4  5  6  7  8
14 15 16 17 18 19 20   12 13 14 15 16 17 18    9 10 11 12 13 14 15
21 22 23 24 25 26 27   19 20 21 22 23 24 25   16 17 18 19 20 21 22
28 29                  26 27 28 29 30         23 24 25 26 27 28 29
                                              30 31

       March                  July                November
 S  M  T  W  T  F  S    S  M  T  W  T  F  S    S  M  T  W  T  F  S
       1  2  3  4  5                1  2       1  2  3  4  5  6  7
 6  7  8  9 10 11 12    3  4  5  6  7  8  9    8  9 10 11 12 13 14
13 14 15 16 17 18 19   10 11 12 13 14 15 16   15 16 17 18 19 20 21
20 21 22 23 24 25 26   17 18 19 20 21 22 23   22 23 24 25 26 27 28
27 28 29 30 31         24 25 26 27 28 29 30   29 30
                       31

       April                  August              December
 S  M  T  W  T  F  S    S  M  T  W  T  F  S    S  M  T  W  T  F  S
                1  2       1  2  3  4  5  6                1  2  3
 3  4  5  6  7  8  9    7  8  9 10 11 12 13    4  5  6  7  8  9 10
10 11 12 13 14 15 16   14 15 16 17 18 19 20   11 12 13 14 15 16 17
17 18 19 20 21 22 23   21 22 23 24 25 26 27   18 19 20 21 22 23 24
24 25 26 27 28 29 30   28 29 30 31            25 26 27 28 29 30 31
```

14 (leap year, January 1 = Saturday)

```
       January                May                 September
 S  M  T  W  T  F  S    S  M  T  W  T  F  S    S  M  T  W  T  F  S
                   1       1  2  3  4  5  6                1  2
 2  3  4  5  6  7  8    7  8  9 10 11 12 13    3  4  5  6  7  8  9
 9 10 11 12 13 14 15   14 15 16 17 18 19 20   10 11 12 13 14 15 16
16 17 18 19 20 21 22   21 22 23 24 25 26 27   17 18 19 20 21 22 23
23 24 25 26 27 28 29   28 29 30 31            24 25 26 27 28 29 30
30 31

       February               June                October
 S  M  T  W  T  F  S    S  M  T  W  T  F  S    S  M  T  W  T  F  S
       1  2  3  4  5             1  2  3    1  2  3  4  5  6  7
 6  7  8  9 10 11 12    4  5  6  7  8  9 10    8  9 10 11 12 13 14
13 14 15 16 17 18 19   11 12 13 14 15 16 17   15 16 17 18 19 20 21
20 21 22 23 24 25 26   18 19 20 21 22 23 24   22 23 24 25 26 27 28
27 28 29              25 26 27 28 29 30       29 30 31

       March                  July                November
 S  M  T  W  T  F  S    S  M  T  W  T  F  S    S  M  T  W  T  F  S
          1  2  3  4                      1             1  2  3  4
 5  6  7  8  9 10 11    2  3  4  5  6  7  8    5  6  7  8  9 10 11
12 13 14 15 16 17 18    9 10 11 12 13 14 15   12 13 14 15 16 17 18
19 20 21 22 23 24 25   16 17 18 19 20 21 22   19 20 21 22 23 24 25
26 27 28 29 30 31      23 24 25 26 27 28 29   26 27 28 29 30
                       30 31

       April                  August              December
 S  M  T  W  T  F  S    S  M  T  W  T  F  S    S  M  T  W  T  F  S
                   1          1  2  3  4  5                1  2
 2  3  4  5  6  7  8    6  7  8  9 10 11 12    3  4  5  6  7  8  9
 9 10 11 12 13 14 15   13 14 15 16 17 18 19   10 11 12 13 14 15 16
16 17 18 19 20 21 22   20 21 22 23 24 25 26   17 18 19 20 21 22 23
23 24 25 26 27 28 29   27 28 29 30 31         24 25 26 27 28 29 30
30                                            31
```

Numbered Days of the Year

Date	Day Number	Date	Day Number	Date	Day Number	Date	Day Number
Jan 1	1	Feb 8	39	Mar 18	77	Apr 25	115
Jan 2	2	Feb 9	40	Mar 19	78	Apr 26	116
Jan 3	3	Feb 10	41	Mar 20	79	Apr 27	117
Jan 4	4	Feb 11	42	Mar 21	80	Apr 28	118
Jan 5	5	Feb 12	43	Mar 22	81	Apr 29	119
Jan 6	6	Feb 13	44	Mar 23	82	Apr 30	120
Jan 7	7	Feb 14	45	Mar 24	83	May 1	121
Jan 8	8	Feb 15	46	Mar 25	84	May 2	122
Jan 9	9	Feb 16	47	Mar 26	85	May 3	123
Jan 10	10	Feb 17	48	Mar 27	86	May 4	124
Jan 11	11	Feb 18	49	Mar 28	87	May 5	125
Jan 12	12	Feb 19	50	Mar 29	88	May 6	126
Jan 13	13	Feb 20	51	Mar 30	89	May 7	127
Jan 14	14	Feb 21	52	Mar 31	90	May 8	128
Jan 15	15	Feb 22	53	Apr 1	91	May 9	129
Jan 16	16	Feb 23	54	Apr 2	92	May 10	130
Jan 17	17	Feb 24	55	Apr 3	93	May 11	131
Jan 18	18	Feb 25	56	Apr 4	94	May 12	132
Jan 19	19	Feb 26	57	Apr 5	95	May 13	133
Jan 20	20	Feb 27	58	Apr 6	96	May 14	134
Jan 21	21	Feb 28	59	Apr 7	97	May 15	135
Jan 22	22	Mar 1	60	Apr 8	98	May 16	136
Jan 23	23	Mar 2	61	Apr 9	99	May 17	137
Jan 24	24	Mar 3	62	Apr 10	100	May 18	138
Jan 25	25	Mar 4	63	Apr 11	101	May 19	139
Jan 26	26	Mar 5	64	Apr 12	102	May 20	140
Jan 27	27	Mar 6	65	Apr 13	103	May 21	141
Jan 28	28	Mar 7	66	Apr 14	104	May 22	142
Jan 29	29	Mar 8	67	Apr 15	105	May 23	143
Jan 30	30	Mar 9	68	Apr 16	106	May 24	144
Jan 31	31	Mar 10	69	Apr 17	107	May 25	145
Feb 1	32	Mar 11	70	Apr 18	108	May 26	146
Feb 2	33	Mar 12	71	Apr 19	109	May 27	147
Feb 3	34	Mar 13	72	Apr 20	110	May 28	148
Feb 4	35	Mar 14	73	Apr 21	111	May 29	149
Feb 5	36	Mar 15	74	Apr 22	112	May 30	150
Feb 6	37	Mar 16	75	Apr 23	113	May 31	151
Feb 7	38	Mar 17	76	Apr 24	114	Jun 1	152

Numbered Days of the Year (continued)

Date	Day Number	Date	Day Number	Date	Day Number	Date	Day Number
Jun 2	153	Jul 10	191	Aug 17	229	Sep 24	267
Jun 3	154	Jul 11	192	Aug 18	230	Sep 25	268
Jun 4	155	Jul 12	193	Aug 19	231	Sep 26	269
Jun 5	156	Jul 13	194	Aug 20	232	Sep 27	270
Jun 6	157	Jul 14	195	Aug 21	233	Sep 28	271
Jun 7	158	Jul 15	196	Aug 22	234	Sep 29	272
Jun 8	159	Jul 16	197	Aug 23	235	Sep 30	273
Jun 9	160	Jul 17	198	Aug 24	236	Oct 1	274
Jun 10	161	Jul 18	199	Aug 25	237	Oct 2	275
Jun 11	162	Jul 19	200	Aug 26	238	Oct 3	276
Jun 12	163	Jul 20	201	Aug 27	239	Oct 4	277
Jun 13	164	Jul 21	202	Aug 28	240	Oct 5	278
Jun 14	165	Jul 22	203	Aug 29	241	Oct 6	279
Jun 15	166	Jul 23	204	Aug 30	242	Oct 7	280
Jun 16	167	Jul 24	205	Aug 31	243	Oct 8	281
Jun 17	168	Jul 25	206	Sep 1	244	Oct 9	282
Jun 18	169	Jul 26	207	Sep 2	245	Oct 10	283
Jun 19	170	Jul 27	208	Sep 3	246	Oct 11	284
Jun 20	171	Jul 28	209	Sep 4	247	Oct 12	285
Jun 21	172	Jul 29	210	Sep 5	248	Oct 13	286
Jun 22	173	Jul 30	211	Sep 6	249	Oct 14	287
Jun 23	174	Jul 31	212	Sep 7	250	Oct 15	288
Jun 24	175	Aug 1	213	Sep 8	251	Oct 16	289
Jun 25	176	Aug 2	214	Sep 9	252	Oct 17	290
Jun 26	177	Aug 3	215	Sep 10	253	Oct 18	291
Jun 27	178	Aug 4	216	Sep 11	254	Oct 19	292
Jun 28	179	Aug 5	217	Sep 12	255	Oct 20	293
Jun 29	180	Aug 6	218	Sep 13	256	Oct 21	294
Jun 30	181	Aug 7	219	Sep 14	257	Oct 22	295
Jul 1	182	Aug 8	220	Sep 15	258	Oct 23	296
Jul 2	183	Aug 9	221	Sep 16	259	Oct 24	297
Jul 3	184	Aug 10	222	Sep 17	260	Oct 25	298
Jul 4	185	Aug 11	223	Sep 18	261	Oct 26	299
Jul 5	186	Aug 12	224	Sep 19	262	Oct 27	300
Jul 6	187	Aug 13	225	Sep 20	263	Oct 28	301
Jul 7	188	Aug 14	226	Sep 21	264	Oct 29	302
Jul 8	189	Aug 15	227	Sep 22	265	Oct 30	303
Jul 9	190	Aug 16	228	Sep 23	266	Oct 31	304

Numbered Days of the Year (continued)

Date	Day Number	Date	Day Number	Date	Day Number	Date	Day Number
Nov 1	305	Nov 17	321	Dec 3	337	Dec 19	353
Nov 2	306	Nov 18	322	Dec 4	338	Dec 20	354
Nov 3	307	Nov 19	323	Dec 5	339	Dec 21	355
Nov 4	308	Nov 20	324	Dec 6	340	Dec 22	356
Nov 5	309	Nov 21	325	Dec 7	341	Dec 23	357
Nov 6	310	Nov 22	326	Dec 8	342	Dec 24	358
Nov 7	311	Nov 23	327	Dec 9	343	Dec 25	359
Nov 8	312	Nov 24	328	Dec 10	344	Dec 26	360
Nov 9	313	Nov 25	329	Dec 11	345	Dec 27	361
Nov 10	314	Nov 26	330	Dec 12	346	Dec 28	362
Nov 11	315	Nov 27	331	Dec 13	347	Dec 29	363
Nov 12	316	Nov 28	332	Dec 14	348	Dec 30	364
Nov 13	317	Nov 29	333	Dec 15	349	Dec 31	365
Nov 14	318	Nov 30	334	Dec 16	350		
Nov 15	319	Dec 1	335	Dec 17	351		
Nov 16	320	Dec 2	336	Dec 18	352		

Note: In leap years, February 29 becomes day 60, and subsequent days are renumbered accordingly.

Julian Day (JD) Count and Before the Present (BP): Measures of Time Used by Scientists

Astronomers use the Julian Day (JD), which counts days instead of years and starts at noon rather than midnight, making recording easier: One night's observations can be noted for one date rather than split over two. The first Julian Day was January 1, 4713 BCE (before the Christian era)—chosen because the Julian calendar, the ancient Roman tax calendar, and the lunar calendar coincided, an event that would not occur for another 7,980 years. The current Julian Period will end at noon on January 1, 3268. Astronomers use conversion tables to convert Gregorian calendar dates to Julian Days. At noon (UTC) on January 1, 1997, Julian Day 2,450,450 will begin.

Archeologists use the abbreviation BP—before the present—to express prehistoric dates to avoid using a calendar based on a religious or cultural dating system. BP is currently defined as being AD 1950.

The First Days of Spring, Summer, Fall, Winter, 1996 through 2000

Year	Vernal Equinox	Summer Solstice	Autumnal Equinox	Winter Solstice
1996	March 20 (0803)	June 21 (0224)	Sept. 22 (1800)	Dec. 21 (1406)
1997	March 20 (1355)	June 21 (0820)	Sept. 22 (2356)	Dec. 21 (2007)
1998	March 20 (1955)	June 21 (1403)	Sept. 23 (0537)	Dec. 22 (0156)
1999	March 21 (0146)	June 21 (1949)	Sept. 23 (1131)	Dec. 22 (0744)
2000	March 20 (0735)	June 21 (0148)	Sept. 22 (1727)	Dec. 21 (1337)

Source: U.S. Naval Observatory.

Note: Numbers in parentheses are hours and minutes expressed in universal time.

Official U.S. Holidays

If holidays with fixed dates—such as Independence Day—fall on a weekend day, the federal government and many other employers give workers the Friday before the holiday off if the holiday falls on a Saturday; they give the Monday after the holiday off if the fixed-date holiday falls on a Sunday.

Holiday	Date or Day of Month
New Year's Day	January 1
Martin Luther King's Birthday	Third Monday in January
Presidents' Day	Third Monday in February
Memorial Day	Last Monday in May
Independence Day	July 4
Labor Day	First Monday in September
Columbus Day	Second Monday in October (or October 12)
Veterans Day	November 11
Thanksgiving Day	Fourth Thursday in November
Christmas Day	December 25

Other U.S. Holidays

Holiday	Date or Day of Month
Inauguration Day	January 20 (every 4 years)
Groundhog Day	February 2
Lincoln's Birthday	February 12
Valentine's Day	February 14
Washington's Birthday	February 22
St. Patrick's Day	March 17
Easter	See table on page 470
April Fool's Day	April 1
Mother's Day	Second Sunday in May
Armed Forces Day	Third Saturday in May
Flag Day	June 14
Father's Day	Third Sunday in June
Halloween	October 31
Election Day	First Tuesday after first Monday in November

Official Canadian Holidays

Holiday	Date or Day of Month
New Year's Day	January 1
Good Friday	Friday before Easter Sunday
Easter Monday	Monday after Easter Sunday
Victoria Day	First Monday before May 25
Canada Day	July 1
Labour Day	First Monday in September
Thanksgiving Day	Second Monday in October
Remembrance Day	November 11
Christmas Day	December 25
Boxing Day	December 26

Other Canadian Holidays

Holiday	Date or Day of Month
Sir John A. MacDonald's Birthday	January 11
Alberta Family Day	Third Monday in February
Mother's Day	Second Sunday in May
Father's Day	Third Sunday in June
Fête Nationale (Quebec)	June 24
Memorial Day (Newfoundland)	July 1
Civic holiday (Manitoba, Northwest Territories, Ontario, Saskatchewan)	First Monday in August
British Columbia Day	First Monday in August
Heritage Day (Alberta)	First Monday in August
Natal Day (Nova Scotia, Prince Edward Island)	First Monday in August (except Halifax)
New Brunswick Day	First Monday in August
Discovery Day (Yukon)	Third Monday in August

Christian Holy Days: Ash Wednesday and Easter Sunday, 1996 Through 2010

Easter is the first Sunday after the first full moon on or after the vernal equinox; it can fall as early as March 22 and as late as April 25. Shrove Tuesday is 1 day before Ash Wednesday. Palm Sunday is 1 week before Easter Sunday; Maundy Thursday is 3 days before Easter Sunday; Good Friday is 2 days before Easter Sunday; Holy Saturday is 1 day before Easter Sunday.

Year	Ash Wednesday	Easter Sunday
1996	February 21	April 7
1997	February 12	March 30
1998	February 25	April 12
1999	February 17	April 4
2000	March 8	April 23
2001	February 28	April 15
2002	February 13	March 31
2003	March 5	April 20
2004	February 25	April 11
2005	February 9	March 27
2006	March 1	April 16
2007	February 21	April 8
2008	February 6	March 23
2009	February 25	April 12
2010	February 17	April 4

Jewish Holy Days, 1996 Through 2010

Year	Purim	1st Day Passover	1st Day Shavuot	1st Day Rosh Hashanah	Yom Kippur	1st Day Sukkot	Simchat Torah	1st Day Hanukkah
1996	Mar 5	Apr 4	May 24	Sept 14	Sept 23	Sept 28	Oct 6	Dec 6
1997	Mar 23	Apr 22	June 11	Oct 2	Oct 11	Oct 16	Oct 24	Dec 24
1998	Mar 12	Apr 11	May 31	Sept 21	Sept 30	Oct 5	Oct 13	Dec 14
1999	Mar 2	Apr 1	May 21	Sept 11	Sept 20	Sept 25	Oct 3	Dec 4
2000	Mar 21	Apr 20	June 9	Sept 30	Oct 9	Oct 14	Oct 22	Dec 22
2001	Mar 9	Apr 8	May 28	Sept 18	Sept 27	Oct 2	Oct 10	Dec 10
2002	Feb 26	Mar 28	May 17	Sept 7	Sept 16	Sept 21	Sept 29	Nov 30
2003	Mar 18	Apr 17	June 6	Sept 27	Oct 6	Oct 11	Oct 19	Dec 20
2004	Mar 7	Apr 6	May 26	Sept 16	Sept 25	Sept 30	Oct 8	Dec 8
2005	Mar 25	Apr 24	June 13	Oct 4	Oct 13	Oct 18	Oct 26	Dec 26
2006	Mar 14	Apr 13	June 2	Sept 23	Oct 2	Oct 7	Oct 15	Dec 16
2007	Mar 4	Apr 3	May 23	Sept 13	Sept 22	Sept 27	Oct 5	Dec 5
2008	Mar 21	Apr 20	June 9	Sept 30	Oct 9	Oct 14	Oct 22	Dec 22
2009	Mar 10	Apr 9	May 29	Sept 19	Sept 28	Oct 3	Oct 11	Dec 12
2010	Feb 28	Mar 30	May 19	Sept 9	Sept 18	Sept 23	Oct 1	Dec 2

Note: Holidays begin at sundown on the previous day.

Islamic Holy Days, 1996 Through 2005

Year	1st Day of Ramadan	1st Day of Shawwal	10th Day of Dhul-Hijja	1st Day of Al-Muharram
1996	Jan 21	Feb 19	Apr 27	May 18
1997	Jan 9 Dec 30	Feb 8	Apr 17	May 7
1998	Dec 19	Jan 29	Apr 6	Apr 27
1999	Dec 8	Jan 18	Mar 27	Apr 16
2000	Nov 26	Jan 7 Dec 26	Mar 16	Apr 5
2001	Nov 17	Dec 17	Mar 5	Mar 26
2002	Nov 6	Dec 6	Feb 23	Mar 15
2003	Oct 27	Nov 26	Feb 12	Mar 5
2004	Oct 15	Nov 14	Feb 2	Feb 22
2005	Oct 4	Nov 3	Jan 21	Feb 10

World Calendars Compared

Type	Days in Year	Days in Month	Explanation
Gregorian	365.2425	30, 31, 28 (29)	Used today for civil purposes almost worldwide, it was developed by Pope Gregory in the 1580s. It has 12 months with 30 or 31 days except February, which has 28 except every fourth, or leap, year, when it has 29. To keep the calendar from gaining extra days over time, February 29 is dropped in century years that cannot be evenly divided by 400 (e.g., 1900 was not a leap year).
Jewish	353–385	29, 30 alternating	Hebrew year designation is Gregorian year plus 3,760. It has 12 months; adds 7 months over 19-year period (1 month inserted in years 3, 6, 8, 11, 14, 17, 19).
Islamic	354–355	29, 30 alternating	Has 12 months; to adjust for lunar year, uses 30-year cycles, in which 19 years have 354-day years and 11 years have extra day each.
Chinese	354	29, 30 alternating	Has 12 months; repeats a month 7 times during each 19-year period. Year starts at second new moon after winter solstice. Gregorian year 1997 is 4634 in the Chinese era.

Why the Year 2000 Is Not the First Year of the 21st Century

A century is defined as a period of 100 years. With that in mind, consider the following:

Year (AD)	Number of Years	Century Name
1–100	100	1st
101–200	100	2nd
....		
1901–2000	100	20th
2001–2100	100	21st

Chinese Calendar

The Chinese calendar is based on cycles lasting 60 years; a new cycle began in 1996. Each year in a cycle is given a name consisting of two terms, one of which is the name of an animal.

Rat	Ox	Tiger	Hare (Rabbit)	Dragon	Snake
1924	1925	1926	1927	1928	1929
1936	1937	1938	1939	1940	1941
1948	1949	1950	1951	1952	1953
1960	1961	1962	1963	1964	1965
1972	1973	1974	1975	1976	1977
1984	1985	1986	1987	1988	1989
1996	1997	1998	1999	2000	2001
2008	2009	2010	2011	2012	2013

Horse	Sheep (Goat)	Monkey	Rooster	Dog	Pig
1930	1931	1932	1933	1934	1935
1942	1943	1944	1945	1946	1947
1954	1955	1956	1957	1958	1959
1966	1967	1968	1969	1970	1971
1978	1979	1980	1981	1982	1983
1990	1991	1992	1993	1994	1995
2002	2003	2004	2005	2006	2007
2014	2015	2016	2017	2018	2019

Chinese New Year, 1996 Through 2010

Year	New Year's Day
1996	February 19
1997	February 7
1998	January 28
1999	February 16
2000	February 5
2001	January 24
2002	February 12
2003	February 1
2004	January 22
2005	February 9
2006	January 29
2007	February 18
2008	February 7
2009	January 26
2010	February 14

Jewish, Islamic, and Hindu Years: Gregorian Year Equivalents

The Jewish year is calculated from the year of the Creation, 3761 BC. The Islamic year is calculated from AD 622, when the Prophet fled from Mecca to Medina. The Hindu year is calculated from AD 78, the start of the Saka era.

Jewish Year (AM)	Islamic Year (AH)	Hindu Year (SE)
5757	1417	1918
Sept 14, 1996–Oct 1, 1997	May 19, 1996–May 8, 1997	Mar 21, 1996–Mar 21, 1997
5758	1418	1919
Oct 2, 1997–Sept 20, 1998	May 9, 1997–Apr 27, 1998	Mar 22, 1997–Mar 21, 1998
5759	1419	1920
Sept 21, 1998–Sept 10, 1999	Apr 28, 1998–Apr 16, 1999	Mar 22, 1998–Mar 21, 1999
5760	1420	1921
Sept 11, 1999–Sept 29, 2000	Apr 17, 1999–Apr 5, 2000	Mar 22, 1999–Mar 21, 2000

Making Calendars Accurate

Societies that used the lunar year of 354 days, rather than the solar year of 365.25 days, found that if adjustments were not made over time, spring holidays began to take place in winter months. The Egyptians were the first to adopt a solar year, and the Romans also eventually came to use it.

Julius Caesar came up with the Julian calendar, which worked well for more than 1,500 years until it became obvious that it had gained about 10 days over actual solar time. To fix this problem, Pope Gregory dropped 10 days from October in 1582 to correct the difference.

Acceptance of this new Gregorian calendar took time. Although much of Europe adopted it immediately, Great Britain didn't use it until 1752, Russia until 1918, and Turkey until 1927. It is still slightly off, using 365.2425 days instead of the true solar year of 365.2422 days—a loss of less than a day over the next 3,300 years.

Other Measures of Time

Words for Time

Word	Meaning
fortnight	2 weeks
biennium	2 years
triennium	3 years
quadrennium, olympiad	4 years
lustrum, quinquennium	5 years
decade, decennium	10 years
score, vicennium	20 years
5 score, century	100 years
millennium, 10 centuries	1,000 years
trice	A very short period of time
diurnal	Happening once a day
semidiurnal	Happening twice a day
annual	Happening once a year
biannual	Happening twice a year
biennial	Happening every second year
biweekly	Happening twice a week *or* every 2 weeks
semiweekly	Happening twice a week
semimonthly	Happening twice a month
bimonthly	Happening twice a month *or* once every 2 months
biyearly	Happening twice a year *or* every 2 years
semiannual	Happening twice a year
triweekly	Happening 3 times a week *or* every 3 weeks
quadrennial	Happening once in 4 years

Words for Time (continued)

Word	Meaning
quindecennial	Happening once every 15 years
quinquennial	5th anniversary
sexennial	6th anniversary
septennial	7th anniversary
decennial	10th anniversary
vicennial	20th anniversary
semicentennial	50th anniversary
centennial	100th anniversary
sesquicentennial	150th anniversary
bicentennial	200th anniversary
tricentennial, tercentenary	300th anniversary
quadricentennial	400th anniversary
quincentennial	500th anniversary

Traditional Anniversary Gifts

Anniversary	Gift	Anniversary	Gift
1st	Cotton	14th	Ivory
2nd	Paper	15th	Glass, crystal
3rd	Leather	20th	China
4th	Silk, books	25th	Silver
5th	Wood	30th	Pearl
6th	Iron	35th	Coral, jade
7th	Wool, copper	40th	Ruby
8th	Bronze	45th	Sapphire
9th	Pottery	50th	Gold
10th	Tin, aluminum	55th	Emerald
11th	Steel	60th	Diamond
12th	Linen	70th	Platinum
13th	Lace		

Astrology: Dates and Signs

Dates	Sign of the Zodiac
Jan 20–Feb 18	Aquarius
Feb 19–Mar 20	Pisces
Mar 21–Apr 19	Aries
Apr 20–May 20	Taurus
May 21–Jun 20	Gemini
Jun 21– Jul 22	Cancer
Jul 23–Aug 22	Leo
Aug 23–Sept 22	Virgo
Sept 23–Oct 22	Libra
Oct 23–Nov 21	Scorpio
Nov 22–Dec 21	Sagittarius
Dec 22–Jan 19	Capricorn

Generations: Names and Dates

In their book, *Generations: The History of America's Future, 1584 to 2069,* William Strauss and Neil Howe list 18 American generations, beginning with the year 1584. The average length of the first 17 generations was 23.4 years, but the average after colonial times dropped to 22.3 years. Here are the 6 U.S. generations living today.

Name	Birth Years
Lost	1883–1900
G.I.	1901–1924
Silent	1925–1942
Boom	1943–1960
13th	1961–1981
Millennial	1982–?

8

Weather

Everyday Weather

Normal Daily High Temperatures, Selected U.S. Cities (in °F)

City	Jan.	Feb.	Mar.	Apr.	May	June	July	Aug.	Sept.	Oct.	Nov.	Dec.	Ann. Avg.
Mobile, AL	59.7	63.6	70.9	78.5	84.6	90.0	91.3	90.5	86.9	79.5	70.3	62.9	77.4
Juneau, AK	29.4	34.1	38.7	47.2	55.1	60.9	63.9	62.7	55.9	47.1	36.7	31.6	46.9
Phoenix, AZ	65.9	70.7	75.5	84.5	93.6	103.5	105.9	103.7	98.3	88.1	74.9	66.2	85.9
Little Rock, AR	49.0	53.9	64.0	73.4	81.3	89.3	92.4	91.4	84.6	75.1	62.7	52.5	72.5
Los Angeles, CA	65.7	65.9	65.5	67.4	69.0	71.9	75.3	76.6	76.6	74.4	70.3	65.9	70.4
Sacramento, CA	52.7	60.0	64.0	71.1	80.3	87.8	93.2	92.1	87.3	77.9	63.1	52.7	73.5
San Diego, CA	65.9	66.5	66.3	68.4	69.1	71.6	76.2	77.8	77.1	74.6	69.9	66.1	70.8
San Francisco, CA	55.6	59.4	60.8	63.9	66.5	70.3	71.6	72.3	73.6	70.1	62.4	56.1	65.2
Denver, CO	43.2	46.6	52.2	61.8	70.8	81.4	88.2	85.8	76.9	66.3	52.5	44.5	64.2
Hartford, CT	33.2	36.4	46.8	59.9	71.6	80.0	85.0	82.7	74.8	63.7	51.0	37.5	60.2
Wilmington, DE	38.7	41.9	52.1	62.6	72.9	81.4	85.6	84.1	77.7	66.6	55.5	43.9	63.6
Washington, DC	42.3	45.9	56.5	66.7	76.2	84.7	88.5	86.9	80.1	69.1	58.3	47.0	66.9
Jacksonville, FL	64.2	67.0	73.0	79.1	84.7	89.3	91.4	90.7	87.2	80.2	73.6	66.8	78.9
Miami, FL	75.2	76.5	79.1	82.4	85.3	87.6	89.0	89.0	87.8	84.5	80.4	76.7	82.8
Atlanta, GA	50.4	55.0	64.3	72.7	79.6	85.8	88.0	87.1	81.8	72.7	63.4	54.0	71.2
Honolulu, HI	80.1	80.5	81.6	82.8	84.7	86.5	87.5	88.7	88.5	86.9	84.1	81.2	84.4
Boise, ID	36.4	44.2	52.9	61.4	71.0	80.9	90.2	88.1	77.0	64.6	48.7	37.7	62.8
Chicago, IL	29.0	33.5	45.8	58.6	70.1	79.6	83.7	81.8	74.8	63.3	48.4	34.0	58.6
Peoria, IL	29.9	34.9	48.1	62.0	72.8	82.2	85.7	83.1	76.9	64.8	49.8	34.6	60.4
Indianapolis, IN	33.7	38.3	50.9	63.3	73.8	82.7	85.5	83.6	77.6	65.8	51.9	38.5	62.1
Des Moines, IA	28.1	33.7	46.9	61.8	73.0	82.2	86.7	84.2	75.6	64.3	48.0	32.6	59.8
Wichita, KS	39.8	45.9	57.2	68.3	76.9	86.8	92.8	90.7	81.4	70.6	55.3	43.0	67.4
Louisville, KY	40.3	44.8	56.3	67.3	76.0	83.5	87.0	85.7	80.3	69.2	56.8	45.1	66.0
New Orleans, LA	60.8	64.1	71.6	78.5	84.4	89.2	90.6	90.2	86.6	79.4	71.1	64.3	77.6
Portland, ME	30.3	33.1	41.4	52.3	63.2	72.7	78.8	77.4	69.3	58.7	47.0	35.1	54.9
Baltimore, MD	40.2	43.7	54.0	64.3	74.2	83.2	87.2	85.4	78.5	67.3	56.5	45.2	65.0
Boston, MA	35.7	37.5	45.8	55.9	66.6	76.3	81.8	79.8	72.8	62.7	52.2	40.4	59.0
Detroit, MI	30.3	33.3	44.4	57.7	69.6	78.9	83.3	81.3	73.9	61.5	48.1	35.2	58.1

Normal Daily High Temperatures, Selected U.S. Cities (in °F) (continued)

City	Jan.	Feb.	Mar.	Apr.	May	June	July	Aug.	Sept.	Oct.	Nov.	Dec.	Ann. Avg.
Sault Ste. Marie, MI	21.1	23.2	32.8	48.0	62.6	70.5	76.3	73.8	65.9	54.3	40.0	26.2	49.6
Duluth, MN	16.2	21.7	32.9	48.2	61.9	71.0	77.1	73.9	63.8	52.3	35.2	20.7	47.9
Minneapolis/ St. Paul, MN	20.7	26.6	39.2	56.5	69.4	78.8	84.0	80.7	70.7	58.8	41.0	25.5	54.3
Jackson, MS	55.6	60.1	69.3	77.4	84.0	90.6	92.4	92.0	88.0	79.1	69.2	59.5	76.4
Kansas City, MO	34.7	40.6	52.8	65.1	74.3	83.3	88.7	86.4	78.1	67.5	52.6	38.8	63.6
St. Louis, MO	37.7	42.6	54.6	66.9	76.1	85.2	89.3	87.3	79.9	68.5	54.7	41.7	65.4
Great Falls, MT	30.6	37.5	43.7	55.3	65.2	74.6	83.3	81.6	69.6	59.3	43.5	33.1	56.4
Omaha, NE	31.3	37.1	49.4	63.8	74.0	83.7	87.9	85.2	76.5	65.6	49.3	34.6	61.5
Reno, NV	45.1	51.7	56.3	63.7	72.9	83.1	91.9	89.6	79.5	68.6	53.8	45.5	66.8
Concord, NH	29.8	33.0	42.8	56.3	68.9	77.3	82.4	79.8	71.6	60.7	47.1	34.2	57.0
Atlantic City, NJ	40.4	42.5	51.6	60.7	71.2	80.0	84.5	83.3	76.6	66.0	55.7	45.3	63.2
Albuquerque, NM	46.8	53.5	61.4	70.8	79.7	90.0	92.5	89.0	81.9	71.0	57.3	47.5	70.1
Albany, NY	30.2	33.2	44.0	57.5	69.7	79.0	84.0	81.4	73.2	61.8	48.7	34.9	58.1
Buffalo, NY	30.2	31.6	41.7	54.2	66.1	75.3	80.2	77.9	70.8	59.4	47.1	35.3	55.8
New York, NY*	37.6	40.3	50.0	61.2	71.7	80.1	85.2	83.7	76.2	65.3	54.0	42.5	62.3
Charlotte, NC	49.0	53.0	62.3	71.2	78.3	85.8	88.9	87.7	81.9	72.0	62.6	52.3	70.4
Raleigh, NC	48.9	52.6	62.1	71.7	78.6	85.0	88.0	86.8	81.1	71.6	62.6	52.7	70.1
Bismarck, ND	20.2	26.4	38.5	54.9	67.8	77.1	84.4	82.7	70.8	58.7	39.3	24.5	53.8
Cincinnati, OH	36.6	40.8	53.0	64.2	74.0	82.0	85.5	84.1	77.9	66.0	53.3	41.5	63.2
Cleveland, OH	31.9	35.0	46.3	57.9	68.6	78.3	82.4	80.5	73.6	62.1	50.0	37.4	58.7
Columbus, OH	34.1	38.0	50.5	62.0	72.3	80.4	83.7	82.1	76.2	64.5	51.4	39.2	61.2
Oklahoma City, OK	46.7	52.1	62.0	71.9	79.1	87.3	93.4	92.5	83.8	73.6	60.4	49.9	71.1
Portland, OR	45.4	51.0	56.0	60.6	67.1	74.0	79.9	80.3	74.6	64.0	52.6	45.6	62.6
Philadelphia, PA	37.9	41.0	51.6	62.6	73.1	81.7	86.1	84.6	77.6	66.3	55.1	43.4	63.4
Pittsburgh, PA	33.7	36.9	49.0	60.3	70.6	78.9	82.6	80.8	74.3	62.5	50.4	38.6	59.9
Providence, RI	36.6	38.3	46.1	57.0	67.3	76.9	82.1	80.7	74.3	64.1	53.0	41.2	59.8
Columbia, SC	55.3	59.3	68.2	76.5	83.5	88.8	91.6	90.1	85.1	76.3	67.8	58.8	75.1
Sioux Falls, SD	24.3	29.6	42.3	59.0	70.7	80.5	86.3	83.3	73.1	61.2	43.4	28.0	56.8
Memphis, TN	48.5	53.5	63.2	73.3	81.0	89.3	92.3	90.8	83.9	74.3	62.3	52.5	72.1
Nashville, TN	45.9	50.8	61.2	70.8	78.8	86.5	89.5	88.4	82.5	72.5	60.4	50.2	69.8
Dallas/Ft. Worth, TX	54.1	58.9	67.8	76.3	82.9	91.9	96.5	96.2	87.8	78.5	66.8	57.5	76.3
El Paso, TX	56.1	62.2	69.9	78.7	87.1	96.5	96.1	93.5	87.1	78.4	66.4	57.5	77.5
Houston, TX	61.0	65.3	71.1	78.4	84.6	90.1	92.7	92.5	88.4	81.6	72.4	64.7	78.6
Salt Lake City, UT	36.4	43.6	52.2	61.3	71.9	82.8	92.2	89.4	79.2	66.1	50.8	37.8	63.6
Burlington, VT	25.1	27.5	39.3	53.6	67.2	75.8	81.2	77.9	69.0	57.0	44.0	30.4	54.0
Norfolk, VA	47.3	49.7	57.9	66.9	75.3	82.9	86.4	85.1	79.6	69.5	61.2	52.2	67.8
Richmond, VA	45.7	49.2	59.5	70.0	77.8	85.1	88.4	87.1	80.9	70.7	61.3	50.2	68.8
Seattle/Tacoma, WA	45.0	49.5	52.7	57.2	63.9	69.9	75.2	75.2	69.3	59.7	50.5	45.1	59.4
Spokane, WA	33.2	40.6	47.7	57.0	65.8	74.7	83.1	82.5	72.0	58.6	41.4	33.8	57.5

Normal Daily High Temperatures, Selected U.S. Cities (in °F) (continued)

City	Jan.	Feb.	Mar.	Apr.	May	June	July	Aug.	Sept.	Oct.	Nov.	Dec.	Ann. Avg.
Charleston, WV	41.2	45.3	56.7	66.8	75.5	83.1	85.7	84.4	78.8	68.2	57.3	46.0	65.8
Milwaukee, WI	26.1	30.1	40.4	52.9	64.3	74.9	79.9	77.8	70.6	58.7	44.7	31.2	54.3
Cheyenne, WY	37.7	40.5	44.9	54.7	64.6	74.4	82.2	80.0	71.1	60.0	46.8	38.8	58.0
San Juan, PR	83.2	83.6	84.4	85.8	87.2	88.6	88.5	88.7	88.8	88.3	85.9	83.8	86.4

Source: U.S. National Oceanic and Atmospheric Administration, *Climatography of the United States*, No. 81.

Note: Airport data except as noted. Based on standard 30-year period, 1961 through 1990.

*City office data.

Fahrenheit/Celsius Equivalents

°F	°C	°F	°C
23.0	−5	64.4	18
24.8	−4	66.2	19
26.6	−3	68.0	20
28.4	−2	69.8	21
30.2	−1	71.6	22
32.0	0	73.4	23
33.8	1	75.2	24
35.6	2	77.0	25
37.4	3	78.8	26
39.2	4	80.6	27
41.0	5	82.4	28
42.8	6	84.2	29
44.6	7	86.0	30
46.4	8	87.8	31
48.2	9	89.6	32
50.0	10	91.4	33
51.8	11	93.2	34
53.6	12	95.0	35
55.4	13	96.8	36
57.2	14	98.6	37
59.0	15	100.4	38
60.8	16	102.2	39
62.6	17	104.0	40

Why Snow and Rain Are Predicted Differently

Snow is forecast in terms of inches expected; rain, in percentage probabilities. The reason for this distinction is that snow and rain come from two different types of weather systems: Snow falls in a broad pattern over many miles, whereas rain is more localized and spotty. It is also difficult to predict how much rain will fall in any given place from an often small, fast-moving thunderstorm.

Normal Daily Low Temperatures, Selected U.S. Cities (in °F)

City	Jan.	Feb.	Mar.	Apr.	May	June	July	Aug.	Sept.	Oct.	Nov.	Dec.	Ann. Avg.
Mobile, AL	40.0	42.7	50.1	57.1	64.4	70.7	73.2	72.9	68.7	57.3	49.1	43.1	57.4
Juneau, AK	19.0	22.7	26.7	32.1	38.9	45.0	48.1	47.3	42.9	37.2	27.2	22.6	34.1
Phoenix, AZ	41.2	44.7	48.8	55.3	63.9	72.9	81.0	79.2	72.8	60.8	48.9	41.8	59.3
Little Rock, AR	29.1	33.2	42.2	50.7	59.0	67.4	71.5	69.8	63.5	50.9	41.5	33.1	51.0
Los Angeles, CA	47.8	49.3	50.5	52.8	56.3	59.5	62.8	64.2	63.2	59.2	52.8	47.9	55.5
Sacramento, CA	37.7	41.4	43.2	45.5	50.3	55.3	58.1	58.0	55.7	50.4	43.4	37.8	48.1
San Diego, CA	48.9	50.7	52.8	55.6	59.1	61.9	65.7	67.3	65.6	60.9	53.9	48.8	57.6
San Francisco, CA	41.8	45.0	45.8	47.2	49.7	52.6	53.9	55.0	55.2	51.8	47.1	42.7	49.0
Denver, CO	16.1	20.2	25.8	34.5	43.6	52.4	58.6	56.9	47.6	36.4	25.4	17.4	36.2
Hartford, CT	15.8	18.6	28.1	37.5	47.6	56.9	62.2	60.4	51.8	40.7	32.8	21.3	39.5
Wilmington, DE	22.4	24.8	33.1	41.8	52.2	61.6	67.1	65.9	58.2	45.7	37.0	27.6	44.8
Washington, DC	26.8	29.1	37.7	46.4	56.6	66.5	71.4	70.0	62.5	50.3	41.1	31.7	49.2
Jacksonville, FL	40.5	43.3	49.2	54.9	62.1	69.1	71.9	71.8	69.0	59.3	50.2	43.4	57.1
Miami, FL	59.2	60.4	64.2	67.8	72.1	75.1	76.2	76.7	75.9	72.1	66.7	61.5	69.0
Atlanta, GA	31.5	34.5	42.5	50.2	58.7	66.2	69.5	69.0	63.5	51.9	42.8	35.0	51.3
Honolulu, HI	65.6	65.4	67.2	68.7	70.3	72.2	73.5	74.2	73.5	72.3	70.3	67.0	70.0
Boise, ID	21.6	27.5	31.9	36.7	43.9	52.1	57.7	56.8	48.2	39.0	31.1	22.5	39.1
Chicago, IL	12.9	17.2	28.5	38.6	47.7	57.5	62.6	61.6	53.9	42.2	31.6	19.1	39.5
Peoria, IL	13.2	17.7	29.8	40.8	50.9	60.7	65.4	63.1	55.2	43.1	32.5	19.3	41.0
Indianapolis, IN	17.2	20.9	31.9	41.5	51.7	61.0	65.2	62.8	55.6	43.5	34.1	23.2	42.4
Des Moines, IA	10.7	15.6	27.6	40.0	51.5	61.2	66.5	63.6	54.5	42.7	29.9	16.1	40.0
Wichita, KS	19.2	23.7	33.6	44.5	54.3	64.6	69.9	67.9	59.2	46.6	33.9	23.0	45.0
Louisville, KY	23.2	26.5	36.2	45.4	54.7	62.9	67.3	65.8	58.7	45.8	37.3	28.6	46.0
New Orleans, LA	41.8	44.4	51.6	58.4	65.2	70.8	73.1	72.8	69.5	58.7	51.0	44.8	58.5
Portland, ME	11.4	13.5	24.5	34.1	43.4	52.1	58.3	57.1	48.9	38.3	30.4	17.8	35.8
Baltimore, MD	23.4	25.9	34.1	42.5	52.6	61.8	66.8	65.7	58.4	45.9	37.1	28.2	45.2
Boston, MA	21.6	23.0	31.3	40.2	49.8	59.1	65.1	64.0	56.8	46.9	38.3	26.7	43.6
Detroit, MI	15.6	17.6	27.0	36.8	47.1	56.3	61.3	59.6	52.5	40.9	32.2	21.4	39.0
Sault Ste. Marie, MI	4.6	4.8	15.3	28.4	38.4	45.5	51.3	51.3	44.3	36.2	25.9	11.8	29.8
Duluth, MN	−2.2	2.8	15.7	28.9	39.6	48.5	55.1	53.3	44.5	35.1	21.5	4.9	29.0
Minneapolis/ St. Paul, MN	2.8	9.2	22.7	36.2	47.6	57.6	63.1	60.3	50.3	38.8	25.2	10.2	35.3
Jackson, MS	32.7	35.7	44.1	51.9	60.0	67.1	70.5	69.7	63.7	50.3	42.3	36.1	52.0

Normal Daily Low Temperatures, Selected U.S. Cities (in °F) (continued)

City	Jan.	Feb.	Mar.	Apr.	May	June	July	Aug.	Sept.	Oct.	Nov.	Dec.	Ann. Avg.
Kansas City, MO	16.7	21.8	32.6	43.8	53.9	63.1	68.2	65.7	56.9	45.7	33.6	21.9	43.7
St. Louis, MO	20.8	25.1	35.5	46.4	56.0	65.7	70.4	67.9	60.5	48.3	37.7	26.0	46.7
Great Falls, MT	11.6	17.2	22.8	31.9	40.9	48.6	53.2	52.2	43.5	35.8	24.3	14.6	33.1
Omaha, NE	10.9	16.7	27.7	39.9	50.9	60.4	65.9	62.9	53.6	41.2	28.7	15.6	39.5
Reno, NV	20.7	24.2	29.2	33.3	40.1	46.9	51.3	49.6	41.3	32.9	26.7	19.9	34.7
Concord, NH	7.4	10.4	22.1	31.5	41.4	51.2	56.5	54.7	46.0	34.9	27.0	14.4	33.1
Atlantic City, NJ	21.4	23.5	31.3	39.3	49.6	58.7	64.8	63.5	55.5	43.7	35.8	26.3	42.8
Albuquerque, NM	21.7	26.4	32.2	39.6	48.6	58.3	64.4	62.6	55.2	43.0	31.2	23.1	42.2
Albany, NY	11.0	13.8	24.5	35.1	45.4	54.6	59.6	57.8	49.4	38.6	30.7	18.2	36.6
Buffalo, NY	17.0	17.4	25.9	36.2	47.0	56.5	61.9	60.1	53.0	42.7	33.9	22.9	39.5
New York, NY*	25.3	26.9	34.8	43.8	53.7	63.0	68.4	67.3	60.1	49.7	41.1	30.7	47.1
Charlotte, NC	29.6	31.9	39.4	47.5	56.4	65.6	69.6	68.9	62.9	50.6	41.5	32.8	49.7
Raleigh, NC	28.8	31.3	38.7	46.2	55.3	63.6	68.1	67.5	61.1	48.4	39.7	32.4	48.4
Bismarck, ND	−1.7	5.1	17.8	31.0	42.2	51.6	56.4	53.9	43.1	32.5	17.8	3.3	29.4
Cincinnati, OH	19.5	22.7	33.1	42.2	51.8	60.0	64.8	62.9	56.6	44.2	35.3	25.3	43.2
Cleveland, OH	17.6	19.3	28.2	37.3	47.3	56.8	61.4	60.3	54.2	43.5	35.0	24.5	40.5
Columbus, OH	18.5	21.2	31.2	40.0	50.1	58.0	62.7	60.8	54.8	42.9	34.3	24.6	41.6
Oklahoma City, OK	25.2	29.6	38.5	48.8	57.7	66.1	70.6	69.6	62.2	50.4	38.6	28.6	48.8
Portland, OR	33.7	36.1	38.6	41.3	47.0	52.9	56.5	56.9	52.0	44.9	39.5	34.8	44.5
Philadelphia, PA	22.8	24.8	33.2	42.1	52.7	61.8	67.2	66.3	58.7	46.4	37.6	28.1	45.1
Pittsburgh, PA	18.5	20.3	29.8	38.8	48.4	56.9	61.6	60.2	53.5	42.3	34.1	24.4	40.7
Providence, RI	19.1	20.9	28.8	37.7	47.3	56.8	63.2	61.9	53.8	43.0	34.9	24.4	41.0
Columbia, SC	32.1	34.2	42.2	49.4	58.2	66.0	70.0	69.2	63.2	50.1	41.5	34.9	50.9
Sioux Falls, SD	3.3	9.7	22.6	34.8	45.9	56.1	62.3	59.4	48.7	36.0	22.6	8.6	34.2
Memphis, TN	30.9	34.8	43.0	52.4	61.2	68.9	72.9	71.1	64.5	51.9	42.7	34.8	52.4
Nashville, TN	26.5	29.9	39.1	47.5	56.6	64.7	68.9	67.7	61.1	48.3	39.6	30.9	48.4
Dallas/Ft. Worth, TX	32.7	36.9	45.6	54.7	62.6	70.0	74.1	73.6	66.9	55.8	45.4	36.3	54.6
El Paso, TX	29.4	33.9	40.2	48.0	56.5	64.3	68.4	66.6	61.6	49.6	38.4	30.7	49.0
Houston, TX	39.7	42.6	50.0	58.1	64.4	70.6	72.4	72.0	67.9	57.6	49.6	42.2	57.3
Salt Lake City, UT	19.3	24.6	31.4	37.9	45.6	55.4	63.7	61.8	51.0	40.2	30.9	21.6	40.3
Burlington, VT	7.5	8.9	22.0	34.2	45.4	54.6	59.7	57.9	48.8	38.6	29.6	15.5	35.2
Norfolk, VA	30.9	32.3	39.3	47.1	56.8	65.2	70.0	69.4	64.2	52.9	43.8	35.4	50.6
Richmond, VA	25.7	28.1	36.3	44.6	54.2	62.7	67.5	66.4	59.0	46.5	37.9	29.9	46.6
Seattle/Tacoma, WA	35.2	37.4	38.5	41.2	46.3	51.9	55.2	55.7	51.9	45.8	40.1	35.8	44.6
Spokane, WA	20.8	25.9	29.6	34.7	41.9	49.2	54.4	54.3	45.8	36.0	28.8	21.7	36.9
Charleston, WV	23.0	25.7	35.0	42.8	51.5	59.8	64.4	63.4	56.5	44.2	36.3	28.0	44.2
Milwaukee, WI	11.6	15.9	26.2	35.8	44.8	55.0	62.0	60.8	52.8	41.8	30.7	17.5	37.9
Cheyenne, WY	15.2	18.1	22.1	30.1	39.4	48.3	54.6	52.8	43.7	33.9	23.7	16.7	33.2
San Juan, PR	70.8	70.6	71.6	72.9	74.5	76.1	76.8	76.7	76.2	75.5	74.0	72.4	74.0

Source: U.S. National Oceanic and Atmospheric Administration, *Climatography of the United States*, No. 81.

Note: Airport data except as noted. Based on standard 30-year period, 1961 through 1990.

*City office data.

How Much Water in Snow?	Average snow: 10 inches = 1 inch of water Heavy, wet snow: 4 to 5 inches = 1 inch of water Dry, powdery snow: 15 inches = 1 inch water

Average Number of Days with Precipitation of 0.01 Inch or More, Selected U.S. Cities

City	Length of Record (years)	Jan.	Feb.	Mar.	Apr.	May	June	July	Aug.	Sept.	Oct.	Nov.	Dec.	Ann. Avg.
Mobile, AL	51	11	10	10	7	9	11	16	14	10	6	8	10	122
Juneau, AK*	48	18	17	18	17	17	15	17	18	20	23	20	21	222
Phoenix, AZ	53	4	4	4	2	1	1	4	5	3	3	3	4	36
Little Rock, AR	50	9	9	10	10	10	8	8	7	7	7	8	9	105
Los Angeles, CA	57	6	6	6	3	1	(a)	1	(a)	1	2	3	5	35
Sacramento, CA	53	10	9	9	5	3	1	(a)	(a)	1	3	7	9	57
San Diego, CA	52	7	6	7	4	2	1	(a)	1	1	2	5	6	42
San Francisco, CA	65	11	10	10	6	3	1	(a)	(a)	1	4	7	10	62
Denver, CO	58	6	6	9	9	11	9	9	9	6	5	6	5	89
Hartford, CT	38	11	10	11	11	12	11	10	10	9	8	11	12	127
Wilmington, DF	45	11	9	11	11	11	10	9	9	8	8	9	10	116
Washington, DC	51	10	9	11	10	11	9	10	9	8	7	9	9	112
Jacksonville, FL	51	8	8	8	7	8	12	14	15	13	8	6	8	116
Miami, FL	50	7	6	6	6	10	15	16	17	17	14	9	6	130
Atlanta, GA	58	11	10	11	9	9	10	12	10	8	6	8	10	115
Honolulu, HI	43	10	9	9	9	7	6	7	6	7	9	9	10	99
Boise, ID	53	12	10	10	8	8	6	2	3	4	6	10	11	90
Chicago, IL	34	11	9	12	13	11	10	10	9	10	9	11	11	126
Peoria, IL	53	9	8	11	12	11	9	9	8	9	8	9	10	114
Indianapolis, IN	53	12	10	13	12	12	10	10	9	8	8	10	12	126
Des Moines, IA	53	7	7	10	11	11	11	9	9	9	8	7	8	107
Wichita, KS	39	5	5	8	8	11	9	7	8	8	6	5	6	86
Louisville, KY	45	11	11	13	12	12	10	11	8	8	8	10	11	125
New Orleans, LA	44	10	9	9	7	8	11	15	13	10	6	7	10	114
Portland, ME	52	11	10	11	12	12	11	10	10	8	9	12	11	128
Baltimore, MD	42	10	9	11	11	11	9	9	10	8	7	9	9	113
Boston, MA	41	11	10	12	11	12	11	9	10	9	9	11	12	126
Detroit, MI	34	13	11	13	13	11	10	9	9	10	10	12	14	136
Sault Ste. Marie, MI	51	19	15	13	11	11	11	10	11	13	13	17	19	166
Duluth, MN	51	12	10	11	10	12	13	11	11	12	9	11	12	134
Minneapolis/ St. Paul, MN	54	9	7	10	10	11	12	10	10	10	8	9	9	114
Jackson, MS	29	11	9	10	8	10	8	10	10	8	6	8	10	110
Kansas City, MO	20	7	7	10	11	11	10	8	9	8	8	8	8	105
St. Louis, MO	35	8	8	11	11	11	9	9	8	8	8	10	9	111

Average Number of Days with Precipitation of 0.01 Inch or More, Selected U.S. Cities (continued)

City	Length of Record (years)	Jan.	Feb.	Mar.	Apr.	May	June	July	Aug.	Sept.	Oct.	Nov.	Dec.	Ann. Avg.
Great Falls, MT	55	9	8	9	9	11	12	7	8	7	6	7	8	101
Omaha, NE	56	6	7	9	10	12	11	9	9	8	6	6	6	99
Reno, NV	50	6	6	6	4	4	3	2	2	2	3	5	6	50
Concord, NH	51	11	10	11	12	12	11	10	10	9	9	11	11	125
Atlantic City, NJ	49	11	10	11	11	10	9	9	9	8	7	9	10	112
Albuquerque, NM	53	4	4	5	3	5	4	9	9	6	5	3	4	61
Albany, NY	46	12	11	12	12	13	11	10	10	10	9	12	12	134
Buffalo, NY	49	20	17	16	14	12	10	10	11	11	12	16	20	169
New York, NY**	123	11	10	11	11	11	10	11	10	8	8	9	10	121
Charlotte, NC	53	10	10	11	9	10	10	11	10	7	7	8	10	111
Raleigh, NC	53	8	7	8	8	10	11	9	8	7	6	6	8	96
Bismarck, ND	48	10	10	10	9	10	9	11	10	8	7	8	9	112
Cincinnati, OH	45	12	11	13	13	12	10	10	9	8	8	11	12	130
Cleveland, OH	51	16	14	15	14	13	11	10	10	10	11	15	16	156
Columbus, OH	53	13	11	14	13	13	11	11	9	8	9	12	13	137
Oklahoma City, OK	53	5	6	7	8	10	9	6	7	7	6	5	6	83
Portland, OR	52	18	16	17	14	12	9	4	5	8	12	18	19	151
Philadelphia, PA	52	11	9	11	11	11	10	9	9	8	8	9	10	117
Pittsburgh, PA	40	16	14	16	14	13	11	11	10	9	10	13	16	153
Providence, RI	39	11	10	12	11	11	11	9	9	8	9	11	12	124
Columbia, SC	45	10	10	10	8	9	10	12	11	8	6	7	9	109
Sioux Falls, SD	47	6	6	9	9	11	11	10	9	8	6	6	6	97
Memphis, TN	42	10	10	11	10	9	8	9	8	7	6	9	10	106
Nashville, TN	51	11	11	12	11	11	9	10	9	8	7	9	11	118
Dallas/Ft. Worth, TX	39	7	6	7	8	9	7	5	5	7	6	6	7	79
El Paso, TX	53	4	3	2	2	3	3	8	8	5	4	3	4	49
Houston, TX	23	11	9	9	7	8	9	10	9	9	7	9	9	106
Salt Lake City, UT	64	10	9	10	9	8	5	5	6	5	6	8	9	90
Burlington, VT	49	14	12	13	12	14	12	12	13	12	12	14	15	154
Norfolk, VA	44	11	10	11	10	10	9	11	11	8	8	8	9	115
Richmond, VA	55	10	9	11	9	11	9	11	10	8	7	8	9	113
Seattle/Tacoma, WA	48	19	16	17	14	10	9	5	6	9	13	18	19	155
Spokane, WA	45	14	11	11	9	9	8	4	5	6	8	13	15	113
Charleston, WV	45	15	14	15	14	13	11	13	11	10	9	12	14	151
Milwaukee, WI	52	11	10	12	12	12	11	10	9	9	9	11	11	125
Cheyenne, WY	57	6	6	9	10	12	11	11	10	7	6	6	6	100
San Juan, PR	37	17	13	12	13	16	15	19	18	17	17	18	19	196

Source: U.S. National Oceanic and Atmospheric Administration, *Comparative Climatic Data*, annual.
Note: Airport data except as noted. For period of record through 1992, except as noted.
(a) = less than 1/2 day.
*For period of record through 1989.
**City office data.

Weather Prediction Reliability Times

Here's how well the American Meteorological Society says its forecasters can predict the weather:

Period	Prediction Reliability
Up to 12 hours	"Considerable skill and utility" except for severe, short-lived storms (e.g., tornadoes)
12 to 48 hours	Good for general weather and movement of large extra-tropical storms
3 to 5 days	Good for major events, such as large storms or cold fronts
6 to 10 days	Temperature forecasts are more accurate than precipitation forecasts
30 to 90 days	Only slight skill; for odds only of higher or lower than normal temperatures and precipitation

Percentage of Sunshine, Number of Days Temperature Is Below 32°F, and Average Relative Humidity, Selected U.S. Cities

State	City	Average Percentage of Possible Sunshine (annual)	Minimum Temperature Below 32° F (mean number of days)	Average Relative Humidity (%)			
				Jan.		July	
				a.m.	p.m.	a.m.	p.m.
AL	Mobile[1]	59	2.2	79	61	87	60
AK	Juneau	30	14.1	81	77	87	70
AZ	Phoenix	86	0.8	66	32	45	20
AR	Little Rock	62	6.0	76	61	83	56
CA	Los Angeles[2]	73	(a)	70	59	86	68
	Sacramento	78	1.7	90	70	76	28
	San Diego	68	(a)	70	56	82	66
	San Francisco	NA	0.2	86	66	86	59
CO	Denver	70	15.7	63	49	68	34
CT	Hartford	56	13.5	69	56	82	51
DE	Wilmington	NA	10.0	73	60	83	54
DC	Washington	56	7.0	67	55	77	53
FL	Jacksonville	63	1.5	85	57	88	58
	Miami	72	(a)	81	59	82	63
GA	Atlanta	61	5.3	74	59	85	60
HI	Honolulu	69	0	81	61	73	51
ID	Boise	64	12.4	81	70	54	21
IL	Chicago	54	13.3	75	67	79	56
	Peoria	57	12.9	78	68	82	59
IN	Indianapolis	55	11.8	78	70	84	59
IA	Des Moines	59	13.5	74	67	76	57
KS	Wichita	65	11.1	76	63	67	48
KY	Louisville	56	8.9	72	64	81	58
LA	New Orleans	59	1.3	82	66	89	66

Percentage of Sunshine, Number of Days Temperature Is Below 32°F, and Average Relative Humidity, Selected U.S. Cities (continued)

State	City	Average Percentage of Possible Sunshine (annual)	Minimum Temperature Below 32° F (mean number of days)	Average Relative Humidity (%)			
				Jan.		July	
				a.m.	p.m.	a.m.	p.m.
ME	Portland	57	15.7	74	60	89	59
MD	Baltimore	57	9.7	69	57	81	53
MA	Boston	58	9.8	65	57	77	56
MI	Detroit	53	13.6	78	69	81	53
	Sault Ste. Marie	47	18.1	81	75	90	61
MN	Duluth	52	18.5	74	70	82	59
	Minneapolis/St. Paul	58	15.6	72	67	74	54
MS	Jackson	60	5.0	84	65	90	59
MO	Kansas City	62	11.0	72	63	75	56
	St. Louis	57	10.0	77	66	77	56
MT	Great Falls	61	15.7	66	60	66	29
NE	Omaha[2]	60	14.1	75	65	78	57
NV	Reno	79	17.4	79	50	63	18
NH	Concord	54	17.3	74	58	90	51
NJ	Atlantic City	56	11.0	76	58	87	57
NM	Albuquerque	76	11.9	70	40	60	27
NY	Albany	52	14.9	76	63	84	55
	Buffalo	49	13.2	77	72	79	55
	New York[3]	58	7.9	65	60	74	55
NC	Charlotte	63	6.6	72	56	83	57
	Raleigh	59	7.8	73	55	88	58
ND	Bismarck	59	18.6	75	68	74	46
OH	Cincinnati	52	10.8	75	67	83	57
	Cleveland	49	12.3	75	69	81	57
	Columbus	49	11.9	74	67	82	56
OK	Oklahoma City	NA	7.7	72	59	70	49
OR	Portland	48	4.3	86	75	82	45
PA	Philadelphia	56	9.7	71	59	81	54
	Pittsburgh	46	12.3	73	65	80	54
RI	Providence	58	11.8	69	56	83	56
SC	Columbia	64	6.0	78	54	87	54
SD	Sioux Falls[4]	63	16.8	75	68	75	53
TN	Memphis	64	5.7	75	63	79	57
	Nashville	56	7.6	75	63	85	57
TX	Dallas/Ft. Worth	63	4.0	73	60	67	49
	El Paso	84	6.5	66	35	63	30
	Houston	56	2.1	82	64	86	58
UT	Salt Lake City	66	12.5	79	69	52	22
VT	Burlington	49	15.5	70	63	82	53
VA	Norfolk	61	5.4	72	59	84	59
	Richmond	62	8.5	77	57	88	56

Percentage of Sunshine, Number of Days Temperature Is Below 32°F, and Average Relative Humidity, Selected U.S. Cities (continued)

| State | City | Average Percentage of Possible Sunshine (annual) | Minimum Temperature Below 32° F (mean number of days) | Average Relative Humidity (%) | | | |
| | | | | Jan. | | July | |
				a.m.	p.m.	a.m.	p.m.
WA	Seattle/Tacoma	46	3.1	81	74	82	49
	Spokane	54	13.9	85	78	64	27
WV	Charleston[5]	40	10.0	74	63	90	60
WI	Milwaukee	54	14.1	75	68	80	61
WY	Cheyenne	65	17.2	57	50	70	38
PR	San Juan	66	0	82	64	84	67

Source: U.S. National Oceanic and Atmospheric Administration, *Comparative Climatic Data*, annual.

Note: Airport data except as noted. For period of record through 1991, except as noted.

NA = not available. (a) = less than 1/2 day.

1. Recording site is in Montgomery, AL.
2. Site is not at the airport.
3. City office data.
4. Recording site is Rapid City, SD.
5. Recording site is Elkins, WV.

The Pollen Index   A number often heard on local weather forecasts is the pollen index. The reading given is based on the number of pollen grains in 1 cubic yard of air.

Average Daily High and Low Temperatures and Precipitation, Selected World Cities (in °F and inches of rainfall equivalent)

| City | January | | | April | | | July | | | October | | |
| | Temperature | | Avg. precip. | Temperature | | Avg. precip. | Temperature | | Avg. precip. | Temperature | | Avg. precip. |
	Max.	Min.		Max.	Min.		Max.	Min.		Max.	Min.	
Accra, Ghana	87	73	0.6	88	76	3.2	81	73	1.8	85	74	2.5
Amsterdam, Netherlands	40	34	2.0	52	43	1.6	69	59	2.6	56	48	2.8
Athens, Greece	54	42	2.2	67	52	0.8	90	72	0.2	74	60	1.7
Auckland, New Zealand	73	60	3.1	67	56	3.8	56	46	5.7	63	52	4.0
Baghdad, Iraq	60	39	0.9	85	57	0.5	110	76	Tr	92	61	0.1
Bangkok, Thailand	89	67	0.2	95	78	2.3	90	76	6.9	88	76	9.9
Beirut, Lebanon	62	51	7.5	72	58	2.2	87	73	Tr	81	69	2.0
Berlin, Germany	35	26	1.9	55	38	1.7	74	55	3.1	55	41	1.7
Bogotá, Colombia	67	48	2.3	67	51	5.8	64	50	2.0	66	50	6.3
Bombay, India	88	62	0.1	93	74	Tr	88	75	24.3	93	73	2.5

Average Daily High and Low Temperatures and Precipitation, Selected World Cities (in °F and inches of rainfall equivalent) (continued)

City	January Temperature Max.	January Temperature Min.	January Avg. precip.	April Temperature Max.	April Temperature Min.	April Avg. precip.	July Temperature Max.	July Temperature Min.	July Avg. precip.	October Temperature Max.	October Temperature Min.	October Avg. precip.
Budapest, Hungary	35	26	1.5	62	44	2.0	82	61	2.0	61	45	2.1
Buenos Aires, Argentina	85	63	3.1	72	53	3.5	57	42	2.2	69	50	3.4
Cairo, Egypt	65	47	0.2	83	57	0.1	96	70	0.0	86	65	Tr
Calcutta, India	80	55	0.4	97	76	1.7	90	79	12.8	89	74	4.5
Cape Town, South Africa	78	60	0.6	72	53	1.9	63	45	3.5	70	52	1.2
Caracas, Venezuela	75	56	0.9	81	60	1.3	78	61	4.3	79	61	4.3
Casablanca, Morocco	63	45	2.1	69	52	1.4	79	65	0.0	76	58	1.5
Copenhagen, Denmark	36	29	1.6	50	37	1.7	72	55	2.2	53	42	2.1
Dakar, Senegal	79	64	Tr	81	65	Tr	88	76	3.5	89	76	1.5
Dhaka, Bangladesh	77	56	0.3	92	74	5.4	89	79	13.0	88	75	5.3
Dublin, Ireland	47	35	2.7	54	38	1.9	67	51	2.8	57	43	2.7
Geneva, Switzerland	39	29	1.9	58	41	2.5	77	58	2.9	58	44	3.8
Hanoi, Vietnam	68	58	0.8	80	70	3.6	92	79	11.9	84	72	3.5
Hong Kong	64	56	1.3	75	67	5.4	87	78	15.0	81	73	4.5
Istanbul, Turkey	45	36	3.7	61	45	1.9	81	65	1.7	67	54	3.8
Jakarta, Indonesia	84	74	11.8	87	75	5.8	87	73	2.5	87	74	4.4
Jerusalem, Israel	55	41	5.1	73	50	0.9	87	63	0.0	81	59	0.3
Kabul, Afghanistan	36	18	1.3	66	43	3.3	92	61	0.1	73	42	0.4
Karachi, Pakistan	77	55	0.5	90	73	0.1	91	81	3.2	91	72	0.1
Kinshasa, Zaire	87	70	5.3	89	71	7.7	81	64	0.1	88	70	4.7
Lagos, Nigeria	88	74	1.1	89	77	5.9	83	74	11.0	85	74	8.1
Lima, Peru	82	66	0.1	80	63	Tr	67	57	0.3	71	58	0.1
Lisbon, Portugal	56	46	3.3	64	52	2.4	79	63	0.2	69	57	3.1
London, United Kingdom	44	35	2.0	56	40	1.8	73	55	2.0	58	44	2.3
Madrid, Spain	47	33	1.1	64	44	1.7	87	62	0.4	66	48	1.9
Manila, Philippines	86	69	0.9	93	73	1.3	88	75	17.0	88	74	7.6
Melbourne, Australia	78	57	1.9	68	51	2.3	56	42	1.9	67	48	2.6
Mexico City, Mexico	66	42	0.2	78	52	0.7	74	54	4.5	70	50	1.6
Montreal, Canada	21	6	3.8	50	33	2.6	78	61	3.7	54	40	3.4
Moscow, Russia	21	9	1.5	47	31	1.9	76	55	3.0	46	34	2.7
Nairobi, Kenya	77	54	1.5	75	58	8.3	69	51	0.6	76	55	2.1
New Delhi, India	71	43	0.9	97	68	0.3	95	80	7.1	93	64	0.4
Osaka, Japan	47	32	1.7	65	47	5.2	87	70	5.9	72	55	5.1
Oslo, Norway	30	20	1.7	50	34	1.6	73	56	2.9	49	37	2.9
Paris, France	42	32	1.5	60	41	1.7	76	55	2.1	59	44	2.2
Prague, Czech Republic	34	25	0.9	55	40	1.5	74	58	2.6	54	44	1.2

Average Daily High and Low Temperatures and Precipitation, Selected World Cities (in °F and inches of rainfall equivalent) (continued)

City	January Temperature Max.	Min.	Avg. precip.	April Temperature Max.	Min.	Avg. precip.	July Temperature Max.	Min.	Avg. precip.	October Temperature Max.	Min.	Avg. precip.
Rio de Janeiro, Brazil	84	73	4.9	80	69	4.2	75	63	1.6	77	66	3.1
Riyadh, Saudi Arabia	70	46	0.1	89	64	1.0	107	78	0.0	94	61	0.0
Rome, Italy	54	39	3.3	68	46	2.0	88	64	0.4	73	53	4.3
St. Petersburg, Russia	23	12	1.0	45	31	1.0	71	57	2.5	45	37	1.8
Santiago, Chile	85	53	0.1	74	45	0.5	59	37	3.0	72	45	0.6
São Paulo, Brazil	77	63	8.8	73	59	2.2	66	53	1.5	68	57	4.6
Seoul, South Korea	32	15	1.2	62	41	3.0	84	70	14.8	67	45	1.6
Shanghai, China	47	32	1.9	67	49	3.6	91	75	5.8	75	56	2.9
Singapore	86	73	9.9	88	75	7.4	88	75	6.7	87	74	8.2
Stockholm, Sweden	31	23	1.5	45	32	1.5	70	55	2.8	48	39	2.1
Sydney, Australia	78	65	3.5	71	58	5.3	60	46	4.6	71	56	2.8
Tahiti, French Polynesia	89	72	13.2	89	72	6.8	86	68	2.6	87	70	3.4
Taipei, Taiwan	66	53	3.8	77	64	5.3	92	76	8.8	80	68	5.5
Tehran, Iran	45	27	1.8	71	49	1.4	99	72	0.1	76	53	0.3
Tokyo, Japan	47	29	1.9	63	46	5.3	83	70	5.6	69	55	8.2
Toronto, Canada	30	16	2.7	50	34	2.5	79	59	3.0	56	40	2.4
Vienna, Austria	34	26	1.5	57	41	2.0	75	59	3.0	55	44	2.0
Warsaw, Poland	30	21	1.2	54	38	1.5	75	56	3.0	54	41	1.7

Source: U.S. Department of Commerce, *Climates of the World*, 1991.

Tr = trace.

Windiest U.S. Locations

Place	Mean Wind Speed (in mph)
St. Paul Island, AK	18.3
Cold Bay, AK	16.9
Amarillo, TX	13.7
Boston, MA	12.4
Wichita, KS	12.4
Buffalo, NY	12.1
Honolulu, HI	11.6
Minneapolis/St. Paul, MN	10.5
Chicago, IL	10.3
New York, NY	9.4
Miami, FL	9.2
Atlanta, GA	9.1
Anchorage, AK	6.8
Los Angeles, CA	6.2

Wind Chill

Wind can make cold temperatures feel even colder. Wind also makes cold weather more dangerous; wind chill factors of −25°F and below are considered to be dangerous even to properly clothed persons. Wind speeds greater than 45 mph have little additional chilling effect.

Wind Speed (mph)	Air Temperature (°F)																
	35	30	25	20	15	10	5	0	−5	−10	−15	−20	−25	−30	−35	−40	−45
	Wind Chill Temperature (°F)																
5	33	27	21	16	12	7	0	−5	−10	−15	−21	−26	−31	−36	−42	−47	−52
10	22	16	10	3	−3	−9	−15	−22	−27	−34	−40	−46	−52	−58	−64	−71	−77
15	16	9	2	−5	−11	−18	−25	−31	−38	−45	−51	−58	−65	−72	−78	−85	−92
20	12	4	−3	−10	−17	−24	−31	−39	46	−53	−60	−67	−74	−81	−88	−95	−103
25	8	1	−7	−15	−22	−29	−36	−44	−51	−59	−66	−74	−81	−88	−96	−103	−110
30	6	−2	−10	−18	−25	−33	−41	−49	−56	−64	−71	−79	−86	−93	−101	−109	−116
35	4	−4	−12	−20	−27	−35	−43	−52	−58	−67	−74	−82	−89	−97	−105	−113	−120
40	3	−5	−13	−21	−29	−37	−45	−53	−60	−69	−76	−84	−92	−100	−107	−115	−123
45	2	−6	−14	−22	−30	−38	−46	−54	−62	−70	−78	−85	−93	−102	−109	−117	−125

Apparent Room Temperature

A room's air during winter months will feel warmer as the moisture content increases:

Room Temperature (°F)	Relative Humidity (%)										
	0	10	20	30	40	50	60	70	80	90	100
	Apparent Room Temperature (°F)										
75	68	69	71	72	74	75	76	76	77	78	79
74	66	68	69	71	72	73	74	75	76	77	78
73	65	67	68	70	71	72	73	74	75	76	77
72	64	65	67	68	70	71	72	73	74	75	76
71	63	64	66	67	68	70	71	72	73	74	75
70	63	64	65	66	67	68	69	70	71	72	73
69	62	63	64	65	66	67	68	69	70	71	72
68	61	62	63	64	65	66	67	68	69	70	71
67	60	61	62	63	64	65	66	67	68	68	69
66	59	60	61	62	63	64	65	66	67	67	68
65	59	60	61	61	62	63	64	65	65	66	67
64	58	59	60	60	61	62	63	64	64	65	66
63	57	58	59	59	60	61	62	62	63	64	64
62	56	57	58	58	59	60	61	61	62	63	63
61	56	57	57	58	59	59	60	60	61	61	62
60	55	56	56	57	58	58	59	59	60	60	61

Relative Humidity Defined The term "relative humidity" refers to the percentage of water molecules the air contains. A reading of 100 percent means that the air is completely saturated—it contains as many water molecules as it can hold. A reading of 50 percent means that the air is halfway saturated with water, a level that is beneficial for humans and other living things.

The Heat Index

Air temperature and humidity combine to affect how hot it feels outside. The table below gives the apparent temperatures.

Air Temperature (°F)	Relative Humidity (%)									
	10	20	30	40	50	60	70	80	90	100
	Apparent Temperature (°F)									
80	75	77	73	79	81	90	85	86	88	91
85	80	82	84	86	88	90	93	97	102	108
90	85	87	90	93	96	100	106	113	122	
95	90	93	96	101	107	114	124	136		
100	95	99	104	110	120	132	144			
105	100	105	113	123	135	149				

Weather Information on the Radio

The National Oceanic and Atmospheric Administration broadcasts continuously on these radio frequencies:

162.400 MHz

162.475 MHz

162.550 MHz

The Coast Guard NAVTEX broadcasts on 518 kHz, and the Coast Guard VHF (Channel 22A) broadcasts on 157.1 MHz.

Storms

Saffir-Simpson Hurricane Scale

Category	Maximum Sustained Wind Speed (mph)	Damage
1 (weak)	74–95	Above normal; no real damage to building structures*
2 (moderate)	96–110	Some roofing, door, and window damage; considerable damage to vegetation, mobile homes, and piers
3 (strong)	111–130	Some buildings damaged; mobile homes destroyed
4 (very strong)	131–156	Complete roof failure on small residences; major erosion of beach areas; major damage to lower floors of structures near shore
5 (devastating)	>156	Complete roof failure on many residences and industrial buildings; some complete building failures

*Actual storm surge values will vary considerably depending on coastal configurations and other factors.

Hurricane Names

There are 6 groups of names used for hurricanes in the Atlantic Ocean, Caribbean Sea, and Gulf of Mexico; they are rotated over a 6-year cycle. If a storm causes major damage, its name is retired; for example, a new name beginning with A will be assigned for 1998 to replace 1992's devastating Hurricane Andrew.

Hurricane Facts   Hurricane Hugo, which hit South Carolina in 1989, measured 4 on the Saffir-Simpson scale, with winds up to 135 mph. Hurricane Andrew, which struck Florida in 1992, also measured 4 on that scale, with winds up to 145 mph. Andrew was the costliest natural disaster in U.S. history at that time; it caused 54 deaths and up to $30 billion in damage, destroying 25,524 homes and damaging 101,241 others.

Fujita Scale (F Scale) for Rating Tornado Intensity

F Scale	Wind Velocity (mph)	Damage
F0	40–72	Light—tree branches, chimneys
F1	73–112	Moderate—mobile homes, autos pushed aside
F2	113–157	Considerable—roofs torn off houses, large trees uprooted
F3	158–206	Severe—houses torn apart, trees uprooted, cars lifted off the ground
F4	207–260	Devastating—houses leveled, cars thrown (less than 2 percent of all tornadoes)
F5	261–318	Incredible—structures lifted off foundations, cars become missiles
F6	318	Maximum tornado wind speed

How Far Away Is the Storm?   When a storm is accompanied by lightning, you can estimate its distance by counting the number of seconds that elapse from the time you see the lightning until the time you hear the thunder. Divide by 5 to get the approximate distance in miles to the storm (or divide by 3 to get kilometers).

Tornado Facts   Tornadoes are most likely to occur between 3 p.m. and 7 p.m. The average tornado is 300 to 400 yards wide and has a path 4 miles long. Average speeds are 25 to 40 mph. Tornadoes occur in all 50 states.

Beaufort Wind Scale: On Land

The Beaufort Wind Scale allows observers to estimate wind speed by noting the effects the wind has on land and over water.

Beaufort Number	Force	Miles per Hour	Effects
0	Calm	<1	Smoke rises vertically; no perceptible movement of anything
1	Light air	1–3	Smoke drift shows wind direction; tree leaves barely move
2	Light breeze	4–7	Wind felt on face; leaves rustle; small twigs move
3	Gentle breeze	8–12	Leaves and small twigs in constant motion; dry leaves blow up from ground
4	Moderate breeze	13–18	Small branches move; dust and paper raised and driven along
5	Fresh breeze	19–24	Large branches and small trees in leaf begin to sway; crested wavelets form on inland water
6	Strong breeze	25–31	Large branches in continuous motion
7	Moderate gale	32–38	Whole trees in motion; inconvenience in walking
8	Fresh gale	39–46	Twigs and small branches break; difficulty in walking
9	Strong gale	47–54	Bricks on chimneys loosen; roofing slates blow off; ground littered with broken branches
10	Whole gale	55–63	Trees uprooted; considerable structural damage
11	Storm	64–75	Widespread damage
12	Hurricane	>75	Severe and extensive damage

Beaufort Wind Scale: On Open Sea

Beaufort Number	Knots (nautical mph)	Description (probable wave height in meters)
0	<1	Glassy sea; smoke rises vertically
1	1–3	Small ripples
2	4–6	Light breeze; wavelets (0.15)
3	7–10	Gentle breeze; crests of large wavelets break occasionally (0.6)
4	11–16	Moderate breeze; small waves with breaking crests (1)
5	17–21	Fresh breeze; long waves, spray from breaking crests (1.8)
6	22–27	Strong breeze; large waves with extensive foamy crests (3)
7	28–33	Near gale; sea heaps up, white foam from crests blows in streaks downwind (4)
8	34–40	Gale; longer, higher waves, spindrift off crests, streaky foam (5.5)
9	41–47	Strong gale; high waves, crests topple (7)
10	48–55	Storm; very high waves, long curved crests, sea white from large foam patches (9)
11	56–63	Violent storm; huge waves, poor visibility (11.3)
12	>64	Hurricane; air full of foam and spray (13.7)

Storm and Cloud Sizes

Here are some approximate sizes (width or length) of typical storms, clouds, and other weather phenomena:

Storm system	600 miles
Hurricane	60 miles
Large thunderstorm	6 miles
Large cumulus cloud	1.2 miles
Fair-weather cumulus cloud	0.03 mile
Tornado funnel	18 miles
Dust devil	0.5 mile

9

People and
Places

World and U.S. Population

Numbers to Know: World Population

According to the U.S. Bureau of the Census,

- Half the world's people are under age 25.

- In 1994, children age 4 and under outnumbered persons age 60 and over; by 2020, the number of elderly will exceed the number of young children.

- Between 1994 and 2020, the number of women of childbearing age will double to more than 400 million in Africa and the Near East.

- In 1994, it was expected that births in India would exceed those in all 50 sub-Saharan countries combined.

- Population growth rates are highest, at 2.9 percent, in sub-Saharan Africa, and are above 2 percent in the Near East and Africa; they are well below 2 percent in Asia and Latin America.

- The average population growth rate for the world as a whole is 2.5 percent; for developing countries it is 2.8 percent; and for developed countries it is 1.4 percent.

World Population by Region

The world's population was 5.64 billion in 1994; that number is expected to reach 7.9 billion by 2020, according to the U.S. Bureau of the Census.

Countries Ranked by Land Area and by Population, 1994

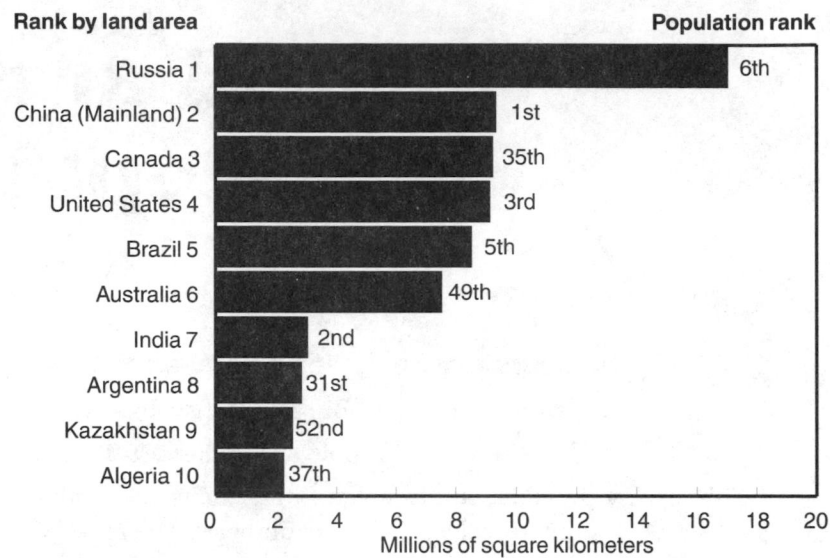

Source: U.S. Bureau of the Census, *World Population Profile: 1994.*

Countries Ranked by Population and by Land Area, 1994

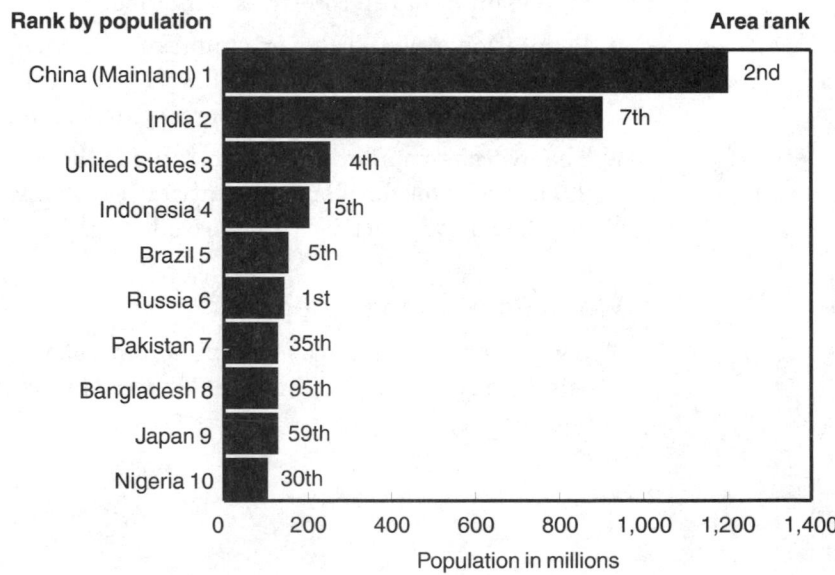

Source: U.S. Bureau of the Census, *World Population Profile: 1994.*

World Population by Region

Area	Area (million sq. miles)	1994 Population (millions)	% of World Population
North America	9.4	289	5.1
South America, Latin America, Caribbean	6.9	474	8.4
Europe	3.9	509	9.0
Asia (excluding former USSR)	17.2	3,344	59.3
Africa	11.7	701	12.4
Former USSR	8.6	296	5.2
Oceania (incl. Australia, New Zealand)	3.3	28	0.5
Antarctica	5.1	—	—

The World Population Boom

The world population almost quadrupled between 1500 and 1900.

Year	World Population
1500	420 million
1700	615 million
1800	900 million
1900	1.625 billion

Distribution of World Population by the Six Most Populous Countries

Half of the world's population lives in just 6 countries: Mainland China, India, the United States, Indonesia, Brazil, and Russia—a total of 2.88 billion people. China is home to 21 percent of the world's population; India, to 16 percent.

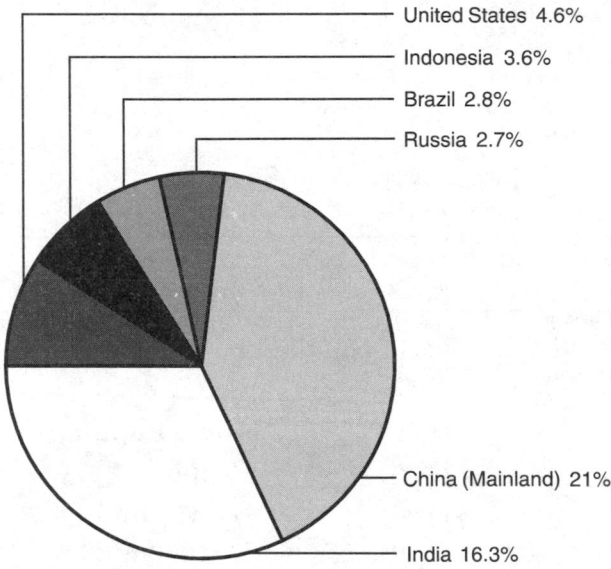

United States 4.6%

Indonesia 3.6%

Brazil 2.8%

Russia 2.7%

China (Mainland) 21%

India 16.3%

New Residents of the Planet Every hour, the world population increases by 100,000 persons. One out of every 2 births occurs in India, China, Bangladesh, Pakistan, Indonesia, or Nigeria. These countries combined make up 47 percent of the world's population.

World Population by Country

Country	Midyear Population (in thousands)				Population per Square Mile 1994
	1994	2000 (est.)	2010 (est.)	2020 (est.)	
World	**5,643,920**	**6,165,079**	**(NA)**	**(NA)**	**112**
Afghanistan	16,903	25,725	32,889	41,518	68
Albania	3,374	3,610	4,016	4,424	319
Algeria	27,895	31,743	38,186	44,096	30
Andorra	64	73	79	78	367
Angola	9,804	11,513	14,982	19,272	20
Antigua/Barbuda	65	68	74	80	381
Argentina	33,913	36,202	39,947	43,190	32
Armenia	3,522	3,685	3,854	3,959	306
Australia	18,077	19,386	21,151	22,724	6
Austria	7,955	8,108	8,259	8,329	249
Azerbaijan	7,684	8,243	8,995	9,689	230
Bahamas	273	298	332	356	70
Bahrain	586	687	849	1,008	2,451
Bangladesh	125,149	143,548	176,902	210,248	2,421
Barbados	256	260	272	284	1,541
Belarus	10,405	10,576	10,864	11,347	130
Belgium	10,063	10,144	10,135	10,015	862
Belize	209	242	299	356	24
Benin	5,342	6,517	8,955	11,920	125
Bhutan	1,739	1,996	2,474	3,035	96
Bolivia	7,719	8,801	10,671	12,547	18
Bosnia/Herzegovina	4,651	4,828	5,039	5,117	236
Botswana	1,359	1,554	1,871	2,187	6
Brazil	158,739	169,543	183,742	197,466	49
Brunci	285	331	410	491	140
Bulgaria	8,800	8,742	8,757	8,642	206
Burkina Faso	10,135	11,871	14,478	18,123	96
Burundi	6,125	6,939	8,382	10,734	618
Cambodia	10,265	12,098	15,679	20,208	151
Cameroon	13,132	15,677	21,165	28,329	72
Canada	28,114	29,867	32,265	34,347	8
Cape Verde	423	503	646	812	272
Central African Republic	3,142	3,511	3,898	4,561	13
Chad	5,467	6,221	7,680	9,396	11
Chile	13,951	15,207	17,266	19,225	48

World Population by Country (continued)

Country	Midyear Population (in thousands)				Population per Square Mile 1994
	1994	2000 (est.)	2010 (est.)	2020 (est.)	
China: Mainland	1,190,431	1,260,154	1,348,429	1,424,725	331
Taiwan	21,299	22,448	24,092	25,122	1,710
Colombia	35,578	39,172	44,504	49,266	89
Comoros	530	656	919	1,249	633
Congo	2,447	2,784	3,219	3,775	19
Costa Rica	3,342	3,797	4,537	5,257	171
Croatia	4,698	4,717	4,729	4,647	215
Cuba	11,064	11,617	12,274	12,755	258
Cyprus	730	768	829	883	205
Czech Republic	10,408	10,607	10,892	10,991	215
Denmark	5,188	5,255	5,311	5,307	317
Djibouti	413	454	588	751	49
Dominica	88	95	107	118	302
Dominican Republic	7,826	8,644	9,931	11,153	419
Ecuador	10,677	11,945	13,990	15,894	100
Egypt	60,765	67,957	82,478	97,434	158
El Salvador	5,753	6,459	7,603	8,763	719
Equatorial Guinea	410	478	615	783	38
Eritrea	(NA)	(NA)	(NA)	(NA)	(NA)
Estonia	1,617	1,670	1,776	1,880	93
Ethiopia	58,710	70,340	94,496	124,294	138
Fiji	764	823	933	1,037	108
Finland	5,069	5,153	5,246	5,283	43
France	57,840	59,354	61,001	61,793	275
Gabon	1,139	1,244	1,445	1,675	11
Gambia	959	1,154	1,561	2,073	248
Georgia	5,681	5,925	6,253	6,506	211
Germany	81,088	82,239	82,837	82,385	600
Ghana	17,225	20,608	27,305	35,877	194
Greece	10,565	10,878	10,920	10,689	209
Grenada	94	98	115	141	719
Guatemala	10,721	12,408	15,284	18,131	256
Guinea	6,392	7,372	9,303	11,664	67
Guinea-Bissau	1,098	1,263	1,579	1,925	102
Guyana	729	710	767	833	10
Haiti	6,491	7,102	8,121	9,499	610

World Population by Country (continued)

Country	Midyear Population (in thousands)				Population per Square Mile 1994
	1994	2000 (est.)	2010 (est.)	2020 (est.)	
Honduras	5,315	6,192	7,643	9,042	123
Hungary	10,319	10,372	10,477	10,449	289
Iceland	264	277	293	306	7
India	919,903	1,018,105	1,173,621	1,320,746	801
Indonesia	200,410	219,496	250,033	276,474	284
Iran	65,615	78,350	107,676	143,624	104
Iraq	19,890	24,731	34,545	46,260	119
Ireland	3,539	3,627	3,846	4,034	133
Israel	5,051	5,507	6,241	6,934	643
Italy	58,138	58,865	59,089	57,844	512
Ivory Coast	14,296	17,371	22,924	29,705	116
Jamaica	2,555	2,746	3,110	3,446	611
Japan	125,107	127,554	129,361	126,062	821
Jordan	3,961	4,814	6,213	7,595	112
Kazakhstan	17,268	17,886	18,794	19,404	16
Kenya	28,241	32,479	37,990	44,240	128
Kyrgyzstan	4,698	5,119	5,810	6,490	61
Kiribati	78	87	95	98	281
Kuwait	1,819	2,494	3,220	4,091	264
Laos	4,702	5,557	7,168	8,923	53
Latvia	2,749	2,833	3,009	3,194	110
Lebanon	3,620	4,115	4,973	5,748	917
Lesotho	1,944	2,242	2,771	3,314	166
Liberia	2,973	3,620	4,903	6,449	80
Libya	5,057	6,294	8,913	12,391	7
Liechtenstein	30	32	34	36	487
Lithuania	3,848	4,007	4,263	4,505	153
Luxembourg	402	415	428	436	403
Madagascar	13,428	16,232	22,064	29,362	60
Malawi	9,732	11,045	13,233	16,697	268
Malaysia	19,283	21,953	26,589	31,681	152
Maldives	252	310	423	554	2,176
Mali	9,113	10,911	14,966	20,427	19
Malta	367	382	404	420	2,959
Mauritania	2,193	2,653	3,630	4,859	6
Mauritius	1,117	1,194	1,322	1,428	1,565

World Population by Country (continued)

Country	Midyear Population (in thousands)				Population per Square Mile 1994
	1994	2000 (est.)	2010 (est.)	2020 (est.)	
Mexico	92,202	102,912	120,115	136,096	124
Moldova	4,473	4,565	4,738	4,880	344
Monaco	31	32	33	34	40,505
Mongolia	2,430	2,826	3,545	4,309	4
Morocco	28,559	32,189	38,112	43,701	166
Mozambique	17,346	20,868	27,381	35,240	57
Myanmar (Burma)	44,277	49,300	57,720	35,240	174
Namibia	1,596	1,957	2,705	3,638	5
Nauru	10	11	11	12	1,236
Nepal	21,042	24,364	30,783	37,767	398
Netherlands	15,368	15,801	16,140	16,222	1,173
New Zealand	3,389	3,476	3,543	3,586	33
Nicaragua	4,097	4,759	5,864	6,945	88
Niger	8,972	10,985	14,652	20,166	18
Nigeria	98,091	118,620	161,969	215,893	279
North Korea	23,067	25,491	28,491	30,969	496
Norway	4,315	4,387	4,424	4,446	36
Oman	1,701	2,098	2,991	4,175	21
Pakistan	128,856	148,540	195,108	251,330	429
Panama	2,630	2,934	3,422	3,886	90
Papua New Guinea	4,197	4,812	5,925	7,044	24
Paraguay	5,214	6,104	7,730	9,474	34
Peru	23,651	26,258	30,483	34,340	48
Philippines	69,809	77,747	90,316	101,530	606
Poland	38,655	39,531	41,332	42,474	329
Portugal	10,524	10,744	10,997	11,038	297
Qatar	513	572	645	713	121
Romania	23,181	23,383	23,950	24,337	261
Russia	149,609	151,460	155,933	159,263	23
Rwanda	8,374	9,715	11,755	15,006	869
St. Kitts/Nevis	41	43	50	57	293
St. Lucia	145	151	169	193	615
St. Vincent/the Grenadines	115	122	136	152	882
San Marino	24	25	26	27	1,040
São Tomé and Príncipe	137	159	196	232	369
Saudi Arabia	18,197	20,070	30,494	42,085	22

World Population by Country (continued)

Country	Midyear Population (in thousands)				Population per Square Mile 1994
	1994	2000 (est.)	2010 (est.)	2020 (est.)	
Senegal	8,731	10,533	14,318	19,127	118
Seychelles	72	75	81	86	410
Sierra Leone	4,630	5,421	7,041	9,036	167
Singapore	2,859	3,025	3,206	3,335	11,867
Slovakia	5,404	5,585	5,883	6,078	112
Slovenia	1,972	1,998	2,025	2,008	252
Solomon Islands	386	470	620	767	36
Somalia	6,667	9,176	12,588	16,832	28
South Africa	43,931	51,334	65,850	82,502	93
South Korea	45,083	47,861	51,677	54,014	1,189
Spain	39,303	39,972	40,682	40,421	204
Sri Lanka	18,130	19,377	20,972	22,463	725
Sudan	29,420	35,236	46,167	58,090	32
Suriname	423	465	534	598	7
Swaziland	936	1,137	1,566	2,128	141
Sweden	8,778	8,994	9,228	9,469	55
Switzerland	7,040	7,268	7,519	7,696	458
Syria	14,887	18,519	25,768	34,309	209
Tajikistan	5,995	6,956	8,619	10,429	109
Tanzania	27,986	32,254	38,651	48,526	82
Thailand	59,510	63,620	64,181	62,941	301
Togo	4,255	5,263	7,401	10,146	203
Tonga	105	110	119	128	378
Trinidad and Tobago	1,328	1,420	1,583	1,722	670
Tunisia	8,727	9,599	10,937	12,144	145
Turkey	62,154	69,624	81,790	93,362	209
Turkmenistan	3,995	4,474	5,277	6,116	21
Tuvalu	10	11	12	15	979
Uganda	19,122	21,358	26,997	34,106	248
Ukraine	51,847	51,931	52,280	52,337	222
United Arab Emirates	2,791	3,582	4,873	6,080	86
United Kingdom	58,135	58,951	59,617	60,042	623
United States	260,714	275,327	298,621	323,113	74
Uruguay	3,199	3,344	3,594	3,822	48
Uzbekistan	22,609	25,467	30,380	35,422	131
Vanuatu	170	193	230	266	30

World Population by Country (continued)

Country	Midyear Population (in thousands)				Population per Square Mile 1994
	1994	2000 (est.)	2010 (est.)	2020 (est.)	
Venezuela	20,562	23,196	27,407	31,312	60
Vietnam	73,104	80,533	91,729	102,359	582
Western Samoa	204	235	288	341	186
Yemen	11,105	13,603	18,985	25,907	54
Yugoslavia (the former)	(NA)	(NA)	11,625	11,881	(NA)
Zaire	42,684	51,413	69,079	91,860	49
Zambia	9,188	10,625	12,614	15,828	32
Zimbabwe	10,975	12,013	12,990	14,620	74

Source: U.S. Bureau of the Census, *World Population Profile: 1994.*
NA = not available.

The World's 75 Most Populous Cities, 1992 to 2000

City and Country	Estimated Midyear Population (in thousands)			Area (sq. mi.)
	1992	1995	2000	
1. Tokyo/Yokohama, Japan	27,540	28,447	29,971	1,089
2. Mexico City, Mexico	21,615	23,913	27,872	522
3. São Paulo, Brazil	19,373	21,539	25,354	451
4. Seoul, South Korea	17,334	19,065	21,976	342
5. **New York, United States**	**14,628**	**14,638**	**14,648**	**1,274**
6. Osaka/Kobe/Kyoto, Japan	13,919	14,060	14,287	495
7. Bombay, India	12,450	13,532	15,357	95
8. Calcutta, India	12,137	12,885	14,088	209
9. Rio de Janeiro, Brazil	12,009	12,786	14,169	260
10. Buenos Aires, Argentina	11,743	12,232	12,911	535
11. Manila, Philippines	10,554	11,342	12,846	188
12. Moscow, Russia	10,526	10,769	11,121	379
13. Cairo, Egypt	10,372	11,155	12,512	104
14. Jakarta, Indonesia	10,185	11,151	12,804	76
15. Tehran, Iran	10,102	11,681	14,251	112
16. **Los Angeles, United States**	**10,072**	**10,414**	**10,714**	**1,110**
17. Delhi, India	9,243	10,105	11,849	138
18. London, United Kingdom	9,168	8,897	8,574	874
19. Paris, France	8,589	8,764	8,803	432

The World's 75 Most Populous Cities, 1992 to 2000 (continued)

City and Country	Estimated Midyear Population (in thousands)			Area (sq. mi.)
	1992	1995	2000	
20. Lagos, Nigeria	8,487	9,799	12,528	56
21. Karachi, Pakistan	8,174	9,350	11,299	190
22. Essen, Germany	7,506	7,364	7,239	704
23. Lima, Peru	7,028	7,853	9,241	120
24. Shanghai, China	7,000	7,194	7,540	78
25. Istanbul, Turkey	6,937	7,624	8,875	165
26. Taipei, Taiwan	6,924	7,477	8,516	138
27. **Chicago, United States**	**6,493**	**6,541**	**6,568**	**762**
28. Bogotá, Colombia	6,176	6,801	7,935	79
29. Bangkok, Thailand	6,088	6,657	7,587	102
30. Madras, India	5,998	6,550	7,384	115
31. Beijing, China	5,791	5,865	5,993	151
32. Hong Kong	5,762	5,841	5,956	23
33. Santiago, Chile	5,484	5,812	6,294	128
34. Pusan, South Korea	5,161	5,748	6,700	54
35. Bangalore, India	5,080	5,644	6,764	50
36. Nagoya, Japan	4,909	5,017	5,303	307
37. Tianjin, China	4,857	5,041	5,298	49
38. Milan, Italy	4,718	4,795	4,839	344
39. St. Petersburg, Russia	4,645	4,694	4,738	139
40. Dacca, Bangladesh	4,640	5,296	6,492	32
41. Madrid, Spain	4,577	4,772	5,104	66
42. Lahore, Pakistan	4,475	4,986	5,864	57
43. Baghdad, Iraq	4,358	4,566	5,239	97
44. Shenyang, China	4,323	4,457	4,684	39
45. Barcelona, Spain	4,221	4,492	4,834	87
46. **San Francisco, United States**	**4,005**	**4,104**	**4,214**	**428**
47. Kinshasa, Zaire	3,997	4,520	5,646	57
48. Manchester, United Kingdom	3,984	3,949	3,827	357
49. **Philadelphia, United States**	**3,970**	**3,988**	**3,979**	**471**
50. Belo Horizonte, Brazil	3,920	4,373	5,125	79
51. Ahmadabad, India	3,826	4,200	4,837	32
52. Hyderabad, India	3,787	4,149	4,765	88
53. Ho Chi Minh City, Vietnam	3,725	4,064	4,481	31
54. Athens, Greece	3,613	3,670	3,866	116
55. Sydney, Australia	3,528	3,619	3,708	338

The World's 75 Most Populous Cities, 1992 to 2000 (continued)

City and Country	Estimated Midyear Population (in thousands)			Area (sq. mi.)
	1992	1995	2000	
56. Guadalajara, Mexico	3,525	3,839	4,451	78
57. **Miami, United States**	**3,522**	**3,679**	**3,894**	**448**
58. Surabaya, Indonesia	3,327	3,428	3,632	43
59. Guangzhou, China	3,314	3,485	3,652	79
60. Caracas, Venezuela	3,247	3,338	3,435	54
61. Wuhan, China	3,231	3,325	3,495	65
62. Porto Alegre, Brazil	3,220	3,541	4,109	231
63. Toronto, Canada	3,182	3,296	3,296	154
64. Casablanca, Morocco	3,136	3,327	3,795	35
65. Monterrey, Mexico	3,084	3,385	3,974	77
66. Rome, Italy	3,028	3,079	3,129	69
67. Greater Berlin, Germany	3,020	3,018	3,006	274
68. Ankara, Turkey	3,000	3,263	3,777	55
69. Naples, Italy	2,996	3,051	3,134	62
70. Alexandria, Egypt	2,981	3,114	3,304	35
71. Montreal, Canada	2,933	2,996	3,071	164
72. **Detroit, United States**	**2,890**	**2,865**	**2,735**	**468**
73. Vangon, Myanmar	2,876	3,075	3,332	47
74. Melbourne, Australia	2,865	2,946	2,968	327
75. **Dallas, United States**	**2,856**	**2,972**	**3,257**	**419**

Source: U.S. Bureau of the Census.

U.S. Population Abroad, by Country

Country	Resident U.S. Citizens*	U.S. Tourists
Argentina	13,000	11,000
Australia	62,000	56,000
Canada	296,000	495,000
Costa Rica	23,000	21,000
Dominican Republic	97,000	36,000
Egypt	17,000	11,000
France	59,000	89,000
Germany	354,000	1,060,000
Greece	32,000	60,000
Hong Kong	24,000	23,000
Ireland	46,000	86,000
Israel	155,000	81,000
Italy	104,000	105,000
Mexico	539,000	1,088,000
Netherlands	19,000	11,000
Panama	36,000	2,000
Portugal	26,000	28,000
Saudi Arabia	40,000	6,000
South Korea	30,000	2,000
Spain	79,000	155,000
Switzerland	27,000	29,000
United Kingdom	255,000	377,000
Venezuela	24,000	2,000

Source: U.S. Department of State, 1993, unpublished data.

*Totals represent broad estimates and may include some non-U.S. citizens as well as dual nationals.

U.S. Population by State, 1970 to 1993
(in thousands)

State	1970	1980	1990	1993
U.S.	**203,302**	**226,546**	**248,710**	**257,908**
Alabama	3,444	3,894	4,041	4,187
Alaska	303	402	550	599
Arizona	1,775	2,718	3,665	3,936
Arkansas	1,923	2,286	2,351	2,424
California	19,971	23,668	29,760	31,211

U.S. Population by State, 1970 to 1993
(in thousands) (continued)

State	1970	1980	1990	1993
Colorado	2,210	2,890	3,294	3,566
Connecticut	3,032	3,108	3,287	3,277
Delaware	548	594	666	700
District of Columbia	757	638	607	578
Florida	6,791	9,746	12,938	13,679
Georgia	4,588	5,463	6,478	6,917
Hawaii	770	965	1,108	1,172
Idaho	713	944	1,007	1,099
Illinois	11,110	11,427	11,431	11,697
Indiana	5,195	5,490	5,544	5,713
Iowa	2,825	2,914	2,777	2,814
Kansas	2,249	2,364	2,478	2,531
Kentucky	3,221	3,661	3,685	3,789
Louisiana	3,645	4,206	4,220	4,295
Maine	994	1,125	1,228	1,239
Maryland	3,924	4,217	4,781	4,965
Massachusetts	5,689	5,737	6,016	6,012
Michigan	8,882	9,262	9,295	9,478
Minnesota	3,806	4,076	4,375	4,517
Mississippi	2,217	2,521	2,573	2,643
Missouri	4,678	4,917	5,117	5,234
Montana	694	787	799	839
Nebraska	1,485	1,570	1,578	1,607
Nevada	489	800	1,202	1,389
New Hampshire	738	921	1,109	1,125
New Jersey	7,171	7,365	7,730	7,879
New Mexico	1,017	1,303	1,515	1,616
New York	18,241	17,558	17,990	18,197
North Carolina	5,084	5,882	6,629	6,945
North Dakota	618	653	639	635
Ohio	10,657	10,798	10,847	11,091
Oklahoma	2,559	3,025	3,146	3,231
Oregon	2,092	2,633	2,842	3,032
Pennsylvania	11,801	11,864	11,882	12,048
Rhode Island	950	947	1,003	1,000
South Carolina	2,591	3,122	3,487	3,643
South Dakota	666	691	696	715

U.S. Population by State, 1970 to 1993
(in thousands) (continued)

State	1970	1980	1990	1993
Tennessee	3,926	4,591	4,877	5,099
Texas	11,199	14,229	16,987	18,031
Utah	1,059	1,461	1,723	1,860
Vermont	445	511	563	576
Virginia	4,651	5,347	6,187	6,491
Washington	3,413	4,132	4,867	5,255
West Virginia	1,744	1,950	1,793	1,820
Wisconsin	4,418	4,706	4,892	5,038
Wyoming	332	470	454	470

Source: U.S. Bureau of the Census, *1990 Census of Population and Housing, Population and Housing Unit Counts* (CPH-2); *Current Population Reports*, P25-1106; and unpublished data.

Centers of U.S. Population, 1790 to 1990

The median center of U.S. population is located at the intersection of two median lines, a north–south line constructed so that half of the nation's population lives east and half lives west of it, and an east–west line selected so that half of the nation's population lives north and half lives south of it. In 1990, the median center was in Crawford County, MO, 10 miles southeast of Steelville. Mean centers are shown below.

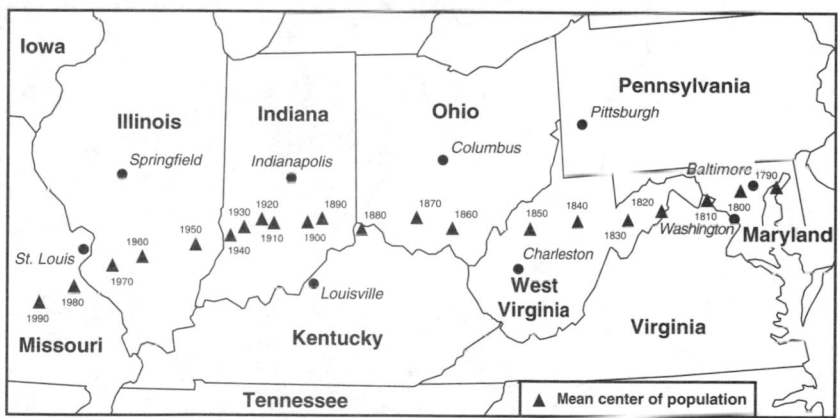

Source: *Statistical Abstract of the United States, 1994.*

Note: The mean center of population is that point at which an imaginary flat, weightless, and rigid map of the United States would balance if weights of identical value were placed on it so that each weight represented the location of one person on the date of the census.

Projected U.S. Population by Age Group, 1995 to 2050 (in thousands)

Age	1995		2000	
	Population	% Distribution	Population	% Distribution
Total	263,433	100.0	276,242	100.0
Under 5 years	20,181	7.7	19,431	7.0
5–13 years	34,262	13.0	36,547	13.2
14–17 years	14,591	5.5	15,811	5.7
18–24 years	25,465	9.7	25,911	9.4
25–34 years	41,670	15.8	38,237	13.8
35–44 years	42,150	16.0	45,123	16.3
45–64 years	51,465	19.5	59,860	21.7
65 years and over	33,649	12.8	35,322	12.8

Age	2010		2050	
	Population	% Distribution	Population	% Distribution
Total	300,431	100.0	392,031	100.0
Under 5 years	20,017	6.7	25,382	6.5
5–13 years	36,213	12.1	45,742	11.7
14–17 years	17,388	5.8	20,630	5.3
18–24 years	30,220	10.1	35,710	9.1
25–34 years	38,179	12.7	49,462	12.6
35–44 years	39,659	13.2	47,739	12.2
45–64 years	78,651	26.2	87,257	22.3
65 years and over	40,104	13.3	80,109	20.4

Source: U.S. Bureau of the Census.

Number of Living Species

Group	Number of Species
Mammals, birds, fish, amphibians, reptiles	4,500
Worms	25,000
Nematodes	80,000
Mollusks	110,000
Arthropods (shrimp, scorpions, mites, spiders, insects)	2,000,000
Plants	268,000

Number of Endangered Species

Group	U.S. Only	U.S. and Foreign	Foreign Only
Animals total	**262**	**51**	**493**
Mammals	36	20	251
Birds	57	16	153
Reptiles	8	8	63
Amphibians	6	0	8
Fishes	60	4	11
Snails	14	0	1
Clams	50	0	2
Crustaceans	11	0	0
Insects	16	3	4
Arachnids	4	0	0
Plants	**378**	**10**	**1**
Total	**640**	**61**	**494**

Source: U.S. Fish and Wildlife Service, 1994.

Evolution of Humans: A Timeline

Scientific Name	Time of Appearance
Homo habilis	2.2 million years ago
Australopithecus boisei	2.0 million years ago
Homo erectus	2.0 million to 90,000 years ago
Australopithecus robustus	1.5 million years ago
Neanderthal man	100,000–35,000 years ago
Archaic *homo sapiens*	90,000 years ago
Homo sapiens (Cro-Magnon)	35,000 years ago

World Languages: Number of Native Speakers

There are about 2,700 world languages in use today. English is on its way to becoming the world language. When native speakers and those who use English as a second language are combined, the total number of speakers is estimated to be close to 1 billion.

Language	Number of Native Speakers
Chinese	1 billion
English	350 million
Spanish	250 million
Hindi	250 million
Arabic	165 million
Bengali	150 million
Russian	150 million
Portuguese	135 million
French	125 million
Japanese	125 million
German	100 million
Urdu	100 million

Note: These numbers are general estimates only.

Word Counts

English has by far the most words of any language—almost 1 million, counting scientific and technical terms; the *Oxford English Dictionary* contains 500,000 words.

The German vocabulary numbers about 185,000 words, and French has fewer than 100,000.

World Religious Affiliations

About 53 percent of the U.S. population in 1990 were Christian adherents; about 2.3 percent were Jewish. Here are the world counts, as of mid-1993:

Christian	1,869,752,000
Roman Catholic	1,042,501,000
Protestant	458,221,000
Orthodox	173,560,000
Other	195,470,000
Muslim	1,014,372,000
Nonreligious	912,874,000
Hindu	751,360,000
Buddhist	334,002,000
Atheist	241,852,000
Chinese folk religions	140,956,000
New religions	123,765,000
Tribal religions	99,736,000
Sikh	19,853,000
Jew	18,153,000
Shaman	10,854,000
Confucian	6,230,000
All other	32,188,800

The Earth

Earth Facts

Age	4.6 billion years
Area	196,950,711 square miles
Mass	6.6 sextillion tons
Land surface	57,000,000 square miles (29% of total area)
Water surface	140,000,000 square miles (71% of total area)
Average elevation of land	2,056 feet
Average depth of oceans	2.28 miles
Mean radius	3,958.7 miles
Circumference (equator)	24,901.6 miles
Circumference (meridian)	24,860 miles
Diameter (equator)	7,927 miles
Diameter (pole to pole)	7,900 miles
Thickness of crust	5 to 25 miles
Thickness of inner core	800 miles
Temperature at core	4,500°C
Distance traveled around sun	595 million miles a year

The Tides

High tides occur twice a day, or every 24 hours and 50 minutes, equally spaced every 12 hours and 25 minutes. They are separated by two low tides. The range of tides—that is, the difference between high and low waters—can vary depending on location and weather conditions. Here are averages for a few U.S. cities:

City	Feet	Inches
Baltimore, MD	1	8
Boston, MA	10	4
Eastport, ME	19	4
Galveston, TX	1	5
Mobile, AL	1	6
Newport, RI	3	1
New York, NY	5	1
San Diego, CA	5	9
Seattle, WA	11	4
Tampa, FL	2	10
Washington, DC	3	2

Geological Eras

Era/Period	Epoch	Millions of Years Ago
Cenozoic		
Quaternary	Recent	0.01
	Pleistocene	1.6
Tertiary	Pliocene	5
	Miocene	23
	Oligocene	35
	Eocene	57
	Paleocene	65
Mesozoic		
Cretaceous		146
Jurassic		208
Triassic		245
Paleozoic		
Permian		290
Carboniferous		363
Devonian		409
Silurian		439
Ordovician		510
Cambrian		570

The Earth's Place in Time

The following "year" is often used to demonstrate the relative youth of our planet and the people on it; it assumes that the universe was formed on the first day of the year and that the present time is near the end of that year:

Date	Event
January 1	Universe formed
April 30	Milky Way galaxy began to take shape
September 15	Earth and solar system formed
December 25	Dinosaurs lived
December 31, 6 a.m.	Humans evolved
December 31, 11:59:50 p.m.	First human civilizations
December 31, 11:59:52 p.m.	Pyramids built
December 31, 11:59:56 p.m.	Roman Empire
December 31, 11:59:58 p.m.	Middle Ages
December 31, 11:59:59 p.m.	Present time

Chemical Elements

Element	Chemical Symbol	Atomic Number
Hydrogen	H	1
Helium	He	2
Lithium	Li	3
Beryllium	Be	4
Boron	B	5
Carbon	C	6
Nitrogen	N	7
Oxygen	O	8
Fluorine	F	9
Neon	Ne	10
Sodium	Na	11
Magnesium	Mg	12
Aluminum	Al	13
Silicon	Si	14
Phosphorus	P	15

Chemical Elements (continued)

Element	Chemical Symbol	Atomic Number
Sulfur	S	16
Chlorine	Cl	17
Argon	Ar	18
Potassium	K	19
Calcium	Ca	20
Scandium	Sc	21
Titanium	Ti	22
Vanadium	V	23
Chromium	Cr	24
Manganese	Mn	25
Iron	Fe	26
Cobalt	Co	27
Nickel	Ni	28
Copper	Cu	29
Zinc	Zn	30
Gallium	Ga	31
Germanium	Ge	32
Arsenic	As	33
Selenium	Se	34
Bromine	Br	35
Krypton	Kr	36
Rubidium	Rb	37
Strontium	Sr	38
Yttrium	Y	39
Zirconium	Zr	40
Niobium	Nb	41
Molybdenum	Mo	42
Technetium	Tc	43
Ruthenium	Ru	44
Rhodium	Rh	45
Palladium	Pd	46
Silver	Ag	47
Cadmium	Cd	48
Indium	In	49
Tin	Sn	50
Antimony	Sb	51
Tellurium	Te	52

Chemical Elements (continued)

Element	Chemical Symbol	Atomic Number
Iodine	I	53
Xenon	Xe	54
Cesium	Cs	55
Barium	Ba	56
Lanthanum	La	57
Cerium	Ce	58
Praseodymium	Pr	59
Neodymium	Nd	60
Promethium	Pm	61
Samarium	Sm	62
Europium	Eu	63
Gadolinium	Gd	64
Terbium	Tb	65
Dysprosium	Dy	66
Holmium	Ho	67
Erbium	Er	68
Thulium	Tm	69
Ytterbium	Yb	70
Lutetium	Lu	71
Hafnium	Hf	72
Tantalum	Ta	73
Tungsten	W	74
Rhenium	Re	75
Osmium	Os	76
Iridium	Ir	77
Platinum	Pt	78
Gold	Au	79
Mercury	Hg	80
Thallium	Tl	81
Lead	Pb	82
Bismuth	Bi	83
Polonium	Po	84
Astatine	At	85
Radon	Rn	86
Francium	Fr	87
Radium	Ra	88
Actinium	Ac	89

Chemical Elements (continued)

Element	Chemical Symbol	Atomic Number
Thorium	Th	90
Protactinium	Pa	91
Uranium	U	92
Neptunium	Np	93
Plutonium	Pu	94
Americium	Am	95
Curium	Cm	96
Berkelium	Bk	97
Californium	Cf	98
Einsteinium	Es	99
Fermium	Fm	100
Mendelevium	Md	101
Nobelium	No	102
Lawrencium	Lr	103
Rutherfordium*	Rf	104
Hahnium*	Hn	105
Seaborgium*	Sg	106
Nielsbohrium*	Ns	107
Hassium*	Hs	108
Meitnerium*	Mt	109

*Names proposed and in use by the American Chemical Society. The International Union of Pure and Applied Chemistry has recommended a different slate of names and numbers; a final decision is expected in August 1997.

Measuring Gold

Gold is measured in karats (not in the carats used to measure gems). Pure gold is 24 karats; an alloy that is 75 percent gold by weight is considered to be 18-karat gold; 14-karat gold is 58.5 percent gold; 12-karat gold is 50.5 percent gold; and 10-karat gold is only 42 percent gold. Gold is the 16th rarest element.

The 10 Most Common Elements on Land and in the Sea

In order of abundance on land:
1. Oxygen
2. Silicon
3. Aluminum
4. Iron
5. Calcium
6. Sodium
7. Potassium
8. Magnesium
9. Titanium
10. Hydrogen

In order of abundance in seawater:
1. Oxygen
2. Hydrogen
3. Chlorine
4. Sodium
5. Magnesium
6. Sulfur
7. Calcium
8. Potassium
9. Bromine
10. Carbon

The Mohs Scale: The Relative Hardness of Minerals

Scale	Mineral	Hardness Test
1	Talc	Crushed by fingernail
2	Gypsum	Scratched by fingernail
3	Calcite	Scratched by copper coin
4	Fluorite	Scratched by glass
5	Apatite	Scratched by penknife
6	Orthoclase (feldspar)	Scratched by quartz
7	Quartz	Scratched by steel file
8	Topaz	Scratched by corundum
9	Corundum	Scratched by diamond
10	Diamond	

Zonal Points and Parallels

Point	Latitude
North Pole	90°00'N
Arctic Circle	66°30'N
Tropic of Cancer	23°27'N
Equator	00°00'
Tropic of Capricorn	23°27'S
Antarctic Circle	66°30'S
South Pole	90°00'S

Latitude and Longitude of Selected U.S. and Canadian Cities

	Latitude		Longitude	
City	°	'	°	'
Albany, NY	42	40	73	45
Albuquerque, NM	35	5	106	39
Atlanta, GA	33	45	84	23
Bismarck, ND	46	48	100	47
Boston, MA	42	21	71	5
Calgary, Alberta	51	1	114	1
Charleston, WV	38	21	81	38
Charlotte, NC	35	14	80	50
Chicago, IL	41	50	87	37
Cleveland, OH	41	28	81	37
Dallas, TX	32	46	96	46
Denver, CO	39	45	105	0
Dubuque, IA	42	31	90	40
Duluth, MN	46	49	92	5
Fargo, ND	46	52	96	48
Grand Rapids, MI	42	58	85	40
Helena, MT	46	35	112	2
Honolulu, HI	21	18	157	50
Indianapolis, IN	39	46	86	10
Jacksonville, FL	30	22	81	40
Juneau, AK	58	18	134	24
Lincoln, NE	40	50	96	40

Latitude and Longitude of Selected U.S. and Canadian Cities (continued)

City	Latitude °	Latitude '	Longitude °	Longitude '
Los Angeles, CA	34	3	118	15
Memphis, TN	35	9	90	3
Miami, FL	25	46	80	12
Milwaukee, WI	43	2	87	55
Montgomery, AL	32	21	86	18
Montreal, Quebec	45	30	73	35
New Haven, CT	41	19	72	55
New Orleans, LA	29	57	90	4
New York, NY	40	47	73	58
Oklahoma City, OK	35	26	97	28
Philadelphia, PA	39	57	75	10
Phoenix, AZ	33	29	112	4
Providence, RI	41	50	71	24
Quebec, Quebec	46	49	71	11
Richmond, VA	37	33	77	29
St. Louis, MO	38	35	90	12
San Antonio, TX	29	23	98	33
San Diego, CA	32	42	117	10
San Francisco, CA	37	47	122	26
San Juan, PR	18	30	66	10
Seattle, WA	47	37	122	20
Tampa, FL	27	57	82	27
Toronto, Ontario	43	40	79	24
Victoria, British Columbia	48	25	123	21
Washington, DC	38	53	77	2
Winnipeg, Manitoba	49	54	97	7

Latitude and Longitude of Selected World Cities

City	Lat. °	Lat. '	Long. °	Long. '
Adelaide, Australia	34	55 s	138	36 e
Athens, Greece	37	58 n	23	43 e
Auckland, New Zealand	36	52 s	174	45 e
Bangkok, Thailand	13	45 n	100	30 e
Beijing, China	39	55 n	116	25 e
Berlin, Germany	52	30 n	13	25 e
Bogotá, Colombia	4	32 n	74	15 w
Brussels, Belgium	50	52 n	4	22 e
Buenos Aires, Argentina	34	35 s	58	22 w
Cairo, Egypt	30	2 n	31	21 e
Calcutta, India	22	34 n	88	24 e
Cape Town, South Africa	33	55 s	18	22 e
Caracas, Venezuela	10	28 n	67	2 w
Copenhagen, Denmark	55	40 n	12	34 e
Edinburgh, Scotland	55	55 n	3	10 w
Hammerfest, Norway	70	38 n	23	38 e
Helsinki, Finland	60	10 n	25	0 e
Jakarta, Indonesia	6	16 s	106	48 e
Johannesburg, South Africa	26	12 s	28	4 e
La Paz, Bolivia	16	27 s	68	22 w
Lima, Peru	12	0 s	77	2 w
London, England	51	32 n	0	5 w
Madrid, Spain	40	26 n	3	42 w
Manila, Philippines	14	35 n	120	57 e
Mecca, Saudi Arabia	21	29 n	39	45 e
Mexico City, Mexico	19	26 n	99	7 w
Milan, Italy	45	27 n	9	10 e
Moscow, Russia	55	45 n	37	36 e
Nairobi, Kenya	1	25 s	36	55 e
Odessa, Ukraine	46	27 n	30	48 e
Paris, France	48	48 n	2	20 e
Prague, Czech Republic	50	5 n	14	26 e
Reykjavik, Iceland	64	4 n	21	58 w
Rio de Janeiro, Brazil	22	57 s	43	12 w
Teheran, Iran	35	45 n	51	45 e

Latitude and Longitude of Selected World Cities (continued)

City	Lat. °	Lat. '	Long. °	Long. '
Tokyo, Japan	35	40 n	139	45 e
Vienna, Austria	48	14 n	16	20 e
Wellington, New Zealand	41	17 s	174	47 e
Yangon, Myanmar	16	50 n	96	0 e
Zürich, Switzerland	47	21 n	8	31 e

Reading a Compass

A compass is read clockwise from 0 to 360 degrees, with north being 0 degrees, east 90 degrees, south 180 degrees, and west 270 degrees. A compass points to the magnetic north, not true north. Magnetic north is located about 1,200 miles from the North Pole off the coast of Canada.

Areas and Dimensions of the Continents

Continent	Area (sq. mi.)	Dimensions North to South (mi.)	Dimensions East to West (mi.)
Asia, including islands	17,276,909	5,300	6,000
Africa	11,688,728	5,000	4,700
North America	9,368,446	5,300	4,000
South America	6,970,760	4,750	3,100
Antarctica	5,165,000	—	—
Europe	3,947,441	2,400	3,900
Australia	3,294,866	1,970	2,400

Sizes of the World's Oceans

Ocean	Area (sq. mi.)	Average Depth (ft.)
Pacific	64,200,000	12,925
Atlantic	33,400,000	11,740
Indian	28,400,000	12,598
Arctic	5,100,000	3,407

The 10 Deepest Ocean Trenches

Location/Name	Deepest Point (ft.)
Pacific Ocean	
Mariana Trench	35,840
Tonga Trench	35,433
Philippine Trench	32,995
Kermadec Trench	32,963
Bonin Trench	32,788
Kuril Trench	31,988
Izu Trench	31,808
New Britain Trench	29,331
Yap Trench	27,976
Atlantic Ocean	
Puerto Rico Trench	28,232

The Deepest Lake

The world's deepest lake is Lake Baikal in Siberia, which is 5,315 feet at its deepest point. This lake contains 20 percent of the world's unfrozen fresh water.

The 20 Largest Lakes and Inland Seas

Name/Location	Area (sq. mi.)
Caspian Sea (Asia)	143,244
Superior (N. America)	31,700
Victoria (Africa)	26,828
Aral Sea (Asia)	24,904
Huron (N. America)	23,000
Michigan (N. America)	22,300
Tanganyika (Africa)	12,700
Baikal (Asia)	12,162
Great Bear (N. America)	12,096
Malawi (Africa)	11,150
Great Slave (N. America)	11,031
Erie (N. America)	9,910
Winnipeg (N. America)	9,417
Ontario (N. America)	7,550
Balkhash (Asia)	7,115
Ladoga (Asia)	6,835
Chad (Africa)	6,300
Maracaibo (S. America)	5,217
Onega (Asia)	3,710
Eyre (Australia)	3,600

Water, Water Everywhere

Here's the percentage distribution of water on our planet:

Oceans	97.3
Glaciers and polar ice	2.1
Underground aquifers	0.6
Lakes and rivers	0.01
Atmosphere	0.001

The water we drink is 3 billion years old.

The 20 Longest Rivers

Name	Length (mi.)
Nile (Africa)	4,160
Amazon (S. America)	4,000
Chang Jiang (Asia)	3,964
Ob-Irtysh (Asia)	3,362
Huang (Asia)	2,903
Congo (Africa)	2,900
Amur (Asia)	2,744
Lena (Asia)	2,734
Missouri (N. America)	2,700
Mackenzie (N. America)	2,635
Mekong (Asia)	2,600
Niger (Africa)	2,590
Yenisey (Asia)	2,543
Parana (S. America)	2,485
Mississippi (N. America)	2,348
Murray-Darling (Australia)	2,310
Volga (Europe)	2,194
Purus (S. America)	2,100
Madeira (S. America)	2,013
São Francisco (S. America)	1,988

The 10 Highest Waterfalls

Name	Longest Single Drop (ft.)
Angel (Venezuela)	2,648
Itatinga (Brazil)	2,060
Cuquenán (Guyana/Venezuela)	2,000
Ormeli (Norway)	1,847
Tysse (Norway)	1,749
Pilão (Brazil)	1,719
Ribbon (United States)	1,612
Vestre Mardola (Norway)	1,535
Roraima (Guyana)	1,500
Cleve-Garth (New Zealand)	1,476

The 10 Largest Deserts

Deserts make up one-seventh of the earth's land area.

Name	Approximate Area (sq. mi.)
Sahara (N. Africa)	3,320,000
Arabian (Arabian Peninsula)	900,000
Gobi (Mongolia/China)	450,000
Patagonian (Argentina)	260,000
Great Victoria (Australia)	250,000
Great Basin (United States)	190,000
Chihuahuan (Mexico)	175,000
Great Sandy (Australia)	150,000
Sonoran (United States)	120,000
Kyzyl-Kum (Kazakhstan/ Uzbekistan)	115,000

Mountains: The Highest Point on Each Continent

Mountain	Feet
Everest (Asia)	29,029
Aconcagua (S. America)	22,834
McKinley (N. America)	20,321
Kilimanjaro (Africa)	19,340
Vinson Massif (Antarctica)	16,066
Blanc (Europe)	15,771
Kosciusko (Australia)	7,316

Mountain Facts

The tallest mountain as measured from the ocean floor is Mauna Kea, Hawaii, at 33,476 feet (13,796 above sea level); Mt. Everest is the tallest mountain above sea level.

Mountain climbers should note that for every 1,000 feet in height, the temperature falls by 3.5°F.

Major Active Volcanoes

Name	Location	Date of Last Eruption
Asia/Oceania		
Agung	Bali	1964
Alaid	Kuril Islands	1972
Asama	Japan	1973
Azuma	Japan	1978
Bagana	Solomon Islands	1960
Batur	Bali	1968
Dukono	Indonesia	1971
Keli Mutu	Indonesia	1968
Kerintiji	Sumatra	1968
Klyuchevskaya	Russia	1994
Koryakskaya	Russia	1957
Lamington	Papua New Guinea	1952
Lokon-Empung	Celebes	1970
Lopevi	New Hebrides	1960
Mayon	Philippines	1978
Merapi	Java	1994
Nasu	Japan	1977
Niigata Yakeyama	Japan	1974
Rabaul	Papua New Guinea	1995
Rindjani	Indonesia	1966
Ruapehu	New Zealand	1995
Semeru	Java	1976
Sundoro	Java	1906
Suwanosezima	Japan	1979
Taal	Philippines	1977
Tambora	Indonesia	1913
Tiatia	Kuril Islands	1973
Ulawun	New Britain	1973
Unzen	Japan	1995
Usu	Japan	1978
White Island	New Zealand	1979
North America		
Augustine	Alaska (U.S.)	1986
Cleveland	Aleutian Islands (U.S.)	1951
Colima	Mexico	1975

Major Active Volcanoes (continued)

Name	Location	Date of Last Eruption
Katmai	Alaska (U.S.)	1931
Martin	Alaska (U.S.)	1912
Paricutin	Mexico	1952
Popocatepetl	Mexico	1994
St. Helens	Washington (U.S.)	1980
Seguam	Alaska (U.S.)	1977
Shishaldin	Aleutian Islands (U.S.)	1979
Spurr	Alaska (U.S.)	1992
Trident	Alaska (U.S.)	1963
Westdahl	Aleutian Islands (U.S.)	1978
South America		
Alcedo	Galápagos Islands	1954
Cotopaxi	Ecuador	1975
Fernandía	Galápagos Islands	1995
Galeras	Colombia	1992
Guallatiri	Chile	1960
Hudson	Chile	1973
Puracé	Colombia	1977
Tolima	Colombia	1943
Antarctica		
Big Ben	Heard Island	1960
Damley	South Sandwich Islands	1956
Deception Island	South Shetland Islands	1970
Erebus	Ross Island	1975
Mid-Pacific		
Kilauea	Hawaii (U.S.)	1995
Mauna Loa	Hawaii (U.S.)	1984
Central America/Caribbean		
Acatenango	Guatemala	1972
Arenal	Costa Rica	1978
Fuego	Guatemala	1979
Irazú	Costa Rica	1967
Izalco	El Salvador	1966
Masaya	Nicaragua	1978

Major Active Volcanoes (continued)

Name	Location	Date of Last Eruption
Pacaya	Guatemala	1995
Poás	Costa Rica	1978
Rincón de la Vieja	Costa Rica	1995
San Miguel	El Salvador	1976
Santa María (Santiaguito)	Guatemala	1993
Soufrière	St. Vincent	1979
Europe		
Etna	Italy	1993
Santorini	Greece	1950
Stromboli	Italy	1995
Vesuvius	Italy	1944
Mid-Atlantic		
Askja	Iceland	1961
Beerenberg	Jan Mayen Island	1970
Helgatell	Iceland	1973
Krafla	Iceland	1978
Leirhnukur	Iceland	1975
Tristan da Cunha	Tristan da Cunha	1962
Africa		
Erta-Ale	Ethiopia	1973
Fogo	Cape Verde Islands	1995
Karthala	Comoros	1977
Nyamuragira	Zaire	1977
Nyirangongo	Zaire	1977
Ol Doinyo Lengai	Tanzania	1960
Palma	Canary Islands	1971
Piton de la Fournaise	Réunion Island	1992
Teide	Canary Islands	1909

Volcano Facts There are 850 known active volcanoes, 80 of which are under ocean and 200 of which are in Indonesia. The largest active volcano is Mauna Loa, in Hawaii. It is 13,677 feet high, with a crater 6 miles wide.

Earthquake Facts

The most destructive earthquake in history occurred in 1923 in Tokyo and Yokohama, Japan; it destroyed 575,000 homes and killed 200,000 people. The worst death rate from an earthquake was in Shaanxi, China, in 1556; that earthquake killed an estimated 830,000 people. In 1976, an earthquake in Tangshan, China, killed 242,000 people. Every year, about 500,000 earthquakes take place, of which about 1,000 cause some damage.

Major Earthquakes Since 1980

Year	Date	Location	Deaths	Magnitude on Richter Scale
1980	Oct. 10	Northwestern Algeria	4,500	7.3
1980	Nov. 23	Southern Italy	4,800	7.2
1982	Dec. 13	Northern Yemen	2,800	6.0
1983	May 26	Northern Honshu, Japan	81	7.7
1983	Oct. 30	Eastern Turkey	1,300	7.1
1985	Mar. 3	Chile	146	7.8
1985	Sept. 19	Mexico City	25,000	8.1
1987	Mar. 5–6	Northeastern Ecuador	4,000	7.3
1988	Aug. 20	India/Nepal border	1,000	6.5
1988	Nov. 6	China/Burma border	1,000	7.3
1988	Dec. 7	Northwestern Armenia	25,000	6.9
1989	Oct. 17	San Francisco Bay	67	7.1
1990	May 30	Northern Peru	115	6.3
1990	June 21	Northwestern Iran	110,000	7.7
1990	July 16	Luzon, Philippines	1,600	7.7
1991	Feb. 1	Pakistan/Afghanistan border	1,200	6.8
1992	Mar. 13	Eastern Turkey	4,000	6.0
1992	June 28	Southern California	1	6.6
1992	Dec. 12	Flores Island, Indonesia	2,500	7.5
1993	July 12	Off Hokkaido, Japan	200+	7.7
1993	Sept. 29	Maharashtra, India	9,700	6.4
1994	Jan. 17	Northridge, California	61	6.8
1994	Feb. 15	Southern Sumatra, Indonesia	215	7.0
1994	June 6	Cauca, Southwestern Colombia	1,000	6.8
1994	June 9	La Paz, Bolivia	NA	8.2

Note: NA = not available.

The Richter Scale and the Modified Mercalli Earthquake Intensity Scale

The Richter scale, developed in 1935, is an estimate of the amount of energy released in an earthquake. It rates earthquakes on an open-ended logarithmic scale, with each increase of 1 on the scale reflecting a 32-fold increase in released energy. On this scale, an earthquake rated as 2 is barely perceptible; one rated 5 or more is destructive. Scientists today prefer to use the modified Mercalli scale, which measures intensity rather than amount of energy released. This scale uses roman numerals to avoid confusion with the arabic numerals of the Richter scale.

Intensity on Mercalli Scale	Description
I	Not felt except by a very few under especially favorable conditions.
II	Felt only by a few persons at rest, especially on upper floors of buildings. Delicately suspended objects may swing.
III	Felt quite noticeably by persons indoors, especially on upper floors of buildings. Not recognized as an earthquake by many people. Standing automobiles may rock slightly. Vibration similar to that of the passing of a truck. Duration estimated.
IV	Felt indoors by many, outdoors by few during the day. At night, some awakened. Dishes, windows, doors disturbed; walls make cracking sound. Sensation like that of heavy truck striking building. Standing automobiles rocked noticeably.
V	Felt by nearly everyone; many awakened. Some dishes, windows broken. Unstable objects overturned. Pendulum clocks may stop.
VI	Felt by all; many frightened. Some heavy furniture moved; a few instances of fallen plaster. Damage slight.
VII	Damage negligible in buildings of good design and construction; slight to moderate in well-built ordinary structures; considerable in poorly built or badly designed structures. Some chimneys broken.
VIII	Damage slight in specially designed structures; considerable in ordinary substantial buildings, with partial collapse; great in poorly built structures. Fall of chimneys, factory stacks, columns, monuments, walls. Heavy furniture overturned.
IX	Damage considerable in specially designed structures; well-designed frame structures thrown out of plumb. Damage great in substantial buildings, with partial collapse. Buildings shifted off foundations.
X	Some well-built wooden structures destroyed; most masonry and frame structures destroyed with foundations. Rails bent.
XI	Few if any masonry structures remain standing. Bridges destroyed. Rails bent greatly.
XII	Damage total. Lines of sight and level distorted. Objects thrown into the air.

Earthquake Magnitudes, Ground Motion, and Energy

Magnitude on Richter Scale	Energy Equivalence and Effect
–2	100-watt light bulb left on for a week
–1	Smallest earthquakes detected
0	Seismic waves from 1 pound of explosives
1	A 2-ton truck traveling 75 miles per hour
2	Not felt but recorded
3	Smallest earthquake commonly felt
4	Seismic waves from 1,000 tons of explosives
5 6 7	Damage varies from slight to great, depending on quality of construction
8	1906 San Francisco earthquake (estimated)
9	Largest recorded earthquake (magnitude = 8.9); destruction nearly total
10	Approximately all the energy used in the United States in one year

Source: U.S. Geological Survey.

The Atmosphere

The earth's atmosphere extends about 500 miles high. All the air in the world weighs about 5.7 trillion tons, or 1.25 ounces per cubic foot. One ton of air presses down on each of us at all times.

Layer	Thickness	Temperature (°F)
Troposphere	6–10 miles	–112
Stratosphere	20 miles	+28
Mesosphere	20 miles	–165
Thermosphere	Begins at 50 miles	+3,600

Pollutant Standards Index (PSI) Values

		Pollutant Levels					
Index Value	Air Quality Level	Particulate Matter (PM$_{10}$) (24-hour) µg/m^3	Sulfur Dioxide (SO$_2$) (24-hour) µg/m^3	Carbon Monoxide (CO) (8-hour) ppm	Ozone (O$_3$) (1-hour) ppm	Nitrogen Dioxide (NO$_2$) (1-hour) ppm	Health Effect Descriptor	PSI Colors
500	Significant harm	600	2,620	50	0.6	2.0	—	—
400	Emergency	500	2,100	40	0.5	1.6	Hazardous	Red
300	Warning	420	1,600	30	0.4	1.2	Very unhealthful	Orange
200	Alert	350	800	15	0.2	0.6	Unhealthful	Yellow
100	NAAQS[a]	150	365	9	0.12	[b]	Moderate	Green
50	50% of NAAQS	50	80	4.5	0.06	[b]	Good	Blue
0		0	0	0	0	[b]	—	—

a. Annual primary National Ambient Air Quality Standards.

b. No index values reported at concentration levels below those specified by "Alert Level" criteria.

Air Pollution in the United States: Number of PSI Days Greater Than 100, 1984 through 1993

Metropolitan Statistical Area (MSA)	No. of Trend Sites	1984	1985	1986	1987	1988	1989	1990	1991	1992	1993
Albany/Schenectady/Troy, NY	3	1	2	0	0	1	0	0	0	0	0
Atlanta	7	8	9	17	19	15	3	16	5	4	14
Baltimore	13	46	21	24	28	41	7	12	20	4	12
Birmingham	13	5	3	7	10	16	2	5	0	2	5
Boston	19	7	3	2	5	12	2	1	3	1	3
Charlotte/Gastonia/Rock Hill, NC/SC	8	19	6	10	10	17	3	4	2	0	4
Chicago	30	12	8	9	15	21	3	3	8	6	1
Cincinnati	18	4	4	6	10	23	2	6	7	0	1
Cleveland/Lorain/Elyria	20	4	1	2	7	20	4	1	5	0	1
Dallas	9	11	15	5	8	3	3	5	0	2	4
Denver	14	62	38	47	36	19	11	7	7	7	3
Detroit	24	7	2	5	9	17	10	3	7	0	2
Hartford, CT	9	31	17	7	18	26	8	7	14	9	9
Honolulu	6	1	0	0	0	0	0	0	0	0	0
Houston	22	49	48	45	55	48	34	48	40	30	26
Indianapolis	15	5	2	2	1	7	2	1	0	1	0
Kansas City	14	12	3	4	3	3	2	2	1	1	2

Air Pollution in the United States: Number of PSI Days Greater Than 100, 1984 through 1993 (continued)

Metropolitan Statistical Area (MSA)	No. of Trend Sites	1984	1985	1986	1987	1988	1989	1990	1991	1992	1993
Los Angeles	17	204	194	211	184	226	212	163	157	169	131
Louisville	13	14	4	9	3	20	3	4	5	0	5
Memphis	10	18	11	12	10	8	4	6	1	1	4
Miami, FL	7	2	5	4	4	4	4	1	2	0	0
Milwaukee/Waukesha, WI	13	11	5	9	12	19	8	2	10	0	0
Minneapolis/St. Paul	11	21	21	13	7	1	5	1	0	1	0
New Orleans	9	8	1	3	4	2	1	0	0	1	2
New York	12	96	61	53	41	43	16	17	22	4	6
Newark	11	20	21	19	22	32	4	8	10	5	2
Oklahoma City	8	14	6	4	6	0	2	1	0	0	0
Philadelphia	29	31	25	21	36	34	19	11	24	3	20
Phoenix/Mesa	16	107	84	85	40	22	30	8	4	8	6
Pittsburgh	23	15	9	8	13	23	9	11	3	1	5
St. Louis	34	22	10	13	14	17	12	8	6	2	5
Salt Lake City/Ogden	15	20	20	36	7	10	15	6	18	9	3
San Antonio	7	2	0	1	1	1	0	0	0	0	0
San Diego	12	51	54	45	41	49	61	39	25	19	14
San Francisco	10	2	5	4	1	1	0	1	0	0	0
San Jose	8	25	27	8	9	7	13	5	9	1	1
Seattle/Bellevue/Everett	10	4	24	10	10	6	4	2	0	0	0
Tampa/St. Petersburg/Clearwater	13	3	5	5	5	0	1	3	0	1	0
Washington, DC	26	30	15	12	25	36	7	5	16	2	12

10

The Universe

Planets, Moons, and Meteors

Planet Facts

Planet	Mean Distance from Sun (mi.)	Diameter (mi.)*	Period of Revolution (earth time)	Length of Planetary Day (earth days)
Mercury	36,000,000	3,031	88 days	58.82
Venus	67,240,000	7,521	224.70 days	224.59
Earth	92,960,000	7,927	365.25 days	1.00
Mars	141,640,000	4,217	686.98 days	1.03
Jupiter	483,640,000	86,696	11.86 years	0.41
Saturn	887,000,000	72,354	29.46 years	0.43
Uranus	1,783,000,000	29,168	84.01 years	0.45
Neptune	2,795,000,000	28,232	164.78 years	0.66
Pluto	3,666,000,000	2,174	248.35 years	6.41

* Figures vary among sources.

If the Earth Were a Grain of Sand... Astronomer Robert Jastrow used this analogy to describe the relative sizes and distances of objects in space: "If the sun were an orange, the earth would be a grain of sand 30 feet away; a block away would be a cherry pit, representing Jupiter, and a block beyond that, another cherry pit, representing Saturn. Pluto would be a grain of sand 10 blocks away. Two thousand miles away is another orange...."

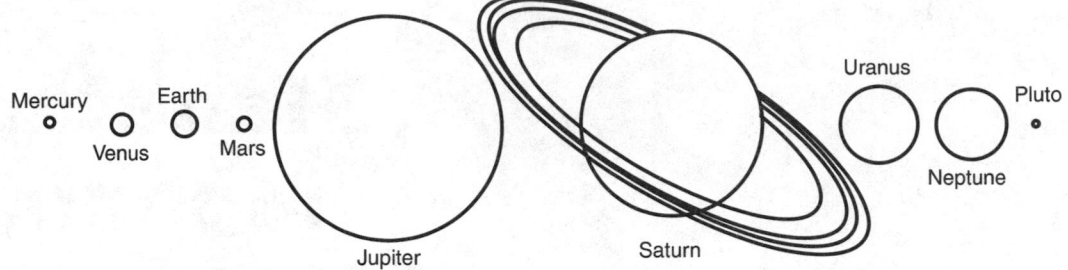

Relative Sizes of the Planets

Rule of Thumb: Observing the Stars	When observing the night sky, keep in mind that a 1-month difference in date is equal to a 2-hour difference in time of night. In other words, the stars you observed at 9:00 p.m. on July 15 will be at that same location at 7:00 p.m. on August 15.

Bode's Law: Planets by the Numbers

A German astronomer, Johann Bode, in 1772 published a book on astronomy in which he included an interesting formula put forth by a German mathematician, Johann Titius:

Start with 0 and a Series of Doubled Numbers	Add 4	Divide by 10
0	4	0.4
3	7	0.7
6	10	1.0
12	16	1.6
24	28	2.8
48	52	5.2
96	100	10.0
192	196	19.6
384	388	38.8
768	772	77.2

Bode noted that 6 of the first 7 numbers of the last column were just about the same as the distance to the sun of the 7 known planets at that time, expressed in astronomical units (AUs, where 1 AU is the mean distance from the earth to the sun):

Planet	Bode's Number	Actual Distance (AU)
Mercury	0.4	0.39
Venus	0.7	0.72
Earth	1.0	1.00
Mars	1.6	1.52
?	2.8	2.77
Jupiter	5.2	5.20
Saturn	10.0	9.54
Uranus	19.6	19.18
Neptune	38.8	30.06
Pluto	77.2	39.44

Note that Uranus, Neptune, and Pluto were discovered after Bode's book was published, in 1781, 1846, and 1930, respectively.

The two problems with this formula are the gap between Mars and Jupiter, which was later shown to be the location of the asteroid belt; and the locations of Neptune and Pluto, which do not fit the formula.

The Planets at Their Brightest: Best Dates for Observing

Mercury, Venus, Mars, Jupiter, Saturn, and Uranus can be seen as bright objects in the night sky at various times throughout the year. The list below gives the dates when those planets will be at their brightest, as viewed from the Northern Hemisphere.

Planet	Best Dates for Observing
Mercury Shortly after sunset in the western sky	April 23, 1996 April 6, 1997 March 20, 1998 March 3, 1999 February 15, June 9, 2000 January 28, May 22, 2001 January 11, May 4, 2002 April 16, 2003 March 29, 2004 March 12, 2005
Venus In the evening	April 1, 1996 November 6, 1997 June 11, 1999 January 17, 2001 August 22, 2002 March 29, 2004 November 3, 2005
In the morning	August 20, 1996 March 27, 1998 October 30, 1999 June 8, 2001 January 11, 2003 August 17, 2004
Mars	March 17, 1997 April 24, 1999 June 13, 2001 August 28, 2003 November 7, 2005
Jupiter	July 4, 1996 August 9, 1997 September 16, 1998 October 23, 1999 November 28, 2000 January 1, 2002 February 2, 2003 March 4, 2004 April 3, 2005

The Planets at Their Brightest: Best Dates for Observing (continued)

Saturn	September 26, 1996
	October 10, 1997
	October 23, 1998
	November 6, 1999
	November 19, 2000
	December 3, 2001
	December 17, 2002
	December 31, 2003
	January 13, 2005
Uranus	July 25, 1996
	July 29, 1997
	August 3, 1998
	August 7, 1999
	August 11, 2000
	August 15, 2001
	August 20, 2002
	August 24, 2003
	August 27, 2004
	September 1, 2005

Viewing the Heavens

Many books on astronomy recommend that amateur astronomers purchase a good pair of binoculars rather than a small telescope to view the night sky, largely because the optics tend to be better and the binoculars are easier to use.

Binoculars are classified by their magnification and the aperture of their lenses; the greater the aperture number, the more light is collected. A rating of 7×50 means that the magnification is ×7 and the aperture of each lens is 50 mm across. The 7×50 size is recommended for good general viewing without the need for a tripod.

Transits of Venus Across the Face of the Sun

When Venus crosses the front of the sun, its path is known as a *transit*. Such crossings of Venus occur only twice every 122 years, and when they do, they are always 8 years apart and always take place in June or December. The next transits are coming up soon, relatively speaking:

- June 8, 2004 (visible in Europe and South America)
- June 6, 2012 (visible in the eastern Pacific and western Americas)

After that, the next dates for the transits of Venus will be in 2117 and 2125.

Transits of Mercury Across the Face of the Sun

- November 15, 1999 (visible from Antarctica)
- May 7, 2003 (visible from North America)
- November 8, 2006
- May 9, 2016

The Planetary Moons

Planet	Number of Moons	Years of Discovery
Earth	1	—
Mars	2	1877
Jupiter	16	1610–1980
Saturn	18*	1655–1990
Uranus	15	1787–1986
Neptune	8	1846–1989
Pluto	1	1978

*Saturn also has several significantly smaller moons.

Quasars Quasars are the most distant and brightest objects in the universe; the brightest quasar shines 1.5 quadrillion times the brightness of our sun. Quasars—"quasi-stellar radio sources"—are mysterious objects; they are not stars but may be the cores of young galaxies.

Dates of Total Lunar Eclipses
(Eastern Standard Time, 24-Hour Clock)

Date	Time at Mid-Eclipse
April 3, 1996	1911
September 26, 1996	2155
September 16, 1997	1347*
January 20, 2000	2345
July 16, 2000	0857*
January 9, 2001	1522*
May 15, 2003	2241
November 8, 2003	2020
May 4, 2004	1532*
October 27, 2004	2205

*Not visible from North America.

Dates of Partial Lunar Eclipses
(Eastern Standard Time, 24-Hour Clock)

Date	Time at Mid-Eclipse
March 23, 1997	2341
July 28, 1999	0634
July 5, 2001	0957

Men on the Moon

The U.S. space program completed 6 manned missions to the moon:

Apollo 11	July 16–24, 1969
Apollo 12	November 14–24, 1969
Apollo 14	January 31–February 9, 1971
Apollo 15	July 26–August 7, 1971
Apollo 16	April 16–27, 1972
Apollo 17	December 7–19, 1972

Levels of Darkness: A Scale for Lunar Eclipses

The darkness of an eclipse varies depending on the amount of dust present in the earth's atmosphere, some of which is caused by volcanic activity. Astronomers use a scale that rates this level of darkness:

Level	Description
0	Very dark eclipse; moon almost invisible, especially at mid-eclipse
1	Dark eclipse, gray or brown color
2	Deep red or rust-colored eclipse, with very dark center of shadow; outer edge of umbra relatively bright
3	Brick-red eclipse, usually with bright rim, sometimes yellowish, to the shadow
4	Very bright coppery-red or orange eclipse, with bluish very bright shadow rim

The Asteroids

Scientists believe that asteroids, most of which are found between the orbits of Mars and Jupiter, are ancient materials that never formed into a planet. There are about 1 million asteroids, 13,000 of which have been identified and 5,000 of which have been named and numbered. Here are the first 10.

Number	Name	Diameter (mi.)
1	Ceres	485
2	Pallas	304
3	Juno	118
4	Vesta	243
5	Astraea	50
6	Hebe	121
7	Iris	121
8	Flora	56
9	Metis	78
10	Hygeia	40

Numbers to Know: Our Galaxy

The Milky Way galaxy, home galaxy for our solar system, is fairly large compared to other galaxies—it's about 100,000 light-years across and contains several hundred billion stars. We are located about two-thirds of the way out from the Milky Way's center. It takes our solar system 250 million years to complete a full orbit of the Milky Way. Astronomers estimate its age to be about 14 billion years.

Dates of Major Meteor Showers

Peak Date*	Shower Name	Numbers/Positions/Times**
January 3	Quadrantids	At least 60 per hour
April 15	April Fireballs	Very bright; look southeast, late night
April 22	Lyrids	Very bright; look overhead around 11 p.m.
April 25	Mu Virginids	About 7 per hour; look overhead around 1 a.m.
May 4	Eta Aquarids	About 20 per hour
June 30	June Draconids	10–100 per hour; look overhead around 9 p.m.
July 28	Delta Aquarids	About 27 per hour; best seen facing south after midnight
July 30	Capricornids	About 10–35 per hour; look south of overhead after midnight
August 12	Perseids	About 25–45 per hour from 10 p.m. to midnight; 50–100 per hour after midnight
October 9	Draconids	Number varies; 200–500 per hour average year; look overhead anytime after dusk
October 20	Orionids	About 30 per hour; look south of overhead at 5 a.m.
November 5	Taurids	About 10 per hour, increasing after midnight
November 17	Leonids	Will peak in 1999; thousands per hour during peak, 20 per hour other years; midnight to predawn
December 14	Geminids	At least 60 per hour; look overhead at midnight
December 20	Delta Arietids	About 12 per hour; look overhead at 10 p.m.

*Showers usually also occur for several days before and after the date given.

**Local time.

The Sun and Stars

Sun Facts

Item	Measurement
Mean distance from earth	92,960,000 miles
Estimated velocity of rotation	175 miles per second
Period of rotation	27 days (average)
Diameter at equator	864,898 miles
Mass	2 octillion tons
Gravity relative to earth	27.8 times
Temperature at core	27 million °F
Temperature at surface	8,700°F
Supply of hydrogen fuel remaining	At least 5 billion years

The Sunspot Cycle

Sunspots are regions of the sun with very strong magnetic fields. On average, they appear in cycles of 11 years. The mid-1990s are in the low end of a cycle, with the next period of maximum sunspot activity occurring about 2001. For reasons unknown, very few sunspots were seen between 1640 and 1710, which happened to be a period of extremely cold weather on earth.

Dates of Solar Eclipses

Date	Type of Eclipse	Visible from—
April 17, 1996	Partial	New Zealand, South Pacific
October 12, 1996	Partial	Canada (northeastern), Greenland, Europe, North Africa
March 8, 1997	Total	Asia (eastern), Japan, North America (northwestern)
September 1, 1997	Partial	Australia (southern), New Zealand, South Pacific
February 26, 1998	Total	North America (eastern and southern), Central America
August 21, 1998	Annular*	Asia (southern and southeastern), Indonesia, Australasia
February 16, 1999	Annular	Indian Ocean, Antarctica, Indonesia
August 11, 1999	Total	North America (northeastern), the Arctic, Europe, Arabian Peninsula
February 5, 2000	Partial	Antarctica
July 1, 2000	Partial	South Pacific, South America
July 31, 2000	Partial	Asia, Alaska, Canada, the Arctic, Greenland
December 25, 2000	Partial	North America, Gulf of Mexico, Caribbean (southern), Greenland
June 21, 2001	Total	South Atlantic, Africa
December 14, 2001	Annular	North America, Hawaii, Canada (southwestern)
June 10, 2002	Annular	Asia (southeastern and northeastern), North Pacific, Philippines
December 4, 2002	Total	Africa (southern), Australia
May 31, 2003	Annular	Europe (northern), Asia (northern), Canada (northern)
November 23, 2003	Total	Antarctica
April 8, 2005	Annular	Pacific Ocean, Central America, South America (northern)
October 3, 2005	Annular	Atlantic Ocean, Spain, Africa, Indian Ocean

*An annular eclipse is one in which the moon covers all but a bright ring around the circumference of the sun.

Length of an Eclipse The maximum duration of a solar eclipse is 7 minutes and 31 seconds.

Comets

Comets are made of frozen gases and orbit the sun in elliptical and often irregular, long-period orbits. Their tails can be millions of miles long. Here are a few of the major comets.

Name	Orbital Period (earth years)
Encke	3.30
Grigg-Skjellerup	4.91
Tempel-2	5.26
Tempel-1	5.98
Pons-Winnecke	6.30
Giacobini-Zinner	6.41
Whipple	7.46
Schwassmann-Wachmann-1	16.10
Tempel-Tuttle	32.91
Halley	76.04

Photographing Celestial Objects

Object	Film Speed (ISO)	Aperture	Exposure Time
Comets	64	f2–f4	up to 30 min.
Meteors	62–200	f5.6	10–30 min.
Full moon	64	f8	$1/250$–$1/500$ sec.
Lunar eclipse			
Half shadow	200	f4	1 sec.
Near totality	200	f2.8	2 sec.
Stars*	200	f5.6	1 min.–1 hr.

Note: Use a tripod, and refer to this table as a general guide only; bracket exposures for best results.

*Use a driven mount.

Apparent Magnitude Explained

The apparent magnitude is an arbitrary number assigned to describe the brightness of an object in the universe. A first magnitude star is 100 times as bright as a sixth magnitude star; each magnitude is 2.152 (the 5th root of 100) times brighter than the previous one. Barely visible stars might be magnitude 6; very bright objects have negative magnitudes, such as the sun (–26.72) and the full moon (–12). The Hubble Space Telescope can detect the spectrums of 26th magnitude stars.

The 10 Closest Stars

Star	Apparent Magnitude	Distance (light-years)*
Sun	–26.72	0.0000015 (8 light-minutes)
Proxima Centauri	11.05	4.2
Alpha Centauri	–0.01	4.3
Alpha Centauri B	1.33	4.3
Barnard's Star	9.54	5.9
Wolf 359	13.53	7.7
Lalande 21185	7.50	8.2
Luyten 726-8a	12.52	8.4
Luyten 726-8b	13.02	8.4
Sirius A	–1.46	8.8

*Estimates of distance vary among sources.

Star Facts

Small stars—and this category includes our sun and stars up to 8 times larger—will end their lives as shrivelled white dwarfs. A white dwarf is extremely dense; a teaspoonful would weigh 5.5 tons.

Large stars—up to 20 times the size of our sun—end up exploding as supernovas and becoming, perhaps, black holes.

Stars in between those sizes will become supernovas, exploding and then collapsing to form neutron stars. Such a star is only 10 miles in diameter and is so dense as to be unimaginable—a teaspoonful would weigh more than 100 million tons.

The 10 Brightest Stars

Star	Apparent Magnitude	Distance (light-years)
Sirius A	−1.46	8.8
Canopus	−0.72	74*
Arcturus	−0.04	36
Alpha Centauri	−0.01	4.3
Vega	0.03	26
Capella	0.08	42
Rigel	0.12	900*
Procyon	0.38	11
Achernar	0.46	85
Betelgeuse	0.50*	310

*Estimates range widely; these numbers are conservative estimates.

A Directory of the Stars

The Hubble Space Telescope's *Guide Star Catalogue* lists 15,169,873 stars and 3.6 million galaxies.

Astronomical Facts and Figures

Speed of light	186,282.3976 miles per second
Light-year (distance light travels in 1 year)	5.88 trillion miles
Parsec	3.259 light-years
Mean distance, earth to sun	92,960,000 miles
Astronomical unit (AU)	1.0167 of mean distance, earth to sun
Radius of sun	432,449 miles
Mean distance, earth to moon	238,857 miles
Radius of earth (equator)	3,963.34 statute miles
Radius of earth (polar)	3,949.99 statute miles
Mean radius of earth	3,958.89 statute miles
Earth's mean velocity in orbit	18.5 miles per second
Distance to center of Milky Way	32,000 light-years
Distance to Andromeda galaxy	2.2 million light-years
Distance to nearest known quasar	15 billion light-years

The Speed of Light

Light travels amazingly fast—186,282 miles per second in a vacuum, such as that found in outer space. At that speed, it takes 8 minutes for the light from the sun to reach the earth. But the speed of light varies as it travels through different substances:

Substance	Speed of Light (mi./sec.)
Air	186,282
Ice	142,000
Glass	109,000
Diamond	77,000

11
Conversion Tables and Everyday Math

Units of Measurement

U.S. Customary System

Length or Distance

1 foot (ft.)	=	12 inches (in.)
1 yard (yd.)	=	3 feet
	=	36 inches
1 rod (rd.)	=	5$\frac{1}{2}$ yards
	=	16$\frac{1}{2}$ feet
1 furlong (fur.)	=	40 rods
	=	220 yards
	=	660 feet
1 statute mile (mi.)	=	8 furlongs
	=	1,760 yards
	=	5,280 feet
1 nautical mile	=	1.15 statute miles
	=	2,025 yards
	=	6,076 feet

Cubic Measure

1 cubic foot (cu. ft.)	=	1,728 cubic inches (cu. in.)
1 cubic yard (cu. yd.)	=	27 cubic feet

Area

1 square foot (sq. ft.)	=	144 square inches (sq. in.)
1 square yard (sq. yd.)	=	9 square feet
1 square rod (sq. rd.)	=	$30\frac{1}{4}$ square yards
	=	$272\frac{1}{4}$ square feet
1 acre	=	160 square rods
	=	4,840 square yards
	=	43,560 square feet
1 square mile (sq. mi.)	=	640 acres
1 section	=	1 square mile
1 township	=	36 sections
	=	36 square miles

Avoirdupois Weight (used for everyday objects)

1 dram (dr.)	=	$27\frac{11}{32}$ grains (gr.)
1 ounce	=	16 drams
	=	$437\frac{1}{2}$ grains
1 pound (lb.)	=	16 ounces
	=	256 drams
	=	7,000 grains
1 short hundredweight (cwt.)	=	100 pounds
1 short ton*	=	20 short hundredweights
	=	2,000 pounds
1 gross or long hundredweight	=	112 pounds
1 gross or long ton**	=	20 gross or long hundredweights
	=	2,240 pounds

* Used in the United States.
** Used in Great Britain.

Rule of Thumb: Pints and Pounds

The old saying, "A pint's a pound the world around," is true for most liquids—a pint of water weighs just a little over a pound. A pint of fluids such as gasoline, alcohol, and turpentine will weigh slightly less, and those such as glycerin will weigh more.

Troy Weight
(used for precious metals and gems)

1 pennyweight (dwt.)	=	24 grains
1 ounce troy (oz. t.)	=	20 pennyweights
	=	480 grains
1 pound troy (lb. t.)	=	12 ounces troy
	=	240 pennyweights
	=	5,760 grains

Apothecaries' Weight (used for drugs)

1 scruple (s. ap.)	=	20 grains
1 apothecaries' dram (dr. ap.)	=	3 scruples
	=	60 grains
1 apothecaries' ounce (oz. ap.)	=	8 apothecaries' drams
	=	24 scruples
	=	480 grains
1 apothecaries' pound (lb. ap.)	=	12 apothecaries' ounces
	=	96 apothecaries' drams
	=	288 scruples
	=	5,760 grains

Fluid Volume

1 tablespoon (tbsp.)	=	3 teaspoons (tsp.)
	=	0.5 fluid ounce (fl. oz.)
1 cup	=	8 fluid ounces
	=	$\frac{1}{2}$ pint
1 pint (pt.)	=	2 cups
	=	16 fluid ounces
1 quart (qt.)	=	2 pints
	=	4 cups
	=	32 fluid ounces
1 gallon (gal.)	=	4 quarts
	=	8 pints
	=	16 cups
	=	231 cubic inches

Dry Volume

1 quart	=	2 pints
1 peck (pk)	=	8 quarts
	=	16 pints
1 bushel (bu)	=	4 pecks
	=	8 gallons
	=	32 quarts
1 barrel (bbl)	=	105 quarts

Metric System

The basic units in the metric system, or Système International (SI), are as follows:

Length	meter, which is the distance light travels in a vacuum in 1/299,792,458 of a second.
Time	second, which is the duration of 9,192,631,770 cycles of the radiation associated with a specified transition of the cesium atom.
Mass	kilogram, as set by a cylinder of platinum/iridium alloy, the original of which is kept by the International Bureau of Weights and Measures in Paris.
Temperature	Kelvin, which is the fraction 1/273.16 of the thermodynamic temperature at which water forms an interface of solid, liquid, and vapor: 0.01°C and 32.02°F. Absolute zero is 0 K.
Electric current	ampere, which is equal to the flow of 1 coulomb per second.
Luminous intensity	candela, which is the luminous intensity of 1/600,000 of a square meter of a cavity at the temperature of freezing platinum.
Amount of substance	mole, which is the amount of substance of a system that contains as many elementary entities as there are atoms in 0.012 kilogram of carbon 12.

Will the United States Ever Move to Metric?

The metric system—considered superior to all other systems because it contains only a few units, all divisible by 10—has been around since the French proposed its adoption to the world in 1799. The United States is one of the last countries to make the switch, which it officially began to do with the passage of the 1975 Metric Conversion Act. The act specifies voluntary conversion, making it easy to put off making the change.

Metric Prefixes: The Smallest and Largest Numbers

Prefix	Symbol	Meaning (using American names of numbers)
yotta-	Y	10^{24} = 1,000,000,000,000,000,000,000,000 = 1 septillion
zetta-	Z	10^{21} = 1,000,000,000,000,000,000,000 = 1 sextillion
exa-	E	10^{18} = 1,000,000,000,000,000,000 = 1 quintillion
peta-	P	10^{15} = 1,000,000,000,000,000 = 1 quadrillion
tera-	T	10^{12} = 1,000,000,000,000 = 1 trillion
giga-	G	10^{9} = 1,000,000,000 = 1 billion
mega-	M	10^{6} = 1,000,000 = 1 million
kilo-	k	10^{3} = 1,000 = 1 thousand
hecto-	h	10^{2} = 100 = 1 hundred
deka-	da	10^{1} = 10 = ten
deci-	d	10^{-1} = 0.1 = a tenth of a
centi-	c	10^{-2} = 0.01 = a hundredth of a
milli-	m	10^{-3} = 0.001 = a thousandth of a
micro-	μ	10^{-6} = 0.000 001 = a millionth of a
nano-	n	10^{-9} = 0.000 000 001 = a billionth of a
pico-	p	10^{-12} = 0.000 000 000 001 = a trillionth of a
femto-	f	10^{-15} = 0.000 000 000 000 001 = a quadrillionth of a
atto-	a	10^{-18} = 0.000 000 000 000 000 001 = a quintillionth of a
zepto-	z	10^{-21} = 0.000 000 000 000 000 000 001 = a sextillionth of a
yocto-	y	10^{-24} = 0.000 000 000 000 000 000 000 001 = a septillionth of a

Googol and Googolplex
A googol is 1 followed by 100 zeroes, or 10 to the 100th power. Lore has it that the number was named by the then 9-year-old nephew of the American mathematician Edward Kasner (1898–1955). A googolplex is 1 followed by a googol of zeroes.

Length or Distance

1 angstrom (Å)	=	0.1 nanometer
1 nanometer (nm)	=	0.001 micrometer
1 micrometer (µm)	=	0.001 millimeter
		0.0001 centimeter
1 millimeter (mm)	=	0.1 centimeter
	=	0.01 decimeter
	=	0.001 meter
1 dekameter (dam)	=	10 meters
1 hectometer (hm)	=	10 dekameters
	=	100 meters
1 kilometer (km)	=	10 hectometers
	=	100 dekameters
	=	1,000 meters

Area

1 square millimeter (mm²)	=	1,000,000 square micrometers (µm²)
1 square centimeter (cm²)	=	100 square millimeters
1 square meter (m²)	=	10,000 square centimeters
1 are (a)	=	100 square meters
1 hectare (ha)	=	100 ares
	=	10,000 square meters
1 square kilometer (km²)	=	100 hectares
	=	1,000,000 square meters

Cubic Measure

"Cubic centimeter" is sometimes abbreviated "cc" and is used in fluid measure interchangeably with milliliter (mL).

1 cubic centimeter (cm^3)	=	1,000 cubic millimeters (mm^3)
1 cubic decimeter (dm^3)	=	1,000 cubic centimeters
1 cubic meter (m^3)	=	1,000 cubic decimeters
	=	1,000,000 cubic centimeters

Fluid Volume

1 centiliter (cL)	=	10 milliliters (mL)
1 deciliter (dL)	=	10 centiliters
	=	100 milliliters
1 liter (L)	=	10 deciliters
	=	100 centiliters
	=	1,000 milliliters
1 dekaliter (daL)	=	10 liters
1 hectoliter (hL)	=	10 dekaliters
	=	100 liters
1 kiloliter (kL)	=	10 hectoliters
	=	1,000 liters

Mass

1 centigram (cg)	=	10 milligrams (mg)
1 decigram (dg)	=	10 centigrams
	=	100 milligrams
1 gram (g)	=	10 decigrams
	=	100 centigrams
	=	1,000 milligrams
1 kilogram (kg)	=	1,000 grams
1 metric ton (t)	=	1,000 kilograms *or* 1 megagram (Mg)

The Value of a Penny If you are without a scale and need to get a rough idea of the number of grams an object weighs, keep in mind that a U.S. penny weighs 2.5 grams and a nickel, 5 grams.

Conversion Charts: U.S. Customary and Metric System Equivalents

Metric to U.S. Quick Reference

If You Know—	Multiply by—	To Get—
millimeters	0.04	inches
centimeters	0.39	inches
meters	3.28	feet
meters	1.09	yards
kilometers	0.62	miles
square centimeters	0.16	square inches
square meters	10.76	square feet
square meters	1.20	square yards
square kilometers	0.39	square miles
hectares	2.47	acres
cubic meters	35.32	cubic feet
cubic meters	1.31	cubic yards
milliliters	0.20	teaspoons
milliliters	0.06	tablespoons
milliliters	0.034	fluid ounces
liters	4.23	cups
liters	2.12	pints
liters	1.06	quarts
liters	0.26	gallons
grams	0.035	ounces
kilograms	2.20	pounds
metric tons (megagrams)	1.10	short tons

U.S. to Metric Quick Reference

If You Know—	Multiply by—	To Get—
inches	25.40	millimeters
inches	2.54	centimeters
feet	30.48	centimeters
yards	0.91	meters
miles	1.61	kilometers
square inches	6.45	square centimeters
square feet	0.09	square meters
square yards	0.84	square meters
square miles	2.60	square kilometers
acres	0.40	hectares
cubic feet	0.028	cubic meters
cubic yards	0.76	cubic meters
teaspoons	4.93	milliliters
tablespoons	14.79	milliliters
fluid ounces	29.57	milliliters
cups	0.24	liters
pints	0.47	liters
quarts	0.95	liters
gallons	3.79	liters
ounces	28.35	grams
pounds	0.45	kilograms
short tons (2,000 lb.)	0.91	metric tons (megagrams)

Energy/Power

If You Know—	Multiply by—	To Get—
ft.-lbs. per second	0.0018	horsepower
horsepower	550	ft.-lbs. per second
ft.-lbs. per minute	0.0226	watts
watts	44.25	ft.-lbs. per minute
ft.-lbs. per second	1.356	watts
watts	0.7373	ft.-lbs. per second
watts	0.001341	horsepower
horsepower	746	watts
Btu per hour	0.000393	horsepower
horsepower	2,545	Btu per hour

Energy

If You Know—	Multiply by—	To Get—
ergs	0.0000001	joules
joules	10,000,000	ergs
joules	0.2388	gram-calories
joules	0.10198	kg.-m.
joules	0.7375	ft.-lbs.
joules	0.000947	Btu
gram-calories	4.186	joules
gram-calories	0.003968	Btu
kg.-m.	9.8117	joules
kg.-m.	7.233	ft.-lbs.
ft.-lbs.	1.356	joules
ft.-lbs.	0.001285	Btu
ft.-lbs.	0.1383	kg.-m.
Btu	252	gram-calories
Btu	1,055	joules
Btu	778	ft.-lbs.
Btu	0.293	watt-hrs.
watt-hrs.	3.413	Btu

Length or Distance

1 inch	=	2.54 centimeters
	=	0.0254 meter
1 foot	=	30.48 centimeters
	=	0.3048 meter
1 yard	=	0.9144 meter
1 rod	=	5.029 meters
1 mile	=	1,609.344 meters
	=	1.609344 kilometers
1 decimeter	=	3.937 inches
1 centimeter	=	0.3937 inch
1 meter	=	39.37 inches
	=	3.28083 feet
	=	1.093613 yards
1 kilometer	=	3,280.8 feet
	=	0.62137 mile

Area

1 square inch	=	6.4516 square centimeters
1 square foot	=	929.0304 square centimeters
	=	0.09290304 square meter
1 square yard	=	0.83612736 square meter
1 acre	=	4,046.8564 square meters
	=	0.40468564 hectare
1 square mile	=	2,589,988.11 square meters
	=	258.998811 hectares
	=	2.58998811 square kilometers
1 square centimeter	=	0.1550003 square inch
1 square meter	=	1,550.001 square inches
	=	10.76391 square feet
	=	1.195990 square yards
1 hectare	=	107,639.1 square feet
	=	11,959.90 square yards
	=	2.4710538 acres
	=	0.003861006 square mile
1 square kilometer	=	247.10538 acres
	=	0.3861006 square mile

Cubic Measure

1 cubic inch	=	16.387064 cubic centimeters
1 cubic foot	=	28,316.846592 cubic centimeters
	=	0.028316847 cubic meter
1 cubic yard	=	764,554.857984 cubic centimeters
	=	0.764554858 cubic meter
1 cubic centimeter	=	0.06102374 cubic inch
1 cubic meter	=	61,023.74 cubic inches
	=	35.31467 cubic feet
	=	1.307951 cubic yards

Fluid Volume

1 fluid ounce	=	29.573528 milliliters
	=	0.029573528 liter
1 cup	=	236.588 milliliters
	=	0.236588 liter
1 pint	=	473.176 milliliters
	=	0.473176 liter
1 quart	=	946.3529 milliliters
	=	0.9463529 liter
1 gallon	=	3,785.41 milliliters
	=	3.78541 liters
1 milliliter	=	0.033814 fluid ounce
1 liter	=	33.814 fluid ounces
	=	4.2268 cups
	=	2.113 pints
	=	1.0567 quarts
	=	0.264 gallon
1 hectoliter	=	26.418 gallons

Dry Volume

1 pint	=	33.600 cubic inches
	=	0.551 liter
1 quart	=	67.201 cubic inches
	=	1.101 liters

Mass/Weight

1 ounce	=	28.3495 grams
1 pound	=	453.59 grams
	=	0.45359 kilogram
1 short ton	=	907.18 kilograms
	=	0.907 metric ton (megagram)
1 milligram	=	0.00003527 ounce
1 gram	=	0.03527 ounce
1 kilogram	=	35.27 ounces
	=	2.2046 pounds
1 metric ton	=	2,204.6 pounds
	=	1.1023 short tons

Miscellaneous Measurements

Force/Energy/Power

poundal	=	fundamental unit of force
foot-pound	=	work done when force of 1 poundal produces a movement of 1 foot
	=	13,560,000 ergs
horsepower	=	550 foot-pounds per second
British thermal unit (Btu)	=	0.778 foot-pound (the energy required to increase the temperature of 1 pound of water by 1°F)
quad	=	10^{15} Btu
erg	=	2.906×10^{-8} calorie
	=	9.48451×10^{-11} Btu
	=	10^{-7} joule
watt	=	10^7 ergs/second
	=	0.001341 horsepower (U.S.)
	=	0.05688 Btu/minute
joule	=	1 watt-second
	=	10^7 dyne-centimeters
calorie	=	4.1868 joules

Electrical Measurements

Unit	Multiple	Value
volt	kilovolt (kV)	1,000 volts
volt	millivolt (mV)	1/1,000 volt
volt	microvolt (μV)	1/1,000,000 volt
ohm	kilohm (kΩ)	1,000 ohms
ohm	megohm (MΩ)	1,000,000 ohms
ampere	milliampere (mA)	1/1,000 ampere
ampere	microampere (μA)	1/1,000,000 ampere

The original unit of quantity was the coulomb, which was equal to the passage of 6.25×10^{18} electrons past a point in the electrical system.

The ampere is the unit of electrical flow, equal to the flow of 1 coulomb per second. The ampere is the basic unit for electricity in the Système International.

The volt, which is 1 joule per coulomb, measures electrical potential energy.

The watt, equal to 1 joule per second, is the basic unit for measuring electrical power. One kilowatt equals 1,000 watts.

The ohm is the unit for measuring electrical resistance. It is the resistance a circuit offers to the flow of 1 ampere driven by the potential of 1 volt.

Temperature: Fahrenheit, Celsius, Kelvin

In the English system, the difference between the freezing and boiling points of water is divided into 180 units, or degrees Fahrenheit. The metric system divides the difference between freezing and boiling into 100 units, or degrees Celsius (the preferred term over centigrade). The Kelvin temperature scale uses absolute zero (meaning no heat or motion is present) as its starting point. It also uses 100 units to describe the range between the freezing and boiling points of water. The word "degrees" is not used with Kelvin temperatures. Kelvin is a universal measure useful for describing extreme temperatures.

System	Water Boils	Water Freezes	Absolute Zero
Fahrenheit	212°	32°	−459.67°
Celsius	100°	0°	−273.15°
Kelvin	373.15	273.15	0

Fahrenheit/Celsius/Kelvin Quick Reference

If You Know—	To Get—	Do This—
Fahrenheit	Celsius	Subtract 32 from the °F temperature and multiply the difference by 5; divide the product by 9
	Kelvin	Add 459.67 to °F and divide by 1.8
Celsius	Fahrenheit	Multiply the °C temperature by 1.8 and add 32
	Kelvin	Add 273.15 to °C
Kelvin	Fahrenheit	Multiply K times 1.8 and subtract 459.67
	Celsius	Subtract 273.15 from K

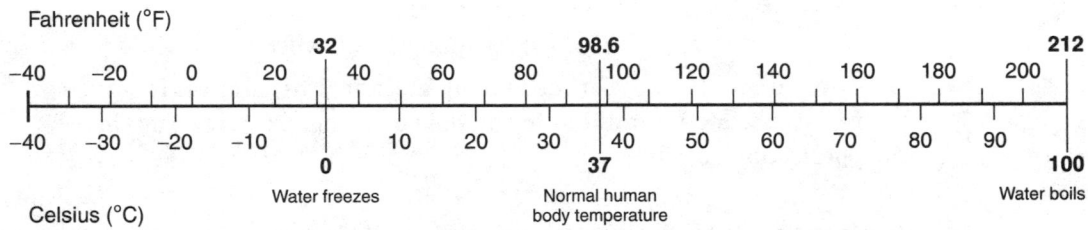

Wood Measures

1 board foot	=	144 cubic inches (1 ft. x 1 ft. x 1 in.)
	=	0.00236 cubic meters
1 cord foot	=	16 cubic feet (4 ft. x 4 ft. x 1 ft.)
	=	0.4531 cubic meters
1 cord	=	8 cord feet (4 ft. x 4 ft. x 8 ft.)
	=	128 cubic feet
	=	3.625 cubic meters

Circular and Angular Measurement

1 second (")	=	1/1,296,000 circle
1 minute (')	=	60 seconds
	=	1/21,600 circle
1 degree (°)	=	60 minutes
	=	1/360 circle
1 quadrant	=	1 right angle
	=	90 degrees
	=	1/4 circle
1 circumference	=	4 quadrants
	=	1 circle
0.017454 radian (rad)	=	1°
1 radian	=	57.2958 degrees
2π radians	=	360 degrees

Surveyors' Chain Measure

1 link	=	7.92 inches
1 chain (ch.)	=	100 links
	=	4 rods
	=	66 feet
1 statute (or survey) mile	=	80 chains
	=	320 rods
	=	5,280 feet

Mariners' Measures

1 fathom	=	6 feet
	=	1.8288 meters
1 cable	=	100 fathoms (approximately)
1 nautical mile	=	10 cables
	=	1,852 meters
	=	1.852 kilometers
	=	1.15 statute miles
	=	2,025 yards
	=	6,076 feet
1 knot	=	1 nautical mile per hour
1 statute mile	=	0.868976 nautical mile
1 kilometer	=	0.539957 nautical mile

How Mark Twain Got His Name

"Mark twain" is a term that was used by riverboat crew members to denote a water depth of 2 fathoms, or 12 feet. "Twain" is an archaic word for "two," from the Middle English word "twayn."

The term, in riverboat parlance, meant a water depth that was barely safe for navigation. The humorist Samuel L. Clemens, a river boat pilot as a young man, is said to have chosen "Mark Twain" as his pseudonym because it suggested a condition that was less than comfortable.

British Measures

1 fluid ounce	=	8 fluid drams
	=	0.961 U.S. fluid dram
	=	28.412 milliliters
1 pint	=	1/2 quart
	=	1.201 U.S. pints
	=	0.5683 liter
1 gill	=	1/2 cup
	=	5 fluid ounces
	=	0.1421 liter
4 gills	=	1 pint
1 quart	=	2 pints
	=	1/4 gallon
	=	1.201 U.S. liquid quarts
	=	1.032 U.S. dry quarts
	=	1.137 liters
1 Imperial gallon	=	160 fluid ounces
	=	8 pints
	=	4 quarts
	=	1.201 U.S. gallons
	=	4.546 liters
1 peck	=	1/4 bushel
	=	8 quarts
	=	1.0314 U.S. peck
	=	9.087 liters
1 bushel	=	4 pecks
	=	1.022 U.S. bushels
	=	36.369 liters

Counting Measures

1 dozen (doz.)	=	12 units
1 gross (gr.)	=	12 dozen
	=	144 units
1 great gross	=	12 gross
	=	144 dozen
	=	1,728 units

Paper Measures

1 quire (qr.)	=	25 sheets
1 ream (rm.)	=	20 quires
	=	500 sheets
1 perfect ream	=	516 sheets
1 bundle (bdl.)	=	2 reams
1 bale	=	5 bundles

Printing Measures

1 point (pt.)	=	approximately 1/72 or 0.013837 inch
	=	0.3514598 millimeter
1 pica	=	12 points
	=	approximately 1/6 or 0.166044 inch
	=	4.2175176 millimeters
1 agate	=	1/14 inch (column length)
1 em	=	square width of any size type (1 em of 12-pt. type = 12 pts.)
1 en	=	1/2 em

Basic Numbers

Fraction/Percentage/Decimal Conversion

Fraction	Percentage	Decimal
1/2	50.00	0.5
1/3	33.33	0.3333
1/4	25.00	0.25
1/5	20.00	0.20
1/6	16.67	0.1667
1/7	14.29	0.1429
1/8	12.50	0.125
1/9	11.11	0.1111
1/10	10.00	0.1
1/11	9.09	0.0909
1/12	8.33	0.0833
1/13	7.69	0.0769
1/14	7.14	0.0714
1/15	6.67	0.0667
1/16	6.25	0.0625
1/32	3.13	0.0313
1/64	1.56	0.0156

Fraction/Percentage/Decimal Conversion (continued)

Fraction	Percentage	Decimal
2/3	66.67	0.6667
2/5	40.00	0.4000
2/7	28.57	0.2857
2/9	22.22	0.2222
2/11	18.18	0.1818
3/4	75.00	0.7500
3/5	60.00	0.6000
3/7	42.86	0.4286
3/8	37.50	0.3750
3/10	30.00	0.3000
3/11	27.27	0.2727
4/5	80.00	0.8000
4/7	57.14	0.5714
4/9	44.44	0.4444
4/11	36.36	0.3636
5/6	83.33	0.8333
5/7	71.43	0.7143
5/8	62.50	0.6250
5/9	55.56	0.5556
5/10	50.00	0.5000
5/11	45.45	0.4545
5/12	41.67	0.4167
6/7	85.71	0.8571
6/10	60.00	0.6000
6/11	54.55	0.5455
6/12	50.00	0.5000
7/8	87.50	0.8750
7/9	77.78	0.7778
7/10	70.00	0.7000
7/11	63.64	0.6364
7/12	58.33	0.5833
8/9	88.89	0.8889
8/10	80.00	0.8000
8/11	72.73	0.7273
8/12	66.67	0.6667
9/10	90.00	0.9000
9/11	81.82	0.8182
9/12	75.00	0.7500
10/11	90.91	0.9091
11/12	91.67	0.9167

The Importance of "and" with Numbers

When speaking of numbers, mathematicians teach that the word "and" should be used only between a whole number and a fraction, or between a whole number and a decimal fraction. For example, "and" is improper in naming years—*nineteen ninety-nine*, not *nineteen and ninety-nine*—or in writing a check—*one hundred two dollars*, not *one hundred and two dollars* (but *one hundred two dollars and 25/100* is correct).

To understand why this distinction is important, consider the difference between these two numbers: *one hundred two thousandths (102/1,000 or 100/2,000)* and *one hundred and two thousandths (100-2/1,000)*.

Very Large Numbers: British versus American Usage

Because British usage is now beginning to change, at least in regard to the use of "billion" to mean 100,000,000 rather than 1,000 million, it's best to use numerals whenever possible to avoid misinterpretation. The highest named number is centillion, which is 1 followed by 303 zeroes in American usage and 600 zeroes in British usage.

Number of Zeroes After 1	American	British
6	million	million
9	billion	1,000 million (rarely, milliard)
12	trillion	billion
15	quadrillion	1,000 billion
18	quintillion	trillion
21	sextillion	1,000 trillion
24	septillion	quadrillion
27	octillion	1,000 quadrillion
30	nonillion	quintillion
33	decillion	1,000 quintillion
36	undecillion	sextillion
39	duodecillion	—
42	tredecillion	septillion
45	quattordecillian	—

Very Large Numbers: British versus American Usage (continued)

Number of Zeroes After 1	American	British
48	quindecillion	octillion
51	sexdecillion	—
54	septendecillion	nonillion
57	octodecillion	—
60	novemdecillion	decillion
63	vigintillion	—
66	—	undecillion
72	—	duodecillion
78	—	tredecillion
84	—	quattordecillion
90	—	quindecillion
96	—	sexdecillion

Roman Numerals

Roman	Arabic	Roman	Arabic
I	1	XIX	19
II	2	XX	20
III	3	XXI	21
IV	4	XXII	22
V	5	XXIII	23
VI	6	XXIV	24
VII	7	XXV	25
VIII	8	XXVI	26
IX	9	XXVII	27
X	10	XXVIII	28
XI	11	XXIX	29
XII	12	XXX	30
XIII	13	XL	40
XIV	14	L	50
XV	15	LX	60
XVI	16	LXX	70
XVII	17	LXXX	80
XVIII	18	XC	90

Roman Numerals (continued)

Roman	Arabic	Roman	Arabic
C	100	$\overline{L}$	50,000
D	500	$\overline{C}$	100,000
M	1,000	$\overline{D}$	500,000
$\overline{V}$	5,000	$\overline{M}$	1,000,000
$\overline{X}$	10,000		

Note: A line over the top of a roman numeral indicates that the symbol's original value is multiplied by 1,000.

The Demise of the Roman Numeral

By the Middle Ages, mathematicians in Europe began to realize that the Hindu-Arabic numbering system had many advantages over the Roman system. The idea of a "place" system, in which the value of the number is determined by its position, was superior to the Roman system, in which the numerals are always the same—an X is 10, a C is 100, and so on. Very simple addition and subtraction are not too difficult in the Roman system, but calculations of large numbers, especially in multiplication and division, are just about impossible.

Another advantage of the Arabic system is that it needs only 10 symbols to express all numbers. It also has the zero, that extremely useful number which the Hindus are credited with inventing. Tools like the abacus, which were necessary for working with Roman numerals, were no longer needed with Arabic numbers. And very large numbers—such as distances to the stars—could be expressed without an extraordinarily long strings of letters.

Prime Numbers from 2 to 1,009

A prime number is any positive integer that is not evenly divisible except by itself and 1. The highest known prime number is $2^{756,839-1}$, which has 227,832 digits. It was discovered in 1992 using a Cray-2 supercomputer.

	2	3	5	7	11	13	17	19	23
29	31	37	41	43	47	53	59	61	67
71	73	79	83	89	97	101	103	107	109
113	127	131	137	139	149	151	157	163	167
173	179	181	191	193	197	199	211	223	227
229	233	239	241	251	257	263	269	271	277
281	283	293	307	311	313	317	331	337	347
349	353	359	367	373	379	383	389	397	401
409	419	421	431	433	439	443	449	457	461
463	467	479	487	491	499	503	509	521	523
541	547	557	563	569	571	577	587	593	599
601	607	613	617	619	631	641	643	647	653
659	661	673	677	683	691	701	709	719	727
733	739	743	751	757	761	769	773	787	797
809	811	821	823	827	829	839	853	857	859
863	877	881	883	887	907	911	919	929	937
941	947	953	967	971	977	983	991	997	1009

Perfect Numbers

A number is referred to as "perfect" if it is equal to the sum of its divisors other than itself. The lowest perfect number is 6 (= 1 + 2 + 3), and the highest of the 32 perfect numbers discovered to date is $2^{756,839-1} \times 2^{756,838}$. This number contains 455,663 digits.

Numbers in American Sign Language

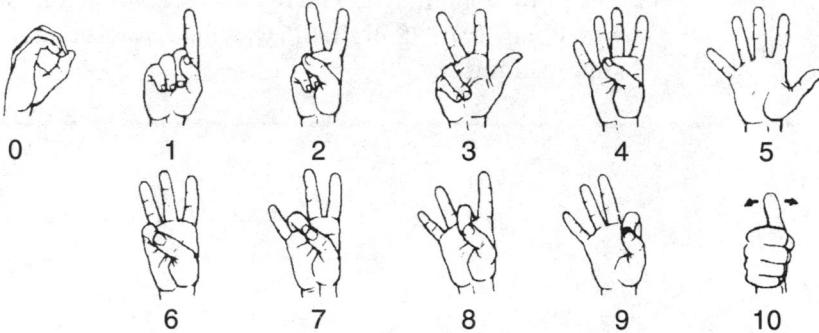

Numbers in Braille

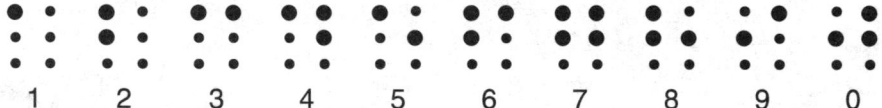

Computer Numbers

1 byte	=	1 character
	=	8 bits
1 kilobyte (1 KB)	=	1,024 bytes*
1 megabyte (1 MB)	=	1,024 kilobytes
	=	1,048,576 bytes
1 gigabyte	=	1 billion bytes

* Where approximate measures are sufficient, 1,000 is often used in place of 1,024 as the multiplier.

Physical Constants

Constant	Value
Speed of light in a vacuum	$2.997925(1) \times 10^8$ m s^{-1}
Electron charge	$1.60210(2) \times 10^{-19}$ C
Avogadro constant	$6.02252(9) \times 10^{26}$ kmol^{-1}
Atomic mass unit (m)	$1.66043(2) \times 10^{-27}$ kg
Electron rest mass	$9.10908(13) \times 10^{-31}$ kg $5.48597(3) \times 10^{-4}$ u
Proton rest mass	$1.67252(3) \times 10^{-27}$ kg $1.00727663(8)$ u
Neutron rest mass	$1.67482(3) \times 10^{-27}$ kg $1.0086654(4)$ u
Electron charge to mass ratio	$1.758796(6) \times 10^{11}$ C kg^{-1}
Planck constant	$6.62559(16) \times 10^{-34}$ J s
Rydberg constant	$1.0973731(1) \times 10^7$ m^{-1}
Gas constant	$8.31434(35)$ J K^{-1} mol^{-1}
Boltzmann constant	$1.38054(6) \times 10^{-23}$ J K^{-1}
Gravitational constant	$6.670(5) \times 10^{-11}$ N m^{2}kg^{-2}
Bohr magneton	$9.2732(2) \times 10^{-24}$ J T^{-1}
Compton wavelength of electron	$2.42621(2) \times 10^{12}$ m
Compton wavelength of proton	$1.321398(13) \times 10^{-15}$ m
Faraday constant	$9.64870(5) \times 10^4$ C mol^{-1}

Numbers in Nature: The Fibonacci Sequence

The Fibonacci sequence works like this: Start with 0 and add the next number—1—to get 1. Then add 1 to itself to get the next number in the sequence: 1 + 1 = 2. Then add this sum to the preceding number to get the next number in the sequence: 2 + 1 = 3. The sequence is thus 0, 1, 1, 2, 3, 5, 8, 13, 21, 34, 55, 89, 144, 233, 377, 610, and so on. This pattern determines a spiral shape in the natural world: It is seen in the nautilus shell, a parrot's beak, the curved horns of a ram, and the sprouting of leaves from the stem of a plant. Flower petals reflect the series; on average, daisies have 21, 34, 55, or 89 petals.

The Fibonacci sequence is reflected in a ratio that the ancient Greeks called the "golden section" or the "divine proportion." After the number 21 in the series, the ratio between any number in the series and the number just before it is 1.618. The Greeks built the Parthenon and other ancient structures according to this ratio; these buildings are 1.618 times as long as they are wide.

One final point of numerical interest: If you calculate the ratios of adjacent numbers in the Fibonacci series—that is, if you divide 1 into 2, 2 into 3, 5 into 8, and so on throughout the sequence—by the 35th number (5,702,887) the ratio stabilizes to 13 decimal places, or 1.6180339887499, for all succeeding numbers in the sequence.

Everyday Math

Rounding Off Numbers

If you need only a rough answer to an everyday math problem—say to leave a tip or to get an idea of how much you have left in your checking account—you don't need a calculator. You can do simple math in your head if you learn how to *round off* numbers. Here's how.

Decide on the level of rounding needed. Depending on how rough an estimate is acceptable for your purposes, round off to the nearest 10, 100, 1,000, or even 1,000,000. Here are a few examples:

Actual Number	Rounded Number
997	1,000
29,985	30,000
92,955,600	93,000,000

For numbers with decimals, rounding to the nearest whole number may be sufficient. For example, here's how to round dollar amounts to the nearest dollar: If the number of cents is less than 50, round the total amount down to the whole-dollar amount. If the number of cents is 50 or more, round the total amount up to the next highest dollar.

Actual Number	Rounded Number
$359.41	$359
$265.62	$266

You can even round off a series of numbers and then add them in your head—for example, when you want to limit your grocery purchases to the amount of cash you have on hand. Here's how:

Item Cost	Rounded Cost
$2.89	$3.00
1.19	1.00
4.56	5.00
.49	—
2.24	2.00
6.73	7.00
1.65	2.00
$19.75	$20.00

In this example, the rounding produced a slightly higher number; the purpose is to get a close answer, not necessarily an accurate one.

Other Tricks for Doing Math in Your Head

To add and subtract. To add a group of numbers quickly, try to make one of the numbers end in 0 to simplify the calculation. For example, to add 47 and 34, add 3 to 47 to make it 50, and subtract that same amount—3—from 34 to make 31. You can then add 50 and 31 to get 81, the correct answer.

To add a long string of numbers, try grouping one or more pairs that are easily added before you try to add all the numbers. For example, you must add 12 + 16 + 8. To simplify the process, look for two numbers that add up to a number ending in 0—in this case, 12 + 8 = 20. Then add the remaining number: 20 + 16 = 36. Add as many groups as you need to simplify the calculation.

To multiply. Again, the idea is to get to 10s, which are easier to work with. For example, if you want to multiply 13 times 21, go through these steps:

13×21 is the same as $(10 + 3) \times 21$

$$10 \times 21 = 210$$
$$3 \times 21 = 63$$
$$63 + 210 = 273$$

To determine percentages. To get a rough idea of various percentage amounts, try dividing by these numbers:

20 percent	=	divide by 5
25 percent	=	divide by 4
33 percent	=	divide by 3
50 percent	=	divide by 2

Business Math Quick Reference

To Calculate—	Do This—
Fraction to percent	Divide the number on the bottom into the number on the top; then multiply by 100
Decimal to percent	Multiply by 100
Percent to decimal	Drop the percent sign; then shift the decimal point 2 places to the left (38.6% = 0.386, 7.33% = 0.0733, 0.21% = 0.0021), inserting a zero or zeroes after the decimal point if necessary
Average	Add all numbers; then divide the sum by the total number of numbers (1 + 3 + 5 + 7 = 16 ÷ 4 = 4)
Mean	Same as an average
Median	List all the numbers in a series from largest to smallest or smallest to largest; the median is the number in the middle of the series
Mode	Find the number that appears most frequently in a series; it may not be the median or the mean
Percentage	Divide the smaller number by the larger number (except for percentage increases; see next entry) and multiply by 100
Percentage increase	Subtract prior amount from current amount; then divide the result by the prior amount and multiply by 100 (move decimal point 2 places to the right)
Current ratio	Divide current assets by current liabilities
Debt/equity ratio	Divide liabilities by tangible net worth
Return on sales	Divide net profit by sales
Return on capital	Divide net profit by tangible net worth
Collection ratio	Divide average accounts receivable by average daily sales; the answer is number of days
Simple interest	Multiply original balance by interest rate (expressed as a decimal) times number of years
Compound interest	$C = P(1 + R)^T$ (C is total cost of loan, P = principal, R = interest rate, T = time period)

Math in Daily Life

Pay Raises. To calculate the dollar amount of your raise—say 4.5 percent:

6. Change 4.5 percent to a decimal: .045.

7. Multiply .045 times your old salary; for example, $28,640:
    $28,640 × .045 = $1,288.80 (dollar amount of raise)

8. To get your new salary, add that result to your old salary:
    $28,640 + $1,288.80 = $29,928.80

To find out your new hourly pay rate, using the example above, divide your new salary by the number of work hours in a year for a full-time job—usually 2,080 (52 weeks × 40 hours/week):

$29,928.80 ÷ 2,080 = $14.39/hour (rounded to the nearest cent)

If you know your hourly rate and want to find out the equivalent annual pay, multiply the rate by 2,080 to get a rough amount (because you rounded your hourly rate):

$14.39 × 2,080 = $29,931

To determine the percentage rate of increase you received, try this. Assume your old salary was $32,000, and your new salary is $35,000.

1. Subtract the old amount from the new amount:
    $35,000 − $32,000 = $3,000

2. Divide the amount of increase by the old salary amount:
    $3,000 ÷ $32,000 = .09375, or about 9.4%

Percentage of Decrease in Investment Value. Two years ago your home was appraised at $187,000; this year's appraisal was $163,000. To determine the percentage decrease—

1. Subtract the more recent amount from the original amount:
    $187,000 − $163,000 = $24,000

2. Divide the amount of decrease by the original amount:
    $24,000 ÷ $187,000 = .01283, or about 13%

Percentage of Increase in Investment Value. You bought your home 10 years ago for $95,000. You just sold it for $157,000. To determine the percentage increase in value,

1. Subtract the original price from the current price:

    $157,000 – $95,000 = $62,000

2. Divide the amount of increase by the original price:

    $62,000 ÷ $95,000 = .6526, or 65%

Percentage of Income Spent on Rent or Mortgage Payment. Financial advisers suggest that you should spend no more than 25 percent of your take-home pay on rent or a mortgage payment. If your take-home pay is $1,864 a month, what is the most you should be paying under this guideline?

$1,864 × .25 = $466 a month

Sales Tax. Here is how to determine the cost of an item with sales tax included.

1. To get the amount of the tax itself, change the percentage of tax to a decimal and multiply the purchase price of the item by the decimal:

    6.0% = .06

    $49.95 × .06 = $2.997, rounded up to $3.00 (tax amount)

2. To get the total amount spent, add the tax amount to the purchase price:

    $49.95 + $3.00 = $52.95

3. The two steps above can be combined into one by placing a 1 in front of the decimal amount of the tax:

    $49.95 × 1.06 = $52.947, rounded up to $52.95

Percentage Discounts. Here are a few quick ways to get rough estimates.

■ For 10 percent discount, divide the original price by 10 to get the amount of the discount:

$45.00 ÷ 10 = $4.50 (discount amount)

Then subtract the discount amount from the original price to get the new price:

$45.00 – $4.50 = $40.50 (new price)

A simpler way to find the amount of a 10 percent discount is to move the decimal point 1 digit to the left: 45 becomes 4.5.

- For 20 percent discount, divide the original price by 5.

    $30 ÷ 5 = $6 (amount of discount)

    $30 − $6 = $24 (new price)

- For 25 percent discount, divide by 4.

    $32 ÷ 4 = $8

    $32 − $8 = $24

- For 33 percent (one-third) discount, divide by 3.

- For 50 percent (one-half) discount, divide by 2.

Percentage Increases. Advertisers' claims of percentage increases in sizes are sometimes misleading; here's how to translate the numbers they use:

Increase Amount	Meaning
100%	Doubled; multiply original number by 2
75%	Increased by 3/4; multiply original number by .75 and add to original number
50%	Increased by 1/2; multiply original number by .5 and add to original number
33%	Increased by 1/3; multiply original number by .33 and add to original number
25%	Increased by 1/4; multiply original number by .25 and add to original number
10%	Increased by 1/10; multiply original number by .1 and add to original number

For example, if a candy bar's size has increased by 33 percent, take its original size—say, 6 ounces—and multiply by .33 to get 1.98 ounces. Then add that number to 6 ounces to get the new size: 7.98 ounces.

Restaurant Tipping. Using the customary 15 percent for restaurant service, you can come up with a rough idea of the tip amount by performing these calculations in your head.

1. Determine 10 percent by taking the total amount of the bill and moving the decimal to the left one place: $36.40 becomes $3.64.

2. Round that number to the closest number ending in zero: $3.60.

3. Divide $3.60 by 2 to get $1.80.

4. Add $3.60 and $1.80 to get the amount of a 15 percent tip: $5.40.

If adding the cent amounts in your head is too difficult, try rounding the numbers to the closest half or whole dollar amount to get a rougher amount; in this example, $3.60 could become $3.50 and $1.80 could be rounded up to $2.00, to get a tip amount of $5.50.

15% Tip Table

Amount	Tip	Amount	Tip	Amount	Tip	Amount	Tip
$ 5.00	$ .75	$29.00	$ 4.35	$53.00	$ 7.95	$77.00	$11.55
6.00	.90	30.00	4.50	54.00	8.10	78.00	11.70
7.00	1.05	31.00	4.65	55.00	8.25	79.00	11.85
8.00	1.20	32.00	4.80	56.00	8.40	80.00	12.00
9.00	1.35	33.00	4.95	57.00	8.55	81.00	12.15
10.00	1.50	34.00	5.10	58.00	8.70	82.00	12.30
11.00	1.65	35.00	5.25	59.00	8.85	83.00	12.45
12.00	1.80	36.00	5.40	60.00	9.00	84.00	12.60
13.00	1.95	37.00	5.55	61.00	9.15	85.00	12.75
14.00	2.10	38.00	5.70	62.00	9.30	86.00	12.90
15.00	2.25	39.00	5.85	63.00	9.45	87.00	13.05
16.00	2.40	40.00	6.00	64.00	9.60	88.00	13.20
17.00	2.55	41.00	6.15	65.00	9.75	89.00	13.35
18.00	2.70	42.00	6.30	66.00	9.90	90.00	13.50
19.00	2.85	43.00	6.45	67.00	10.05	91.00	13.65
20.00	3.00	44.00	6.60	68.00	10.20	92.00	13.80
21.00	3.15	45.00	6.75	69.00	10.35	93.00	13.95
22.00	3.30	46.00	6.90	70.00	10.50	94.00	14.10
23.00	3.45	47.00	7.05	71.00	10.65	95.00	14.25
24.00	3.60	48.00	7.20	72.00	10.80	96.00	14.40
25.00	3.75	49.00	7.35	73.00	10.95	97.00	14.55
26.00	3.90	50.00	7.50	74.00	11.10	98.00	14.70
27.00	4.05	51.00	7.65	75.00	11.25	99.00	14.85
28.00	4.20	52.00	7.80	76.00	11.40		

Amount of Down Payment. Simply multiply the purchase amount by the percentage of down payment required.

$18,900 × 20% (change to decimal = .20) = $3,780

On some calculators, you do not need to convert the percentage to a decimal:

1. Enter $18,900.
2. Press the "×" key.
3. Enter 20 and then press the "%" key.
4. The answer—3,780—will appear.

Square Feet in a House or a Room. Practice varies as to what constitutes the square footage in a house, but generally the number is meant to indicate livable space, not attics, unfinished basements, garages, or closets. The most conservative way is to determine the square footage in each room and hallway of the house and add the numbers together.

To determine a room's square footage, multiply the length by the width.

12 feet × 14 feet = 168 square feet

For odd-shaped rooms—say, a living room/dining room combination—divide the area into multiple rectangles and calculate the square footage of each.

Number of Acres. If you know the number of square feet in your lot, divide it by 43,560, the number of square feet in an acre.

Lot = 28,500 square feet ÷ 43,560 = 0.654 acres, or about two-thirds of an acre.

Driving Times. If you are planning a trip of 250 miles, most of it on an interstate highway at which the posted speed limit is 65 mph, how long will it take, approximately, to get to your destination, assuming perhaps one short stop?

250 miles ÷ 62 mph (average speed) = about 4 hours

**More House
and Garden
Calculations**

See Chapter 5, Daily Life, for help in calculating the following:
- Square yardage needed for floor coverings
- Number of rolls of wallpaper needed
- Gallons of paint needed
- Number of bags of fertilizer needed
- Number of bricks needed
- Number of tiles needed
- Amount of wood flooring needed
- Number of board feet needed
- Amount of slipcover and upholstery yardage needed

Useful Math Formulas

Distance

(rate = r, time = t) $\hfill d = rt$

Length

Perimeter (sides = a, b, etc.) $\hfill p = a + b + c + ...$

Circumference (distance around a circle = C) (diameter = d, or radius $\hfill C = \pi d$ or $C = 2\pi r$
(distance to center of circle) = r) (π = 3.1459)

Area

Rectangle (length = l, width = w) $\hfill A = lw$

Square (length of side = s) $\hfill A = s^2$

Circle (radius = r) $\hfill A = \pi r^2$

Triangle (base = b, height = h) $\hfill A = 1/2\ bh$

Right triangle (lengths a and b of 2 sides that form right angle) $\hfill A = 1/2\ ab$

Parallelogram (base = b, height = h) $\hfill A = bh$

Trapezoid (two bases = B and b, height = h) $\hfill A = 1/2\ h(B + b)$

Regular pentagon (a = one side) $\hfill A = 1.720a^2$

Regular hexagon (a = one side) $\hfill A = 2.598a^2$

Regular octagon (a = one side) $\hfill A = 4.828a^2$

Volume

Cube (length of an edge = e) $\hfill V = e^3$

Prism (area of base = B, height = h) $\hfill V = Bh$

Pyramid (area of base = B, height = h) $\hfill V = 1/3\ Bh$

Sphere (radius = r) $\hfill V = 4/3\ \pi r^3$

Cylinder (radius = r, height = h) $\hfill V = \pi r^2 h$

Cone (radius = r, height = h) $\hfill V = 1/3\ \pi r^2 h$

Odds and Statistics

What the Odds Are

Games and Gambling

Event	Odds It Will Happen
Winning the Powerball lottery jackpot (single winner)	1 in 54 million
Winning instant lottery	
Matching 3 digits in predetermined order	1 in 1,000
Matching 3 digits, any order	1 in 167
String of even numbers in roulette	
2	1 in 4.2
5	1 in 37
8	1 in 319
10	1 in 1,350
15	1 in 50,000
Winning roulette	
On 1 number	1 in 37
On 2 numbers	1 in 18
On odd or even	9 in 10
On a color	9 in 10
Dice throws (with 2 dice)	
2 or 12	1 in 36
3 or 11	1 in 18
4 or 10	1 in 12
5 or 9	1 in 9
6 or 8	5 in 36
7	1 in 6

Games and Gambling (continued)

Event	Odds It Will Happen
Craps	
4 or 10	2 to 1
5 or 9	3 to 2
6 or 8	6 to 5
Against making craps on 1 roll	8 to 1
Against making 11 on 1 roll	17 to 1
Poker	
Royal flush	649,739 to 1
Straight flush	72,192 to 1
4 of a kind	4,164 to 1
Full house	693 to 1
Flush	508 to 1
Straight	254 to 1
3 of a kind	46 to 1
2 pair	20 to 1
1 pair	4 to 3
Bingo	
Against winning jackpot (covering all 24 numbers) with 49 or fewer numbers drawn	212,084 to 1
Against winning jackpot (covering all 24 numbers) with 64 or fewer numbers drawn	102 to 1
Against winning jackpot (covering all 24 numbers) with 70 or fewer numbers drawn	6 to 1
Monopoly	
Against landing on Park Place within 10 turns	3.56 to 1
Against landing on Boardwalk within 10 turns	2.87 to 1
Against going to jail	1.74 to 1
Scrabble	
Drawing an E in 1st draw of 7	6.5 to 1
Drawing an A or I in 1st draw	1.86 to 1
Drawing an O in 1st draw	1.37 to 1
Against drawing a J, K, Q, X, or Z in 1st draw	13 to 1

Calculating Probabilities

To determine the probability of something happening, divide the number of possibilities (assuming they are all equally likely) into 1. For example, to determine the chance of rolling a 6 on a single roll of a die, consider that there are 6 sides to the die and therefore 6 equal chances:

$$1 \div 6 = 0.16666667$$

which is almost 17 out of 100 rolls, or 1.7 out of 10.

Marriage and Family (in the United States, late 20th century)

Event	Odds It Will Happen
Getting married	3 in 4 (over lifetime)
Marriage ending in divorce	2 in 5
Being born on Friday the 13th	1 in 213
Child being classified as a genius	1 in 250
Hair color	
Brown	7 in 10
Blonde	1 in 7
Black	1 in 10
Red	1 in 16
Multiple births	
Any twins	1 in 90
Identical twins	1 in 273
Triplets	1 in 90,000
Siamese twins	1 in 100,000
Quadruplets	1 in 900,000
Quintuplets	1 in 85 million

Marriage and Family (in the United States, late 20th century) (continued)

Event	Odds It Will Happen
Against miscarriage in 1st 12 weeks of pregnancy	9 to 1
Against miscarriage after 12th week	75 to 1
Remaining childless	1 in 5
Having only 1 child	1 in 4
Cohabitating before marriage	9 in 20

Life and Death

Event	Odds It Will Happen
Life on earth being destroyed by a meteorite	1 in 1.2 million (over 50-year period)
Being killed by a lightning bolt	1 in 3.4 million
Dying of cancer	1 in 500
Dying as the result of a vehicle accident	1 in 6,000
Dying as the result of an airplane crash	1 in 258,000
Dying of heart disease	1 in 357
Dying of job-related causes	1 in 13,000
Being murdered	1 in 9,735
Dying in a fire	1 in 63,000
Dying as the result of a fall	1 in 20,500

Note: Over 1-year period, except as noted.

Professional Sports

Event	Odds It Will Happen
Golf	
Hole in one (one round of a PGA event)	3,708 to 1
Major League Baseball	
Home team winning	1.15 to 1
Game being decided by 1 run	5 to 2
Going into extra innings	10 to 1
The World Series lasting 7 games	5 to 3
The World Series lasting only 4 games	5 to 1
Basketball (NBA)	
Home team winning	7 to 3
Going into overtime	23 to 1
Playoff lasting 7 games	5 to 3
Playoff lasting only 4 games	11 to 1
Bowling	
Against pro bowler bowling a perfect 300 game in 1 year	4,000 to 1
Football (NFL)	
Against fumble being recovered and run in for touchdown	7 to 1
Against kickoff being returned for touchdown	45 to 1
Against game with no intercepted passes	18 to 1
Against team that is losing at the end of the 3rd-quarter winning game	5.5 to 1
Against team that scores fewer touchdowns winning the game	32 to 1
Against losing team not scoring	11 to 1
Against missing at least 1 extra point attempt after touchdown	11 to 4
Hockey (NHL)	
Against short-handed team scoring	35 to 1
Against scoring on power play	4 to 1
Against scoring winning goal in last 5 minutes of game	20 to 1
Against a tie game	5 to 1

Coincidences —Or Are They?

"If two strangers sit next to each other on an airplane, more than 99 times out of 100 they will be linked in some way by two or fewer intermediaries," says John Allen Paulos in his book *Beyond Numeracy*. This statistic is based on the assumption that everyone has 1,500 relatives, friends, and acquaintances.

He also mentions the famous birthday "coincidence" that stems from applying probability theory. To ensure that two people in a group share a birthday (the same month and day, not necessarily same year), that group would have to contain 367 people. "But if one is willing to settle for a 50-50 chance of this happening, only 23 people need to be gathered," Paulos says.

Statistics

Nine Ways Numbers Can Lie

When looking at statistics, it pays to be skeptical. Here are a few questions to ask:

1. What is the time period—1 year or a lifetime? This is important when assessing the odds of dying from a particular disease, for example.

2. What group (or individual) is making the claim, and what point might they be trying to make? Bias, whether intentional or not, can affect the way statistics are collected and presented.

3. What is the size of the sample or group being used? For example, in a survey or a medical study, is the sample size 200 or 200,000? The larger the number in the group, the more reliable the claim being made. Another potential problem is the effect of "self-selection." If 1,000 surveys were mailed and only 100 were returned, the sample is said to be self-selected; the results will be skewed in favor of the opinions of those who responded.

4. When applying risks to yourself or someone else, what individual factors come into play? For example, if you do not smoke, are not overweight, and exercise daily, your chances of dying from heart disease are much less than those of someone who smokes and is overweight. If you play golf daily in Florida, your chances of being struck by lightning are greater than those of someone in Arizona who rarely leaves the house.

5. In a poll or survey, how were the questions phrased? This factor alone can make an enormous difference in the results obtained and relates to item 2 above. In her book *Tainted Truth*, Cynthia Crossen described a study done in 1991 that "found" that 1 of every 4 children in the United States was "hungry" or "at risk of hunger." Parents were surveyed, and a "yes" to any of 8 questions resulted in the child being labeled as "hungry" or "at risk." One question read, "Did you ever rely on a limited number of foods to feed your children because you were running out of money to buy food for a meal?"

6. Does the study or survey cited also indicate the size of the sample as well as the expected margin of error?

7. Are percentages backed up with the raw numbers they were drawn from? A 50 percent response may represent only 5 people if the sample size was 10.

8. When comparing statistics gathered over many years or places, is it possible that different methods were used by different people at different times? Look at the possible reasons behind large variations in numbers over a range of years. Related to this problem is the use of "constant dollars" (those adjusted for inflation) versus "current dollars" (today's value). Be sure that monetary figures compared over time use the same measurement.

9. Are the results expressed in averages? An average may not be a true representation; consider a claim that the average price of a home in a particular neighborhood is $122,727. If the neighborhood has 33 homes, 10 of which are worth $100,000, 10 of which are worth $90,000, 10 of which are worth $110,000, and 3 of which are worth $350,000, then only 3 of those homes are worth more than the average price stated—and those 3 are worth much more than the average.

Problems with Percentages

Percentages are frequently misstated and misused in the media. A common problem is lack of information about exactly what is being counted and what it is being compared to. Another is simply incorrect math. In his book *Mathsemantics: Making Numbers Talk Sense,* Edward MacNeal observes,

> John Tesh on *Entertainment Tonight* reported that the PBS series "The Civil War" had an audience of 13% versus the usual 4%, "an increase of more than 300%." The percentage-point increase is 13% minus 4%, or 9%. The base is 4%. Therefore the percentage increase is not more than 300%, but only 9/4, or 225%.

> Failure to allow for shifting bases leads to bad judgments. If sales drop by 20% and then grow by 25%, you're now better off, right? Wrong. You're back where you started. Proof: $100 minus 20% (one fifth) of $100 is $80. $80 plus 25% (one fourth) of $80 is $100. Add three zeros, or six, to each dollar figure for realism. Same conclusion.

> Not all examples are so innocuous. A drop of 70% followed by a rebound of 80% leaves one in deep trouble. Proof: $100 minus 70% is $30. $30 plus 80% is only $54.

> A gain of 80% followed by a loss of 70% is equally bad. $100 plus 80% is $180. $180 minus 70% is again $54.

Problems with Numbers in Charts and Graphs

Charts and graphs can be misleading if they are not prepared
properly; even if the raw numbers are correct, the way the image is
prepared can leave the wrong impression visually.

**Graph A:
Median Family Income**

Adjusted for inflation,
in thousands of dollars

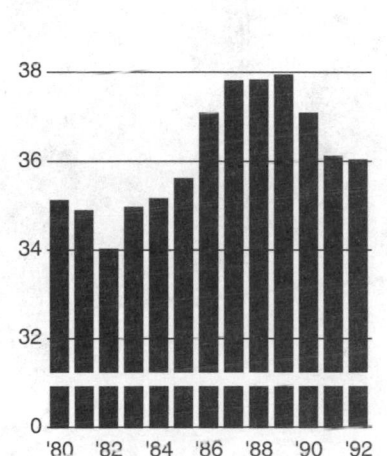

**Graph B:
Median Family Income**

Adjusted for inflation,
in thousands of dollars

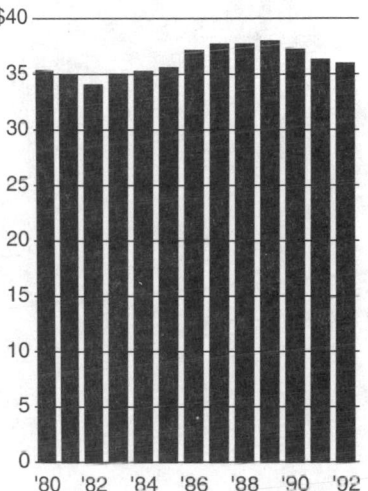

Charts that do not start at 0 can sometimes be misleading; and in a
line or bar chart (as shown in the top graphics on this page), the angle
of the line can vary greatly depending on the numbers used on the
horizontal and vertical axes. Truncation, in which the chart may start
at 0 and then cut off some of the lower or middle numbers (as in
Graph A above), can also give a distorted picture. If the scales are
mixed—for example, if a chart reports figures for the years 1970,
1975, 1980, 1985, 1990, and 1993—the resulting line may not be truly
representative.

In pictographs—like the barrels shown in the graphic below—
artists often take a two-dimensional figure and translate it into a
three-dimensional drawing. This can cause the image area to increase
disproportionately in size, making an increase seem much larger than
it really is.

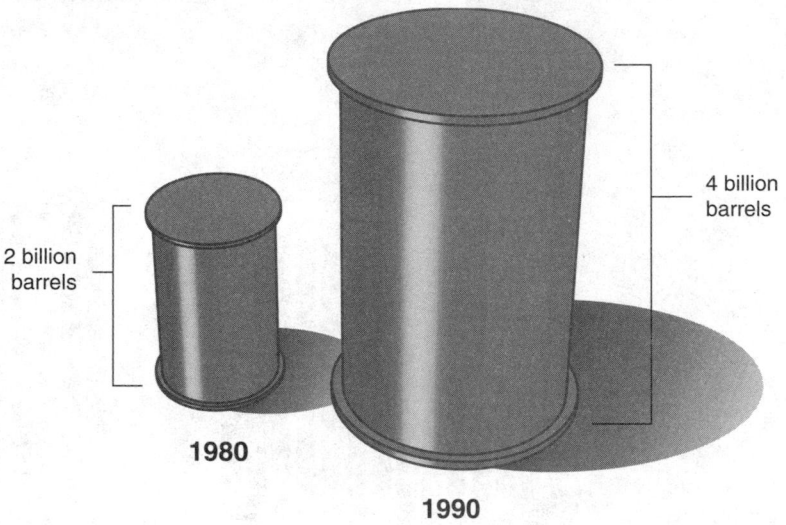

2 billion
barrels

4 billion
barrels

1980

1990

Recommended Resources

Almanacs and Sourcebooks

Information Please Almanac, annual. Houghton Mifflin Company, Boston and New York.

Godin, Seth, ed. *Information Please Business Almanac and Sourcebook*, annual. Houghton Mifflin Company, New York and Boston. Arranged by general business topic (including manufacturing, marketing, finance, human resources, business law and government, international business), this book provides statistics, phone numbers, and rankings.

Lesko, Matthew. *Lesko's Info-Power II*. Information USA, Kensington, MD, 1994. This book, which is also available on CompuServe, lists and describes more than 45,000 sources of low-cost or free information from state and federal government agencies. Topics include education, health, housing, careers, consumer issues, travel, taxes, and financial services.

Traveler's Atlas and World Guide. Rand McNally and Company, 1994.

The Universal Almanac, annual. Andrews and McMeel, Kansas City, MO.

The World Almanac and Book of Facts, annual. Pharos Books, New York.

General References

Lord, John. *Sizes: The Illustrated Encyclopedia*. HarperCollins Publishers, New York, 1995.

The Macmillan Dictionary of Measurement. Macmillan, New York, 1994.

Demographics/Statistics

Bureau of the Census, U.S. Department of Commerce. *Statistical Abstract of the United States*, annual. Subtitled "The Annual Data Book," this volume collects information on a wide range of topics from government and private sources. Topics include population, vital statistics, health, law enforcement, the environment, parks and recreation, banking and finance, business, national defense, energy, transportation, agriculture, manufacturing, and government spending (local, state, and federal). Some international statistics are also included. Available through the Government Printing Office (phone (202) 512-1800) or from the National Technical Information Service (phone (800) 553-6847 or (703) 487-4650).

Bureau of Labor Statistics, U.S. Department of Labor. *Occupational Outlook Handbook,* revised every 2 years. This volume provides information on 250 occupations covering about 104 million jobs. It describes job duties, training and education needed, earnings, working conditions, and future job prospects. Available through the Government Printing Office (phone (202) 512-1800 or from the National Technical Information Service (phone (800) 553-6847 or (703) 487-4650).

National Safety Council, *Accident Facts,* annual. This booklet reports statistics on "unintentional" deaths and injuries in the United States gathered from a variety of sources. Available through the National Safety Council (phone (800) 621-7619).

Office of the Federal Register. *United States Government Manual*. This comprehensive book, published every 2 years, describes agencies and offices, including boards, commissions, and committees, of the federal government, including phone numbers and addresses. Available through the Government Printing Office (phone (202) 512-1800).

Note: The U.S. Government Printing Office (GPO) sells hundreds of publications produced by virtually every federal department and agency and is a good source for unbiased facts and figures; GPO bookstores are located in 24 U.S. cities--call (202) 512-1800 for more information. Trade and professional associations are also good sources of statistical information about a particular topic, if you keep in mind that they are usually supporting their point of view; check the 3-volume *Encyclopedia of Associations* (annual, Gale Research Inc.), available at most libraries.

Mathematics

Huff, Darrell. *How to Lie with Statistics*. W.W. Norton and Co., New York, 1954, 1982. This is the classic work on how statistics can be used to deceive and mislead.

MacNeal, Edward. *Mathsemantics*. Viking, New York. 1994. The author discusses, for the layperson, how to handle the combination of mathematics and semantics; that is, how to talk correctly about numbers. He addresses averages, percentages, survey results, and misunderstandings about math in general.

Paulos, John Allen. *Innumeracy: Mathematical Illiteracy and Its Consequences*. Hill and Wang, New York, 1988. The author defines "innumeracy" as an "inability to deal comfortably with the fundamental notions of numbers and chance," and he explains, in straightforward language, such things as probabilities, coincidences, statistics, and surveys.

Paulos, John Allen. *A Mathematician Reads the Newspaper*. Basic Books, 1995. In this collection of short essays, Paulos dissects daily newspaper articles that, intentionally or unintentionally, serve to mislead readers through the manipulation of numbers and statistics.

Index